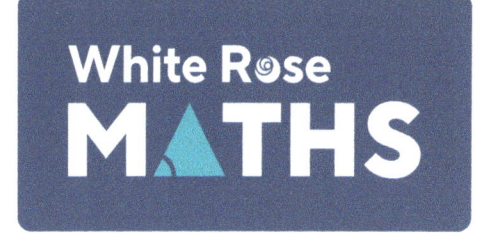

White Rose Maths Edition

Year 3C
A Guide to Teaching for Mastery

Series Editor: Tony Staneff
Lead author: Josh Lury

Contents

Introduction to the author team	4
What is *Power Maths*?	5
What's different in the new edition?	6
Your *Power Maths* resources	7
The *Power Maths* teaching model	10
The *Power Maths* lesson sequence	12
Using the *Power Maths* Teacher Guide	15
Power Maths Year 3, yearly overview	16
Mindset: an introduction	22
The *Power Maths* characters	23
Mathematical language	24
The role of talk and discussion	25
Assessment strategies	26
Keeping the class together	28
Same-day intervention	29
The role of practice	30
Structures and representations	31
Variation helps visualisation	32
Practical aspects of *Power Maths*	33
Working with children below age-related expectation	35
Providing extra depth and challenge with *Power Maths*	37
Using *Power Maths* with mixed age classes	39
List of practical resources	40
Getting started with *Power Maths*	43

Unit 11 – Fractions (2) — 44

Add fractions	46
Subtract fractions	50
Partition the whole	54
Problem solving – add and subtract fractions	58
Unit fractions of a set of objects	62
Non-unit fractions of a set of objects	66
Reason with fractions of an amount	70
Problem solving – fractions of measures	74
End of unit check	78

Unit 12 – Money — 80

Pounds and pence	82
Convert pounds and pence	86
Add money	90
Subtract money	94
Find change	98
End of unit check	102

Unit 13 – Time — 104

Roman numerals to 12	106
Tell the time to 5 minutes	110
Tell the time to the minute	114
Read time on a digital clock	118
Use am and pm	122
Years, months and days	126
Days and hours	130
Hours and minutes – start and end times	134
Hours and minutes – durations	138
Hours and minutes – compare durations	142
Minutes and seconds	146
Solve problems with time	150
End of unit check	154

Unit 14 – Angles and properties of shapes — 156

Turns and angles	158
Right angles in shapes	162
Compare angles	166
Measure and draw accurately	170
Horizontal and vertical	174
Parallel and perpendicular	178
Recognise, draw and describe 2D shapes	182
Recognise and describe 3D shapes	186
Make 3D shapes	190
End of unit check	194

Unit 15 – Statistics — 196

Interpret pictograms (1)	198
Interpret pictograms (2)	202
Draw pictograms	206
Interpret bar charts (1)	210
Interpret bar charts (2)	214
Collect and represent data in a bar chart	218
Simple two-way tables	222
End of unit check	226

Introduction to the author team

Power Maths arises from the work of maths mastery experts who are committed to proving that, given the right mastery mindset and approach, **everyone can do maths**. Based on robust research and best practice from around the world, *Power Maths* was developed in partnership with a group of UK teachers to make sure that it not only meets our children's wide-ranging needs but also aligns with the National Curriculum in England.

Power Maths – White Rose Maths edition

This edition of *Power Maths* has been developed and updated by:

Tony Staneff, Series Editor and Author

Vice Principal at Trinity Academy, Halifax, Tony also leads a team of mastery experts who help schools across the UK to develop teaching for mastery via nationally recognised CPD courses, problem-solving and reasoning resources, schemes of work, assessment materials and other tools.

Josh Lury, Lead Author

Josh is a specialist maths teacher, author and maths consultant with a passion for innovative and effective maths education.

The first edition of *Power Maths* was developed by a team of experienced authors, including:

- **Tony Staneff and Josh Lury**
- **Trinity Academy Halifax** (Michael Gosling CEO, Emily Fox, Kate Henshall, Rebecca Holland, Stephanie Kirk, Stephen Monaghan and Rachel Webster)
- **David Board, Belle Cottingham, Jonathan East, Tim Handley, Derek Huby, Neil Jarrett, Stephen Monaghan, Beth Smith, Tim Weal, Paul Wrangles** – skilled maths teachers and mastery experts
- **Cherri Moseley** – a maths author, former teacher and professional development provider
- **Professors Liu Jian and Zhang Dan**, Series Consultants and authors, and their team of mastery expert authors: **Wei Huinv, Huang Lihua, Zhu Dejiang, Zhu Yuhong, Hou Huiying, Yin Lili, Zhang Jing, Zhou Da and Liu Qimeng**

  Used by over 20 million children, Professor Liu Jian's textbook programme is one of the most popular in China. He and his author team are highly experienced in intelligent practice and in embedding key maths concepts using a C-P-A approach.

- **A group of 15 teachers and maths co-ordinators**

  We consulted our teacher group throughout the development of *Power Maths* to ensure we are meeting their real needs in the classroom.

What is *Power Maths*?

Created especially for UK primary schools, and aligned with the new National Curriculum, *Power Maths* is a whole-class, textbook-based mastery resource that empowers every child to understand and succeed. *Power Maths* rejects the notion that some people simply 'can't do' maths. Instead, it develops growth mindsets and encourages hard work, practice and a willingness to see mistakes as learning tools.

Best practice consistently shows that mastery of small, cumulative steps builds a solid foundation of deep mathematical understanding. *Power Maths* combines interactive teaching tools, high-quality textbooks and continuing professional development (CPD) to help you equip children with a deep and long-lasting understanding. Based on extensive evidence, and developed in partnership with practising teachers, *Power Maths* ensures that it meets the needs of children in the UK.

Power Maths and Mastery

Power Maths makes mastery practical and achievable by providing the structures, pathways, content, tools and support you need to make it happen in your classroom.

To develop mastery in maths, children must be enabled to acquire a deep understanding of maths concepts, structures and procedures, step by step. Complex mathematical concepts are built on simpler conceptual components and when children understand every step in the learning sequence, maths becomes transparent and makes logical sense. Interactive lessons establish deep understanding in small steps, as well as effortless fluency in key facts such as tables and number bonds. The whole class works on the same content and no child is left behind.

Power Maths

- Builds every concept in small, progressive steps
- Is built with interactive, whole-class teaching in mind
- Provides the tools you need to develop growth mindsets
- Helps you check understanding and ensure that every child is keeping up
- Establishes core elements such as intelligent practice and reflection

The *Power Maths* approach

Everyone can!
Founded on the conviction that every child can achieve, *Power Maths* enables children to build number fluency, confidence and understanding, step by step.

Child-centred learning
Children master concepts one step at a time in lessons that embrace a concrete-pictorial-abstract (C-P-A) approach, avoid overload, build on prior learning and help them see patterns and connections. Same-day intervention ensures sustained progress.

Continuing professional development
Embedded teacher support and development offer every teacher the opportunity to continually improve their subject knowledge and manage whole-class teaching for mastery.

Whole-class teaching
An interactive, whole-class teaching model encourages thinking and precise mathematical language and allows children to deepen their understanding as far as they can.

What's different in the new edition?

If you have previously used the first editions of *Power Maths*, you might be interested to know how this edition is different. All of the improvements described below are based on feedback from *Power Maths* customers.

Changes to units and the progression

- The order of units has been slightly adjusted, creating closer alignment between adjacent year groups, which will be useful for mixed age teaching.

- The flow of lessons has been improved within units to optimise the pace of the progression and build in more recap where needed. For key topics, the sequence of lessons gives more opportunities to build up a solid base of understanding. Other units have fewer lessons than before, where appropriate, making it possible to fit in all the content.

- Overall, the lessons put more focus on the most essential content for that year, with less time given to non-statutory content.

- The progression of lessons matches the steps in the new White Rose Maths schemes of learning.

Lesson resources

- There is a Quick recap for each lesson in the Teacher Guide, which offers an alternative lesson starter to the Power Up for cases where you feel it would be more beneficial to surface prerequisite learning than general number fluency.

- In the **Discover** and **Share** sections there is now more of a progression from 1 a) to 1 b). Whereas before, 1 b) was mainly designed as a separate question, now 1 a) leads directly into 1 b). This means that there is an improved whole-class flow, and also an opportunity to focus on the logic and skills in more detail. As a teacher, you will be using 1 a) to lead the class into the thinking, then 1 b) to mould that thinking into the core new learning of the lesson.

- In the **Share** section, for KS1 in particular, the number of different models and representations has been reduced, to support the clarity of thinking prompted by the flow from 1 a) into 1 b).

- More fluency questions have been built into the guided and independent practice.

- Pupil pages are as easy as possible for children to access independently. The pages are less full where this supports greater focus on key ideas and instructions. Also, more freedom is offered around answer format, with fewer boxes scaffolding children's responses; squared paper backgrounds are used in the Practice Books where appropriate. Artwork has also been revisited to ensure the highest standards of accessibility.

New components

480 Individual Practice Games are available in *ActiveLearn* for practising key facts and skills in Years 1 to 6. These are designed in an arcade style, to feel like fun games that children would choose to play outside school. They can be accessed via the Pupil World for homework or additional practice in school – and children can earn rewards. There are Support, Core and Extend levels to allocate, with Activity Reporting available for the teacher. There is a Quick Guide on *ActiveLearn* and you can use the Help area for support in setting up child accounts.

There is also a new set of lesson video resources on the Professional Development tile, designed for in-school training in 10- to 20-minute bursts. For each part of the *Power Maths* lesson sequence, there is a slide deck with embedded video, which will facilitate discussions about how you can take your *Power Maths* teaching to the next level.

Your *Power Maths* resources

Pupil Textbooks

Discover, **Share** and **Think together** sections promote discussion and introduce mathematical ideas logically, so that children understand more easily.

Using a Concrete-Pictorial-Abstract approach, clear mathematical models help children to make connections and grasp concepts.

Appealing scenarios stimulate curiosity, helping children to identify the maths problem and discover patterns and relationships for themselves.

Friendly, supportive characters help children develop a growth mindset by prompting them to think, reason and reflect.

To help you teach for mastery, *Power Maths* comprises a variety of high-quality resources.

The coherent *Power Maths* lesson structure carries through into the vibrant, high-quality textbooks. Setting out the core learning objectives for each class, the lesson structure follows a carefully mapped journey through the curriculum and supports children on their journey to deeper understanding.

Pupil Practice Books

The Practice Books offer just the right amount of intelligent practice for children to complete independently in the final section of each lesson.

Practice questions are finely tuned to move children forward in their thinking and to reveal misconceptions.

The practice questions are for everyone – each question varies one small element to move children on in their thinking.

Calculations are connected so that children think about the underlying concept.

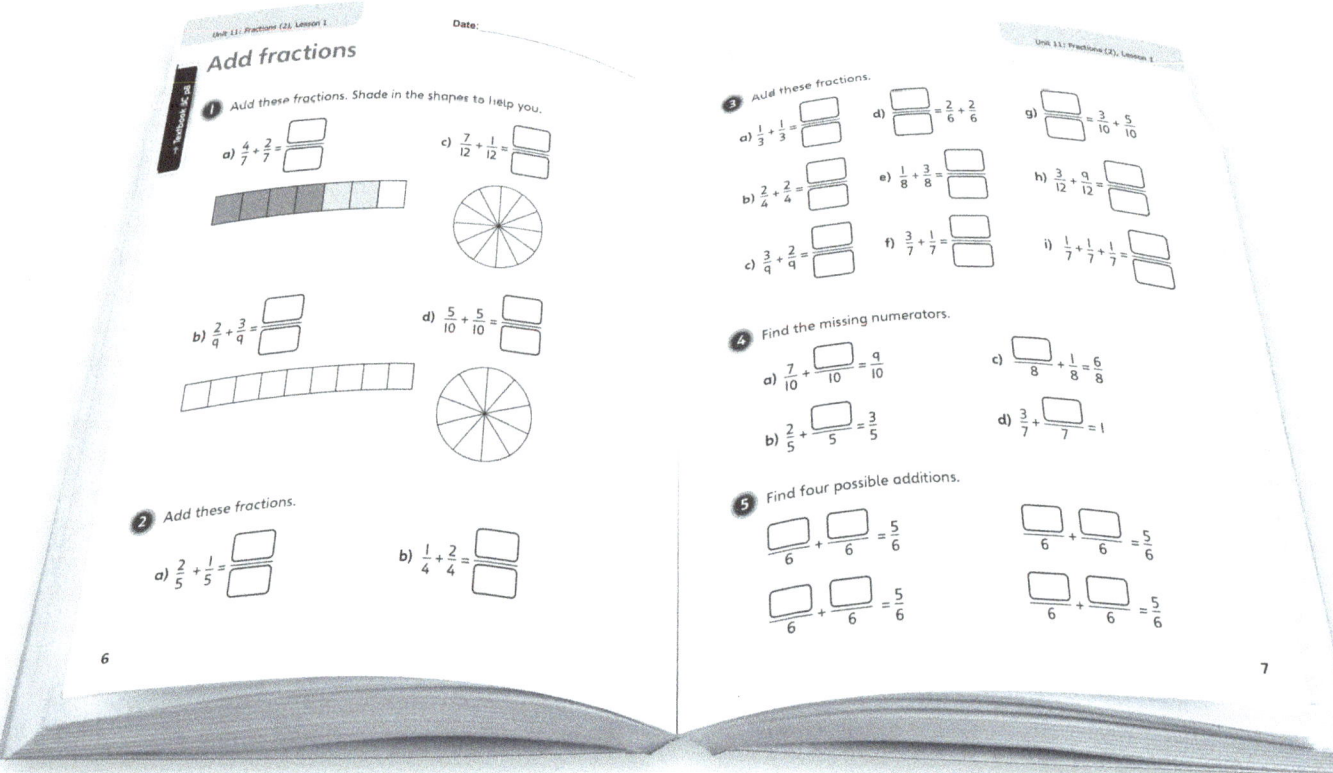

Challenge questions allow children to delve deeper into a concept.

The *Power Maths* characters support and encourage children to think and work in different ways.

Think differently questions encourage children to use reasoning as well as their mathematical knowledge to reach a solution.

Reflect questions reveal the depth of each child's understanding before they move on.

Online subscription

The online subscription will give you access to additional resources and answers from the Textbook and Practice Book.

eTextbooks

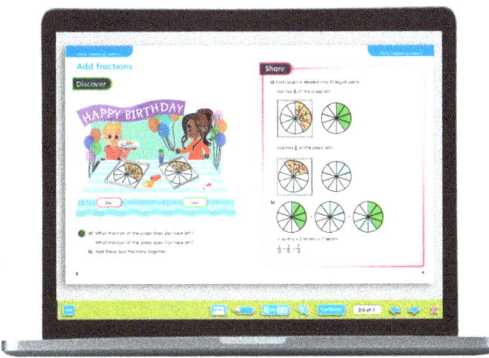

Digital versions of *Power Maths* Textbooks allow class groups to share and discuss questions, solutions and strategies. They allow you to project key structures and representations at the front of the class, to ensure all children are focusing on the same concept.

Teaching tools

Here you will find interactive versions of key *Power Maths* structures and representations.

Power Ups

Use this series of daily activities to promote and check number fluency.

Online versions of Teacher Guide pages

PDF pages give support at both unit and lesson levels. You will also find help with key strategies and templates for tracking progress.

Unit videos

Watch the professional development videos at the start of each unit to help you teach with confidence. The videos explore common misconceptions in the unit, and include intervention suggestions as well as suggestions on what to look out for when assessing mastery in your students.

End of unit Strengthen and Deepen materials

The Strengthen activity at the end of every unit addresses a key misconception and can be used to support children who need it. The Deepen activities are designed to be low ceiling/high threshold and will challenge those children who can understand more deeply. These resources will help you ensure that every child understands and will help you keep the class moving forward together. These printable activities provide an optional resource bank for use after the assessment stage.

Individual Practice Games

These enjoyable games can be used at home or at school to embed key number skills (see page 6).

Professional Development videos and slides

These slides and videos of *Power Maths* lessons can be used for ongoing training in short bursts or to support new staff.

The *Power Maths* teaching model

At the heart of *Power Maths* is a clearly structured teaching and learning process that helps you make certain that every child masters each maths concept securely and deeply. For each year group, the curriculum is broken down into core concepts, taught in units. A unit divides into smaller learning steps – lessons. Step by step, strong foundations of cumulative knowledge and understanding are built.

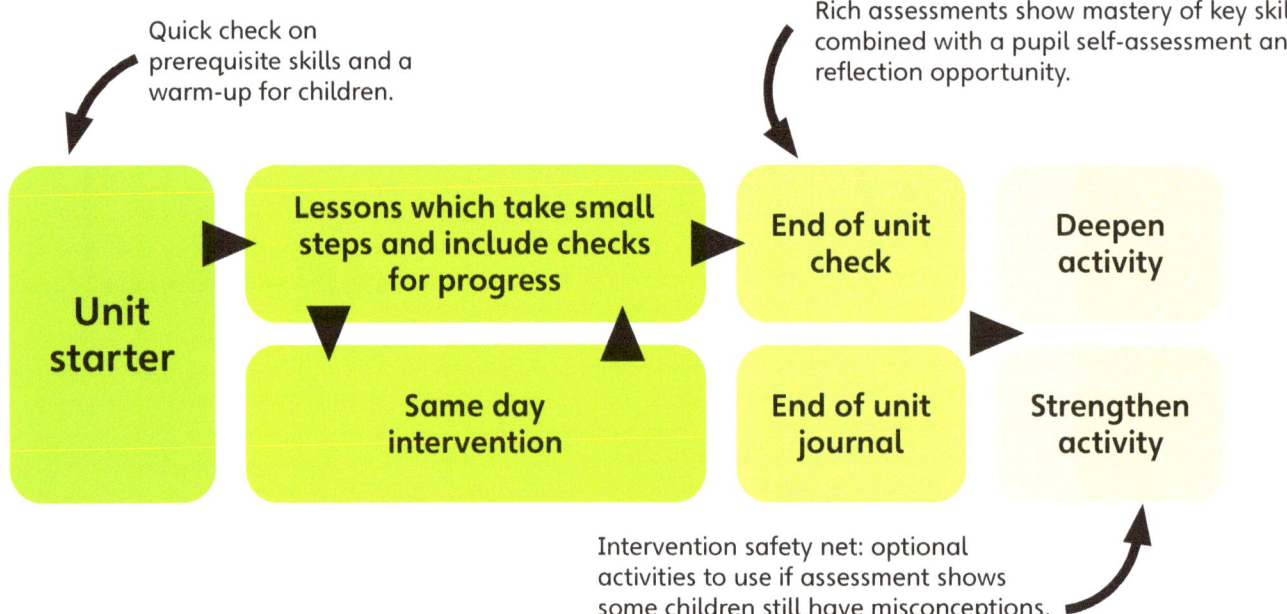

Unit starter

Each unit begins with a unit starter, which introduces the learning context along with key mathematical vocabulary and structures and representations.

- The Textbooks include a check on readiness and a warm-up task for children to complete.
- Your Teacher Guide gives support right from the start on important structures and representations, mathematical language, common misconceptions and intervention strategies.
- Unit-specific videos develop your subject knowledge and insights so you feel confident and fully equipped to teach each new unit. These are available via the online subscription.

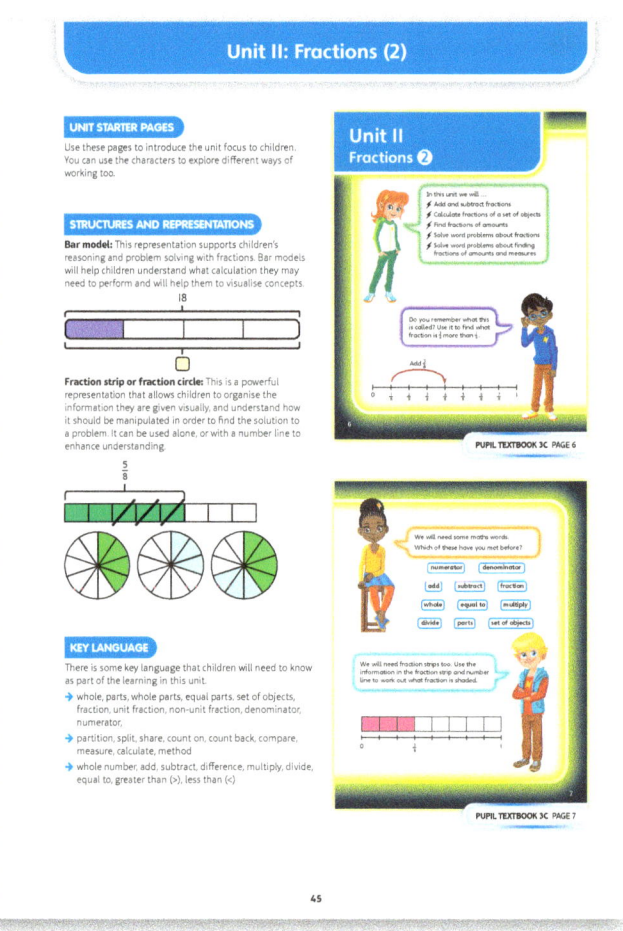

Lesson

Once a unit has been introduced, it is time to start teaching the series of lessons.

- Each lesson is scaffolded with Textbook and Practice Book activities and begins with a Power Up activity (available via online subscription) or the Quick recap activity in the Teacher Guide (see page 15).
- *Power Maths* identifies lesson by lesson what concepts are to be taught.
- Your Teacher Guide offers lots of support for you to get the most from every child in every lesson. As well as highlighting key points, tricky areas and how to handle them, you will also find question prompts to check on understanding and clarification on why particular activities and questions are used.

Same-day intervention

Same-day interventions are vital in order to keep the class progressing together. This can be during the lesson as well as afterwards (see page 29). Therefore, *Power Maths* provides plenty of support throughout the journey.

- Intervention is focused on keeping up now, not catching up later, so interventions should happen as soon as they are needed.
- Practice section questions are designed to bring misconceptions to the surface, allowing you to identify these easily as you circulate during independent practice time.
- Child-friendly assessment questions in the Teacher Guide help you identify easily which children need to strengthen their understanding.

End of unit check and journal

For each unit, the End of unit check in the Textbook lets you see which children have mastered the key concepts, which children have not and where their misconceptions lie. The Practice Books also include an End of unit journal in which children can reflect on what they have learned. Each unit also offers Strengthen and Deepen activities, available via the online subscription.

The Teacher Guide offers different ways of managing the End of unit assessments as well as giving support with handling misconceptions.

The End of unit check presents multiple-choice questions. Children think about their answer, decide on a solution and explain their choice.

The End of unit journal is an opportunity for children to test out their learning and reflect on how they feel about it. Tackling the 'journal' problem reveals whether a child understands the concept deeply enough to move on to the next unit.

In KS2, the End of unit assessment will also include at least one SATs-style question.

The *Power Maths* lesson sequence

At the heart of *Power Maths* is a unique lesson sequence designed to empower children to understand core concepts and grow in confidence. Embracing the National Centre for Excellence in the Teaching of Mathematics' (NCETM's) definition of mastery, the sequence guides and shapes every *Power Maths* lesson you teach.

Flexibility is built into the *Power Maths* programme so there is no one-to-one mapping of lessons and concepts and you can pace your teaching according to your class. While some children will need to spend longer on a particular concept (through interventions or additional lessons), others will reach deeper levels of understanding. However, it is important that the class moves forward together through the termly schedules.

Power Up  5 minutes

Each lesson begins with a Power Up activity (available via the online subscription) which supports fluency in key number facts.

The whole-class approach depends on fluency, so the Power Up is a powerful and essential activity.

The Quick recap is an alternative starter, for when you think some or all children would benefit more from revisiting pre-requisite work (see page 15).

TOP TIP
If the class is struggling with the task, revisit it later and check understanding.

Power Ups reinforce the two key things that are essential for success: times-tables and number bonds.

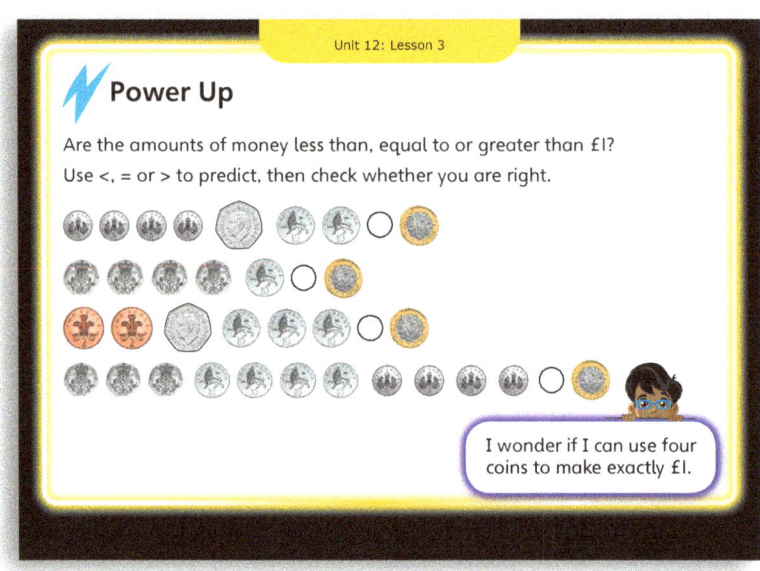

Discover 10 minutes

A practical, real-life problem arouses curiosity. Children find the maths through story telling.

A real-life scenario is provided for the **Discover** section but feel free to build upon these with your own examples that are more relevant to your class, or get creative with the context.

TOP TIP
Discover works best when run at tables, in pairs with concrete objects.

Question ① a) tackles the key concept and question ① b) digs a little deeper. Children have time to explore, play and discuss possible strategies.

Share ⏱ 10 minutes

Teacher-led, this interactive section follows the **Discover** activity and highlights the variety of methods that can be used to solve a single problem.

TOP TIP
Pairs sharing a textbook is a great format for **Share**!

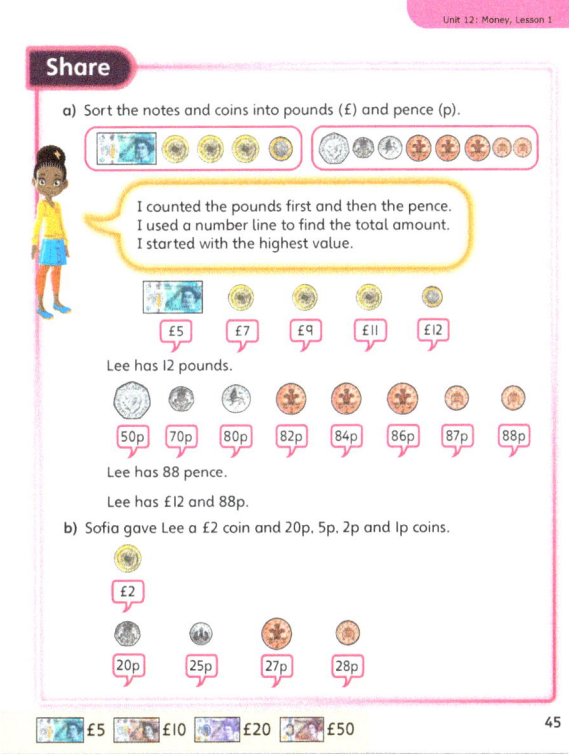

Your Teacher Guide gives target questions for children. The online toolkit provides interactive structures and representations to link concrete and pictorial to abstract concepts.

Bring children to the front to share and celebrate their solutions and strategies.

Think together

⏱ 10 minutes

Children work in groups on the carpet or at tables, using their textbooks or eBooks.

TOP TIP
Make sure children have mini whiteboards or pads to write on if they are not at their tables.

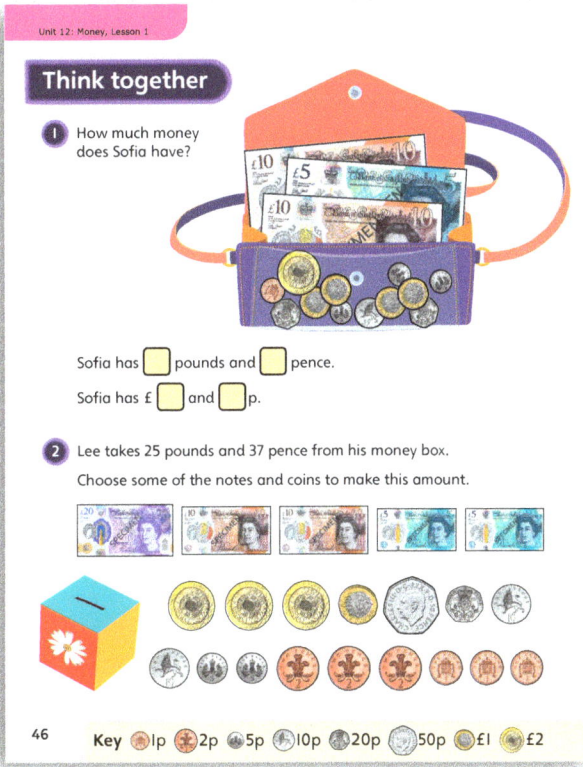

Using the Teacher Guide, model question ❶ for your class.

Question ❷ is less structured. Children will need to think together in their groups, then discuss their methods and solutions as a class.

In question ❸ children try working out the answer independently. The openness of the **Challenge** question helps to check depth of understanding.

Practice ⏱ 15 minutes

Using their Practice Books, children work independently while you circulate and check on progress.

Questions follow small steps of progression to deepen learning.

TOP TIP
Some children could work separately with a teacher or assistant.

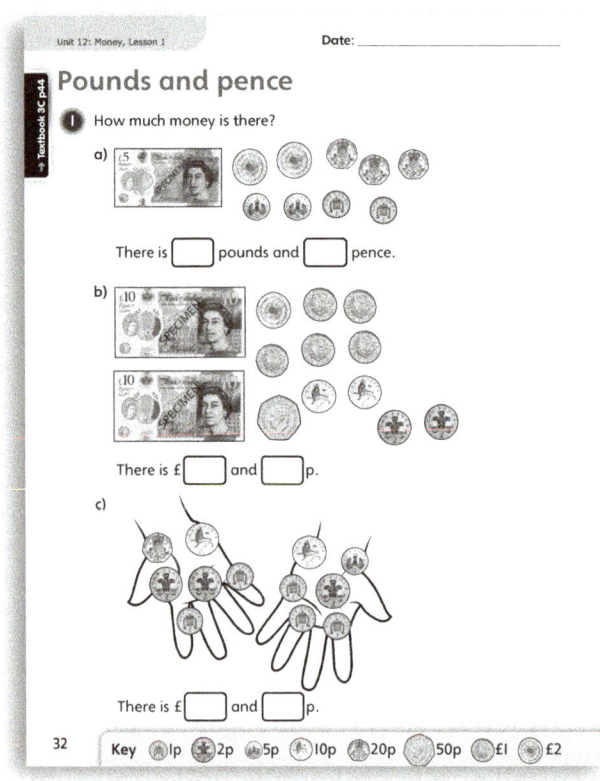

Are some children struggling? If so, work with them as a group, using mathematical structures and representations to support understanding as necessary.

There are no set routines: for real understanding, children need to think about the problem in different ways.

Reflect ⏱ 5 minutes

'Spot the mistake' questions are great for checking misconceptions.

The **Reflect** section is your opportunity to check how deeply children understand the target concept.

The Practice Books use various approaches to check that children have fully understood each concept.

Looking like they understand is not enough! It is essential that children can show they have grasped the concept.

Using the *Power Maths* Teacher Guide

Think of your Teacher Guides as *Power Maths* handbooks that will guide, support and inspire your day-to-day teaching. Clear and concise, and illustrated with helpful examples, your Teacher Guides will help you make the best possible use of every individual lesson. They also provide wrap-around professional development, enhancing your own subject knowledge and helping you to grow in confidence about moving your children forward together.

There is a Teacher Guide per year group for every term, with unit and lesson level guidance and support.

Never feel stuck! You will find ideas for introducing every unit and lesson and questions to encourage teacher reflection before and after each lesson.

Tips and advice on key elements such as C-P-A approaches, misconceptions, language, modelling growth mindsets and same day intervention.

Annotations for every Textbook and Practice Book page, providing prompts for key questions to ask to expose understanding and explanations as to why key questions have been chosen.

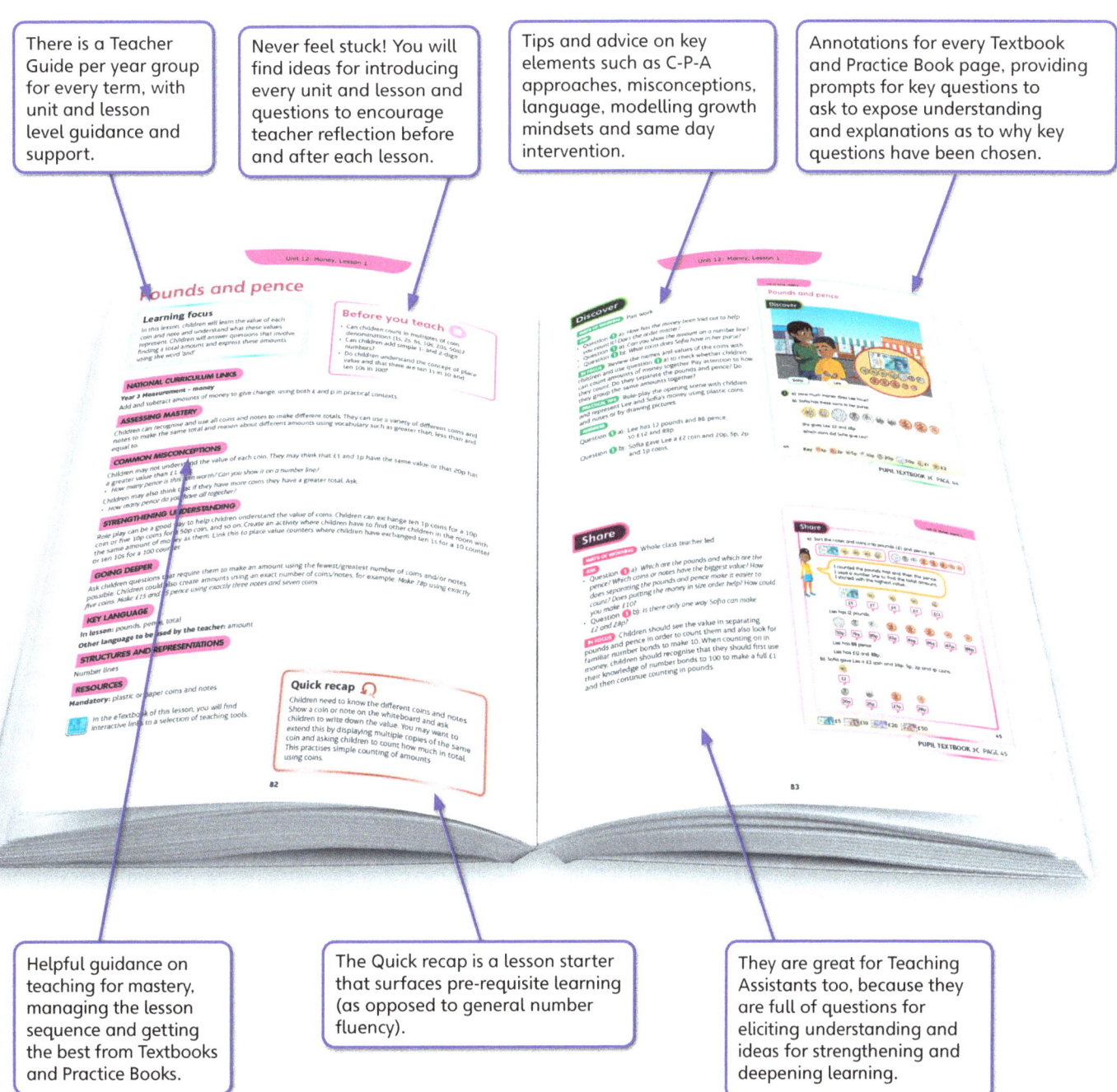

Helpful guidance on teaching for mastery, managing the lesson sequence and getting the best from Textbooks and Practice Books.

The Quick recap is a lesson starter that surfaces pre-requisite learning (as opposed to general number fluency).

They are great for Teaching Assistants too, because they are full of questions for eliciting understanding and ideas for strengthening and deepening learning.

At the end of each unit, your Teacher Guide helps you identify who has fully grasped the concept, who has not and how to move every child forward. This is covered later in the Assessment strategies section.

Power Maths Year 3, yearly overview

Textbook	Strand	Unit		Number of lessons
Textbook A / Practice Workbook A (Term 1)	Number – number and place value	1	Place value within 1,000	13
	Number – addition and subtraction	2	Addition and subtraction (1)	10
	Number – addition and subtraction	3	Addition and subtraction (2)	13
	Number – multiplication and division	4	Multiplication and division (1)	5
	Number – multiplication and division	5	Multiplication and division (2)	13
Textbook B / Practice Workbook B (Term 2)	Number – multiplication and division	6	Multiplication and division (3)	13
	Measurement	7	Length and perimeter	11
	Number – fractions	8	Fractions (1)	10
	Measurement	9	Mass	7
	Measurement	10	Capacity	6
Textbook C / Practice Workbook C (Term 3)	Number – fractions	11	Fractions (2)	8
	Measurement	12	Money	5
	Measurement	13	Time	12
	Geometry – properties of shapes	14	Angles and properties of shapes	9
	Statistics	15	Statistics	7

Power Maths Year 3, Textbook 3C (Term 3) overview

Strand	Unit	Unit title	Lesson number	Lesson title	NC Objective 1	NC Objective 2
Number – fractions	11	Fractions (2)	1	Add fractions	Add and subtract fractions with the same denominator within one whole [for example, $\frac{5}{7} + \frac{1}{7} = \frac{6}{7}$]	
Number – fractions	11	Fractions (2)	2	Subtract fractions	Add and subtract fractions with the same denominator within one whole [for example, $\frac{5}{7} + \frac{1}{7} = \frac{6}{7}$]	
Number – fractions	11	Fractions (2)	3	Partition the whole	Add and subtract fractions with the same denominator within one whole [for example, $\frac{5}{7} + \frac{1}{7} = \frac{6}{7}$]	
Number – fractions	11	Fractions (2)	4	Problem solving – add and subtract fractions	Solve problems that involve all of the above	
Number – fractions	11	Fractions (2)	5	Unit fractions of a set of objects	Recognise, find and write fractions of a discrete set of objects: unit fractions and non-unit fractions with small denominators	
Number – fractions	11	Fractions (2)	6	Non-unit fractions of a set of objects	Recognise, find and write fractions of a discrete set of objects: unit fractions and non-unit fractions with small denominators	
Number – fractions	11	Fractions (2)	7	Reason with fractions of an amount	Recognise, find and write fractions of a discrete set of objects: unit fractions and non-unit fractions with small denominators	
Number – fractions	11	Fractions (2)	8	Problem solving – fractions of measures	Solve problems that involve all of the above	
Measurement	12	Money	1	Pounds and pence	Add and subtract amounts of money to give change, using both £ and p in practical contexts	
Measurement	12	Money	2	Convert pounds and pence	Add and subtract amounts of money to give change, using both £ and p in practical contexts	
Measurement	12	Money	3	Add money	Add and subtract amounts of money to give change, using both £ and p in practical contexts	
Measurement	12	Money	4	Subtract money	Add and subtract amounts of money to give change, using both £ and p in practical contexts	
Measurement	12	Money	5	Find change	Add and subtract amounts of money to give change, using both £ and p in practical contexts	

Strand	Unit	Unit title	Lesson number	Lesson title	NC Objective 1	NC Objective 2
Measurement	13	Time	1	Roman numerals to 12	Tell and write the time from an analogue clock, including using Roman numerals from I to XII, and 12-hour and 24-hour clocks	
Measurement	13	Time	2	Tell the time to 5 minutes	Tell and write the time from an analogue clock, including using Roman numerals from I to XII, and 12-hour and 24-hour clocks	
Measurement	13	Time	3	Tell the time to the minute	Tell and write the time from an analogue clock, including using Roman numerals from I to XII, and 12-hour and 24-hour clocks	Estimate and read time with increasing accuracy to the nearest minute; record and compare time in terms of seconds, minutes and hours; use vocabulary such as o'clock, am/pm, morning, afternoon, noon and midnight
Measurement	13	Time	4	Read time on a digital clock	Estimate and read time with increasing accuracy to the nearest minute; record and compare time in terms of seconds, minutes and hours; use vocabulary such as o'clock, am/pm, morning, afternoon, noon and midnight	Tell and write the time from an analogue clock, including using Roman numerals from I to XII, and 12-hour and 24-hour clocks
Measurement	13	Time	5	Use am and pm	Estimate and read time with increasing accuracy to the nearest minute; record and compare time in terms of seconds, minutes and hours; use vocabulary such as o'clock, am/pm, morning, afternoon, noon and midnight	Tell and write the time from an analogue clock, including using Roman numerals from I to XII, and 12-hour and 24-hour clocks
Measurement	13	Time	6	Years, months and days	Know the number of seconds in a minute and the number of days in each month, year and leap year	
Measurement	13	Time	7	Days and hours	Estimate and read time with increasing accuracy to the nearest minute; record and compare time in terms of seconds, minutes and hours; use vocabulary such as o'clock, am/pm, morning, afternoon, noon and midnight	Tell and write the time from an analogue clock, including using Roman numerals from I to XII, and 12-hour and 24-hour clocks

Strand	Unit	Unit title	Lesson number	Lesson title	NC Objective 1	NC Objective 2
Measurement	13	Time	8	Hours and minutes – start and end times	Estimate and read time with increasing accuracy to the nearest minute; record and compare time in terms of seconds, minutes and hours; use vocabulary such as o'clock, am/pm, morning, afternoon, noon and midnight	Compare durations of events [for example to calculate the time taken by particular events or tasks]
Measurement	13	Time	9	Hours and minutes – durations	Compare durations of events [for example to calculate the time taken by particular events or tasks]	Estimate and read time with increasing accuracy to the nearest minute; record and compare time in terms of seconds, minutes and hours; use vocabulary such as o'clock, am/pm, morning, afternoon, noon and midnight
Measurement	13	Time	10	Hours and minutes – compare durations	Estimate and read time with increasing accuracy to the nearest minute; record and compare time in terms of seconds, minutes and hours; use vocabulary such as o'clock, am/pm, morning, afternoon, noon and midnight	Compare durations of events [for example to calculate the time taken by particular events or tasks]
Measurement	13	Time	11	Minutes and seconds	Estimate and read time with increasing accuracy to the nearest minute; record and compare time in terms of seconds, minutes and hours; use vocabulary such as o'clock, am/pm, morning, afternoon, noon and midnight	
Measurement	13	Time	12	Solve problems with time	Estimate and read time with increasing accuracy to the nearest minute; record and compare time in terms of seconds, minutes and hours; use vocabulary such as o'clock, am/pm, morning, afternoon, noon and midnight	
Geometry – properties of shapes	14	Angles and properties of shapes	1	Turns and angles	Recognise angles as a property of shape or a description of a turn	Identify right angles, recognise that two right angles make a half-turn, three make three quarters of a turn and four a complete turn; identify whether angles are greater than or less than a right angle

Strand	Unit	Unit title	Lesson number	Lesson title	NC Objective 1	NC Objective 2
Geometry – properties of shapes	14	Angles and properties of shapes	2	Right angles in shapes	Recognise angles as a property of shape or a description of a turn	Identify right angles, recognise that two right angles make a half-turn, three make three quarters of a turn and four a complete turn; identify whether angles are greater than or less than a right angle
Geometry – properties of shapes	14	Angles and properties of shapes	3	Compare angles	Identify right angles, recognise that two right angles make a half-turn, three make three quarters of a turn and four a complete turn; identify whether angles are greater than or less than a right angle	Recognise angles as a property of shape or a description of a turn
Geometry – properties of shapes	14	Angles and properties of shapes	4	Measure and draw accurately	Draw 2D shapes and make 3D shapes using modelling materials; recognise 3D shapes in different orientations and describe them	Identify horizontal and vertical lines and pairs of perpendicular and parallel lines
Geometry – properties of shapes	14	Angles and properties of shapes	5	Horizontal and vertical	Identify horizontal and vertical lines and pairs of perpendicular and parallel lines	
Geometry – properties of shapes	14	Angles and properties of shapes	6	Parallel and perpendicular	Identify horizontal and vertical lines and pairs of perpendicular and parallel lines	
Geometry – properties of shapes	14	Angles and properties of shapes	7	Recognise, draw and describe 2D shapes	Draw 2D shapes and make 3D shapes using modelling materials; recognise 3-D shapes in different orientations and describe them	
Geometry – properties of shapes	14	Angles and properties of shapes	8	Recognise and describe 3D shapes	Draw 2D shapes and make 3D shapes using modelling materials; recognise 3-D shapes in different orientations and describe them	
Geometry – properties of shapes	14	Angles and properties of shapes	9	Make 3D shapes	Draw 2D shapes and make 3D shapes using modelling materials; recognise 3-D shapes in different orientations and describe them	
Statistics	15	Statistics	1	Interpret pictograms (1)	Interpret and present data using bar charts, pictograms and tables	Solve one-step and two-step questions [for example, 'How many more?' and 'How many fewer?'] using information presented in scaled bar charts and pictograms and tables
Statistics	15	Statistics	2	Interpret pictograms (2)	Interpret and present data using bar charts, pictograms and tables	

Strand	Unit	Unit title	Lesson number	Lesson title	NC Objective 1	NC Objective 2
Statistics	15	Statistics	3	Draw pictograms	Interpret and present data using bar charts, pictograms and tables	Solve one-step and two-step questions [for example, 'How many more?' and 'How many fewer?'] using information presented in scaled bar charts and pictograms and tables
Statistics	15	Statistics	4	Interpret bar charts (1)	Interpret and present data using bar charts, pictograms and tables	
Statistics	15	Statistics	5	Interpret bar charts (2)	Interpret and present data using bar charts, pictograms and tables	Solve one-step and two-step questions [for example, 'How many more?' and 'How many fewer?'] using information presented in scaled bar charts and pictograms and tables
Statistics	15	Statistics	6	Collect and represent data in a bar chart	Interpret and present data using bar charts, pictograms and tables	
Statistics	15	Statistics	7	Simple two-way tables	Interpret and present data using bar charts, pictograms and tables	

Mindset: an introduction

Global research and best practice deliver the same message: learning is greatly affected by what learners perceive they can or cannot do. What is more, it is also shaped by what their parents, carers and teachers perceive they can do. Mindset – the thinking that determines our beliefs and behaviours – therefore has a fundamental impact on teaching and learning.

Everyone can!

Power Maths and mastery methods focus on the distinction between 'fixed' and 'growth' mindsets (Dweck, 2007).[1] Those with a fixed mindset believe that their basic qualities (for example, intelligence, talent and ability to learn) are pre-wired or fixed: 'If you have a talent for maths, you will succeed at it. If not, too bad!' By contrast, those with a growth mindset believe that hard work, effort and commitment drive success and that 'smart' is not something you are or are not, but something you become. In short, everyone can do maths!

Key mindset strategies

A growth mindset needs to be actively nurtured and developed. *Power Maths* offers some key strategies for fostering healthy growth mindsets in your classroom.

It is okay to get it wrong

Mistakes are valuable opportunities to re-think and understand more deeply. Learning is richer when children and teachers alike focus on spotting and sharing mistakes as well as solutions.

Praise hard work

Praise is a great motivator, and by focusing on praising effort and learning rather than success, children will be more willing to try harder, take risks and persist for longer.

Mind your language!

The language we use around learners has a profound effect on their mindsets. Make a habit of using growth phrases, such as, 'Everyone can!', 'Mistakes can help you learn' and 'Just try for a little longer'. The king of them all is one little word, 'yet'… I can't solve this…yet!' Encourage parents and carers to use the right language too.

Build in opportunities for success

The step-by-small-step approach enables children to enjoy the experience of success. In addition, avoid ability grouping and encourage every child to answer questions and explain or demonstrate their methods to others.

[1] Dweck, C (2007) *The New Psychology of Success*, Ballantine Books: New York

The *Power Maths* characters

The *Power Maths* characters model the traits of growth mindset learners and encourage resilience by prompting and questioning children as they work. Appearing frequently in the Textbooks and Practice Books, they are your allies in teaching and discussion, helping to model methods, alternatives and misconceptions, and to pose questions. They encourage and support your children, too: they are all hardworking, enthusiastic and unafraid of making and talking about mistakes.

Meet the team!

Creative Flo is open-minded and sometimes indecisive. She likes to think differently and come up with a variety of methods or ideas.

Determined Dexter is resolute, resilient and systematic. He concentrates hard, always tries his best and he'll never give up – even though he doesn't always choose the most efficient methods!

'Let's try again.'

'Mistakes are cool!'

'Have I found all of the solutions?'

'Let's try it this way…'

'Can we do it differently?'

'I've got another way of doing this!'

'I'm going to try this!'

'I know how to do that!'

'Want to share my ideas?'

Curious Ash is eager, interested and inquisitive, and he loves solving puzzles and problems. Ash asks lots of questions but sometimes gets distracted.

'What if we tried this…?'

'I wonder…'

'Is there a pattern here?'

Sparks the Cat

Miaow!

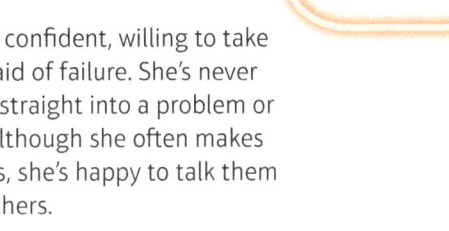

Brave Astrid is confident, willing to take risks and unafraid of failure. She's never scared to jump straight into a problem or question, and although she often makes simple mistakes, she's happy to talk them through with others.

Mathematical language

Traditionally, we in the UK have tended to try simplifying mathematical language to make it easier for young children to understand. By contrast, evidence and experience show that by diluting the correct language, we actually mask concepts and meanings for children. We then wonder why they are confused by new and different terminology later down the line! *Power Maths* is not afraid of 'hard' words and avoids placing any barriers between children and their understanding of mathematical concepts. As a result, we need to be deliberate, precise and thorough in building every child's understanding of the language of maths. Throughout the Teacher Guides you will find support and guidance on how to deliver this, as well as individual explanations throughout the pupil Textbooks.

Use the following key strategies to build children's mathematical vocabulary, understanding and confidence.

Precise and consistent

Everyone in the classroom should use the correct mathematical terms in full, every time. For example, refer to 'equal parts', not 'parts'. Used consistently, precise maths language will be a familiar and non-threatening part of children's everyday experience.

Full sentences

Teachers and children alike need to use full sentences to explain or respond. When children use complete sentences, it both reveals their understanding and embeds their knowledge.

Stem sentences

These important sentences help children express mathematical concepts accurately, and are used throughout the *Power Maths* books. Encourage children to repeat them frequently, whether working independently or with others. Examples of stem sentences are:

'4 is a part, 5 is a part, 9 is the whole.'

'There are .... groups. There are .... in each group.'

Key vocabulary

The unit starters highlight essential vocabulary for every lesson. In the pupil books, characters flag new terminology and the Teacher Guide lists important mathematical language for every unit and lesson. New terms are never introduced without a clear explanation.

Mathematical signs

Mathematical signs are used early on so that children quickly become familiar with them and their meaning. Often, the *Power Maths* characters will highlight the connection between language and particular signs.

The role of talk and discussion

When children learn to talk purposefully together about maths, barriers of fear and anxiety are broken down and they grow in confidence, skills and understanding. Building a healthy culture of 'maths talk' empowers their learning from day one.

Explanation and discussion are integral to the *Power Maths* structure, so by simply following the books your lessons will stimulate structured talk. The following key 'maths talk' strategies will help you strengthen that culture and ensure that every child is included.

Sentences, not words

Encourage children to use full sentences when reasoning, explaining or discussing maths. This helps both speaker and listeners to clarify their own understanding. It also reveals whether or not the speaker truly understands, enabling you to address misconceptions as they arise.

Working together

Working with others in pairs, groups or as a whole class is a great way to support maths talk and discussion. Use different group structures to add variety and challenge. For example, children could take timed turns for talking, work independently alongside a 'discussion buddy', or perhaps play different *Power Maths* character roles within their group.

Think first – then talk

Provide clear opportunities within each lesson for children to think and reflect, so that their talk is purposeful, relevant and focused.

Give every child a voice

Where the 'hands up' model allows only the more confident child to shine, *Power Maths* involves everyone. Make sure that no child dominates and that even the shyest child is encouraged to contribute – and praised when they do.

Assessment strategies

Teaching for mastery demands that you are confident about what each child knows and where their misconceptions lie; therefore, practical and effective assessment is vitally important.

Formative assessment within lessons

The **Think together** section will often reveal any confusions or insecurities; try ironing these out by doing the first **Think together** question as a class. For children who continue to struggle, you or your Teaching Assistant should provide support and enable them to move on.

➤ Performance in practice can be very revealing: check Practice Books and listen out both during and after practice to identify misconceptions.

➤ The **Reflect** section is designed to check on the all-important depth of understanding. Be sure to review how the children performed in this final stage before you teach the next lesson.

End of unit check – Textbook

Each unit concludes with a summative check to help you assess quickly and clearly each child's understanding, fluency, reasoning and problem solving skills. Your Teacher Guide will suggest ideal ways of organising a given activity and offer advice and commentary on what children's responses mean. For example, 'What misconception does this reveal?'; 'How can you reinforce this particular concept?'

For younger children, assess in small, teacher-led groups, giving each child time to think and respond while also consolidating correct mathematical language. Assessment with young children should always be an enjoyable activity, so avoid one-to-one individual assessments, which they may find threatening or scary. If you prefer, the End of unit check can be carried out as a whole-class group using whiteboards and Practice Books.

End of unit check – Practice Book

The Practice Book contains further opportunities for assessment, and can be completed by children independently whilst you are carrying out diagnostic assessment with small groups. Your Teacher Guide will advise you on what to do if children struggle to articulate an explanation – or perhaps encourage you to write down something they have explained well. It will also offer insights into children's answers and their implications for next learning steps. It is split into three main sections, outlined below.

My journal is designed to allow children to show their depth of understanding of the unit. It can also serve as a way of checking that children have grasped key mathematical vocabulary. The question children should answer is first presented in the Textbook in the Think! section. This provides an opportunity for you to discuss the question first as a class to ensure children have understood their task. Children should have some time to think about how they want to answer the question, and you could ask them to talk to a partner about their ideas. Then children should write their answer in their Practice Book, using the word bank provided to help them with vocabulary.

The **Power check** allows pupils to self-assess their level of confidence on the topic by colouring in different smiley faces. You may want to introduce the faces as follows:

Each unit ends with either a Power play or a Power puzzle. This is an activity, puzzle or game that allows children to use their new knowledge in a fun, informal way.

Progress Tests

There are *Power Maths* Progress Tests for each half term and at the end of the year, including an Arithmetic test and Reasoning test in each case. You can enter results in the online markbook to track and analyse results and see the average for all schools' results. The tests use a 6-step scale to show results against age-related expectation.

How to ask diagnostic questions

The diagnostic questions provided in children's Practice Books are carefully structured to identify both understanding and misconceptions (if children answer in a particular way, you will know why). The simple procedure below may be helpful:

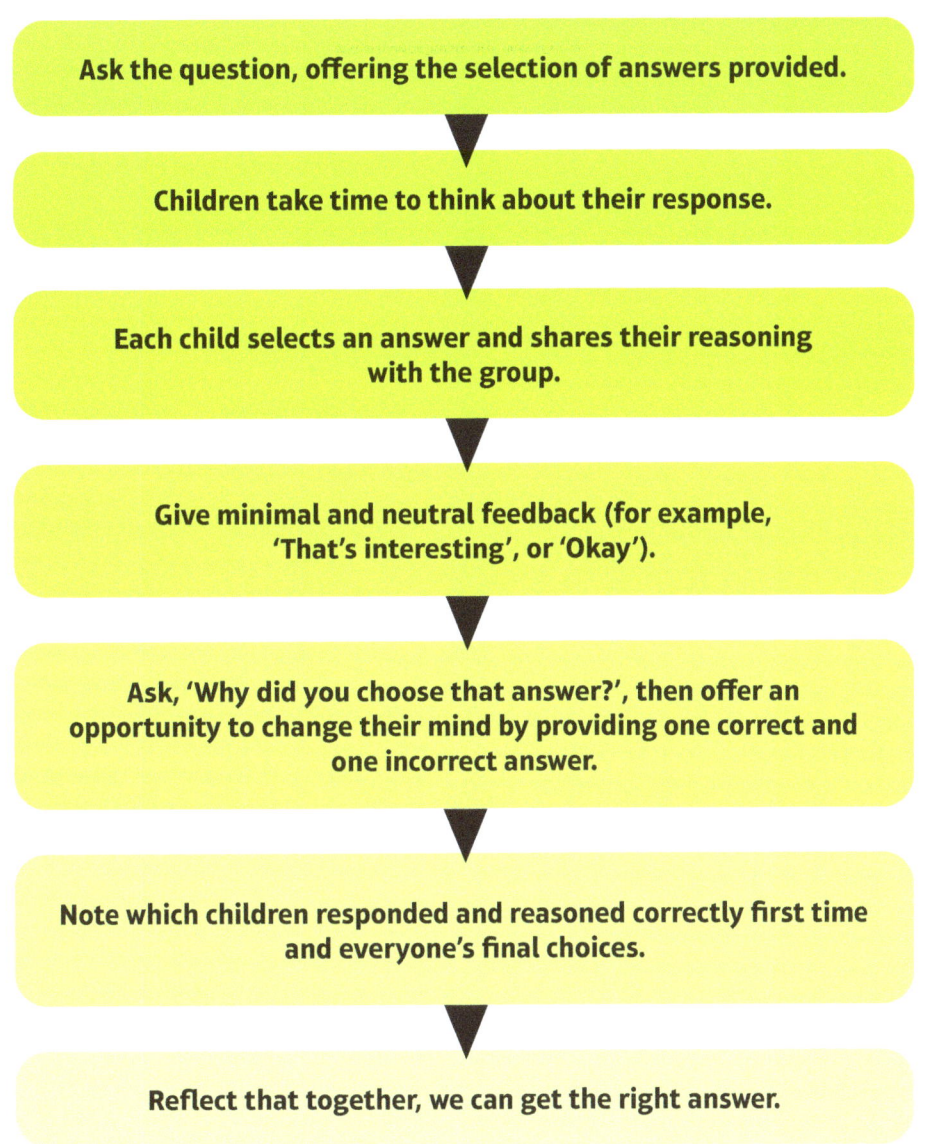

Keeping the class together

Traditionally, children who learn quickly have been accelerated through the curriculum. As a consequence, their learning may be superficial and will lack the many benefits of enabling children to learn with and from each other.

By contrast, *Power Maths'* mastery approach values real understanding and richer, deeper learning above speed. It sees all children learning the same concept in small, cumulative steps, each finding and mastering challenge at their own level. Remember that when you teach for mastery, EVERYONE can do maths! Those who grasp a concept easily have time to explore and understand that concept at a deeper level. The whole class therefore moves through the curriculum at broadly the same pace via individual learning journeys.

For some teachers, the idea that a whole class can move forward together is revolutionary and challenging. However, the evidence of global good practice clearly shows that this approach drives engagement, confidence, motivation and success for all learners, and not just the high flyers. The strategies below will help you keep your class together on their maths journey.

Mix it up

Do not stick to set groups at each table. Every child should be working on the same concept, and mixing up the groupings widens children's opportunities for exploring, discussing and sharing their understanding with others.

Recycling questions

Reuse the Textbook and Practice Book questions with concrete materials to allow children to explore concepts and relationships and deepen their understanding. This strategy is especially useful for reinforcing learning in same-day interventions.

Strengthen at every opportunity

The next lesson in a *Power Maths* sequence always revises and builds on the previous step to help embed learning. These activities provide golden opportunities for individual children to strengthen their learning with the support of Teaching Assistants.

Prepare to be surprised!

Children may grasp a concept quickly or more slowly. The 'fast graspers' won't always be the same individuals, nor does the speed at which a child understands a concept predict their success in maths. Are they struggling or just working more slowly?

Same-day intervention

Since maths competence depends on mastering concepts one by one in a logical progression, it is important that no gaps in understanding are ever left unfilled. Same-day interventions – either within or after a lesson – are a crucial safety net for any child who has not fully made the small step covered that day. In other words, intervention is always about keeping up, not catching up, so that every child has the skills and understanding they need to tackle the next lesson. That means presenting the same problems used in the lesson, with a variety of concrete materials to help children model their solutions.

We offer two intervention strategies below, but you should feel free to choose others if they work better for your class.

Within-lesson intervention

The **Think together** activity will reveal those who are struggling, so when it is time for practice, bring these children together to work with you on the first practice questions. Observe these children carefully, ask questions, encourage them to use concrete models and check that they reach and can demonstrate their understanding.

After-lesson intervention

You might like to use the **Think together** questions to recap the lesson with children who are working behind expectations during assembly time. Teaching Assistants could also work with these children at other convenient points in the school day. Some children may benefit from revisiting work from the same topic in the previous year group. Note also the suggestion for recycling questions from the Textbook and Practice Book with concrete materials on page 28.

The role of practice

Practice plays a pivotal role in the *Power Maths* approach. It takes place in class groups, smaller groups, pairs, and independently, so that children always have the opportunities for thinking as well as the models and support they need to practise meaningfully and with understanding.

Intelligent practice

In *Power Maths*, practice never equates to the simple repetition of a process. Instead we embrace the concept of intelligent practice, in which all children become fluent in maths through varied, frequent and thoughtful practice that deepens and embeds conceptual understanding in a logical, planned sequence. To see the difference, take a look at the following examples.

Traditional practice

- Repetition can be rote – no need for a child to think hard about what they are doing
- Praise may be misplaced
- Does this prove understanding?

Intelligent practice

- Varied methods – concrete, pictorial and abstract
- Equation expressed in different ways, requiring thought and understanding
- Constructive feedback

All practice questions are designed to move children on and reveal misconceptions.

Simple, logical steps build onto earlier learning.

C-P-A runs throughout – different ways of modelling and understanding the same concept.

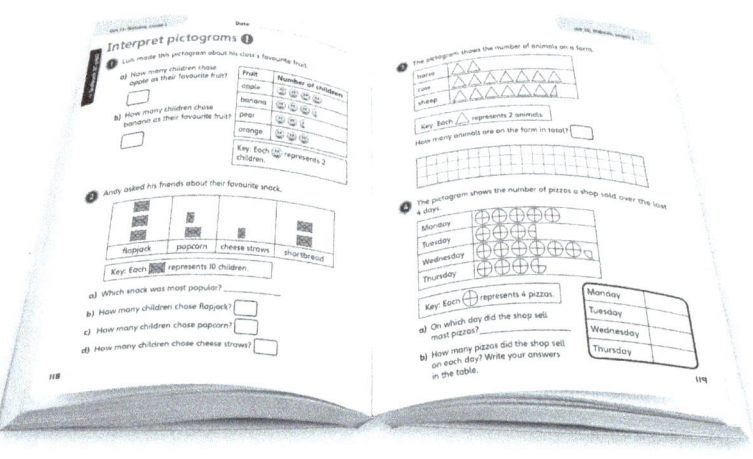

Conceptual variation – children work on different representations of the same maths concept.

Friendly characters offer support and encourage children to try different approaches.

A carefully designed progression

The Practice Books provide just the right amount of intelligent practice for children to complete independently in the final sections of each lesson. It is really important that all children are exposed to the practice questions, and that children are not directed to complete different sections. That is because each question is different and has been designed to challenge children to think about the maths they are doing. The questions become more challenging so children grasping concepts more quickly will start to slow down as they progress. Meanwhile, you have the chance to circulate and spot any misconceptions before they become barriers to further learning.

Homework and the role of parents and carers

While *Power Maths* does not prescribe any particular homework structure, we acknowledge the potential value of practice at home. For example, practising fluency in key facts, such as number bonds and times-tables, is an ideal homework task. You can share the Individual Practice Games for homework (see page 6), or parents and carers could work through uncompleted Practice Book questions with children at either primary stage.

However, it is important to recognise that many parents and carers may themselves lack confidence in maths, and few, if any, will be familiar with mastery methods. A Parents' and Carers' evening that helps them understand the basics of mindsets, mastery and mathematical language is a great way to ensure that children benefit from their homework. It could be a fun opportunity for children to teach their families that everyone can do maths!

Structures and representations

Unlike most other subjects, maths comprises a wide array of abstract concepts – and that is why children and adults so often find it difficult. By taking a concrete-pictorial-abstract (C-P-A) approach, *Power Maths* allows children to tackle concepts in a tangible and more comfortable way.

Non-linear stages

Concrete

Replacing the traditional approach of a teacher working through a problem in front of the class, the concrete stage introduces real objects that children can use to 'do' the maths – any familiar object that a child can manipulate and move to help bring the maths to life. It is important to appreciate, however, that children must always understand the link between models and the objects they represent. For example, children need to first understand that three cakes could be represented by three pretend cakes, and then by three counters or bricks. Frequent practice helps consolidate this essential insight. Although they can be used at any time, good concrete models are an essential first step in understanding.

Pictorial

This stage uses pictorial representations of objects to let children 'see' what particular maths problems look like. It helps them make connections between the concrete and pictorial representations and the abstract maths concept. Children can also create or view a pictorial representation together, enabling discussion and comparisons. The *Power Maths* teaching tools are fantastic for this learning stage, and bar modelling is invaluable for problem solving throughout the primary curriculum.

Abstract

Our ultimate goal is for children to understand abstract mathematical concepts, symbols and notation and of course, some children will reach this stage far more quickly than others. To work with abstract concepts, a child must be comfortable with the meaning of and relationships between concrete, pictorial and abstract models and representations. The C-P-A approach is not linear, and children may need different types of models at different times. However, when a child demonstrates with concrete models and pictorial representations that they have grasped a concept, we can be confident that they are ready to explore or model it with abstract symbols such as numbers and notation.

Use at any time and with any age to support understanding

Variation helps visualisation

Children find it much easier to visualise and grasp concepts if they see them presented in a number of ways, so be prepared to offer and encourage many different representations.

For example, the number six could be represented in various ways:

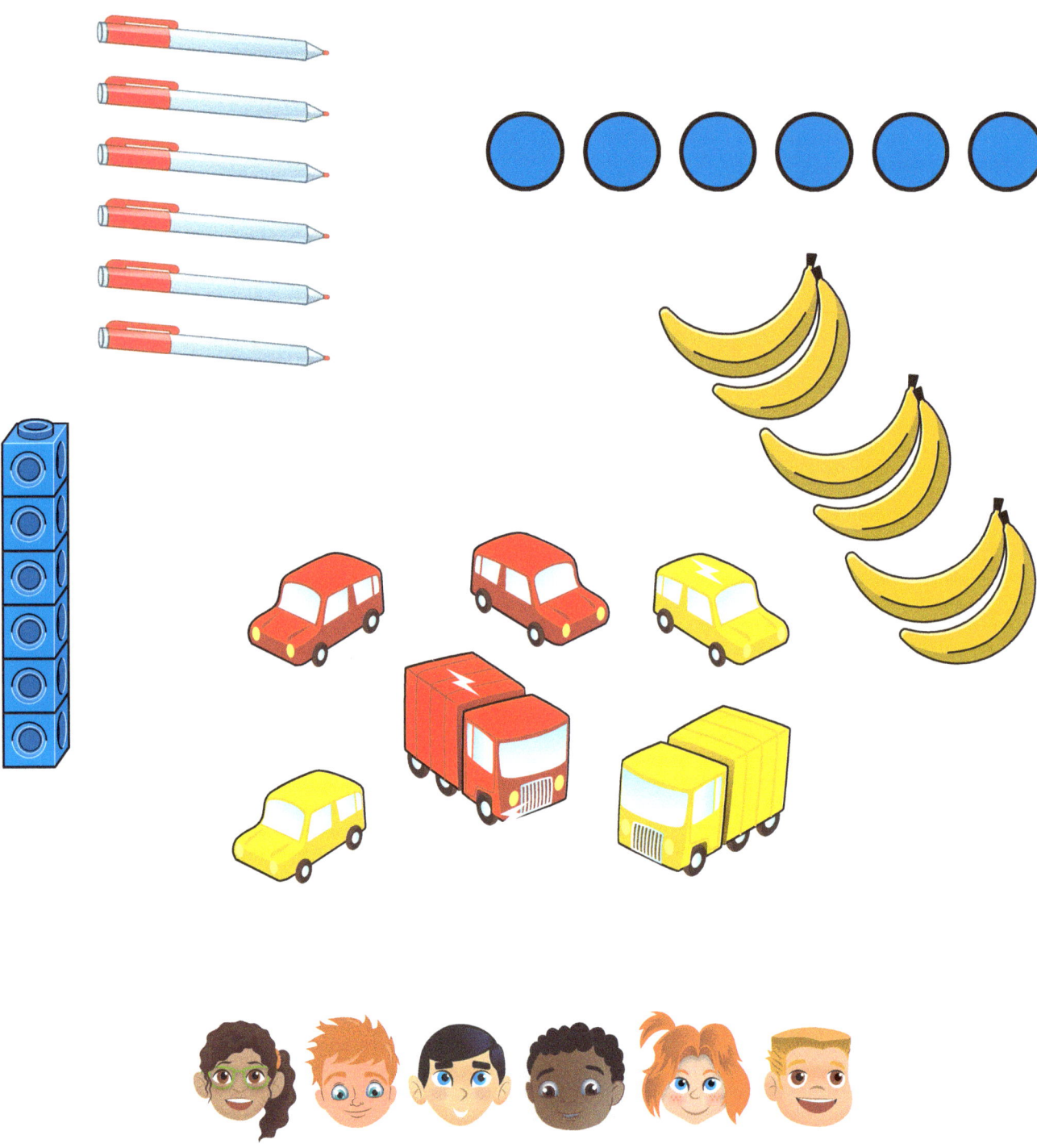

Practical aspects of *Power Maths*

One of the key underlying elements of *Power Maths* is its practical approach, allowing you to make maths real and relevant to your children, no matter their age.

Manipulatives are essential resources for both key stages and *Power Maths* encourages teachers to use these at every opportunity, and to continue the Concrete-Pictorial-Abstract approach right through to Year 6.

The Textbooks and Teacher Guides include lots of opportunities for teaching in a practical way to show children what maths means in real life.

Discover and Share

The **Discover** and **Share** sections of the Textbook give you scope to turn a real-life scenario into a practical and hands-on section of the lesson. Use these sections as inspiration to get active in the classroom. Where appropriate, use the **Discover** contexts as a springboard for your own examples that have particular resonance for your children – and allow them to get their hands dirty trying out the mathematics for themselves.

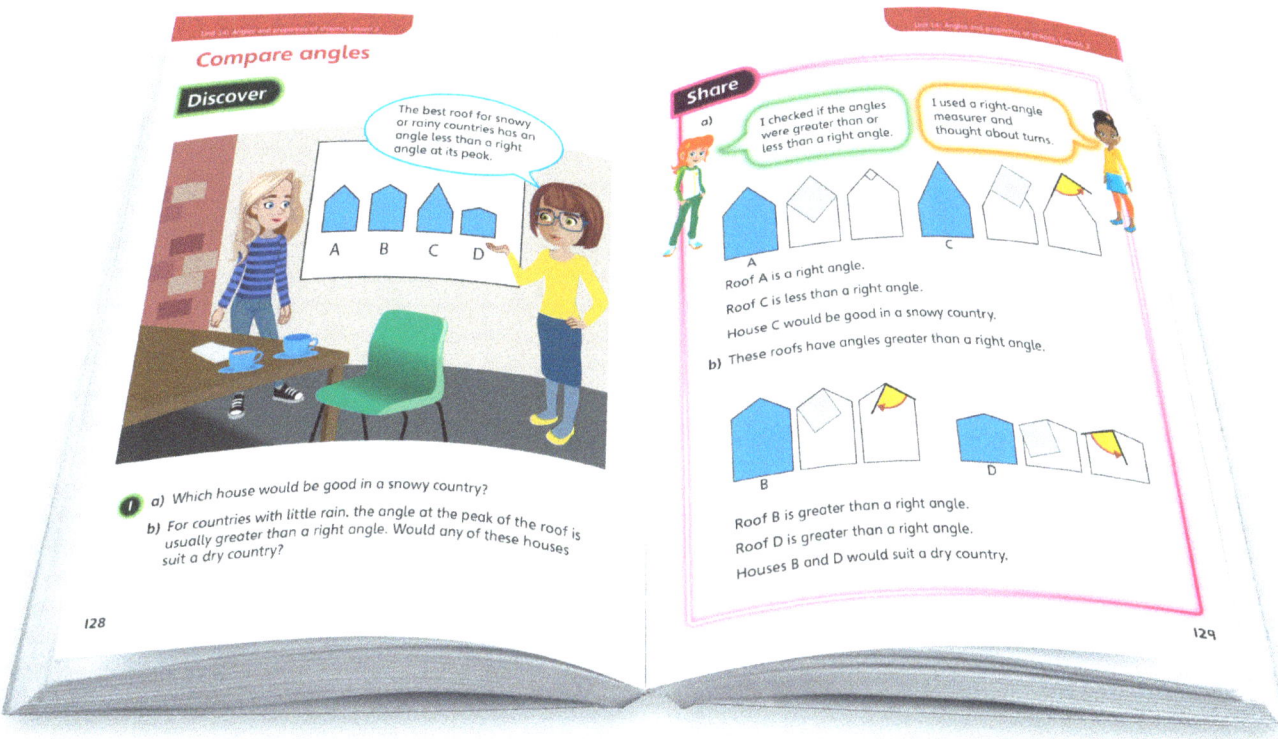

Unit videos

Every term has one unit video which incorporates real-life classroom sequences.

These videos show you how the reasoning behind mathematics can be carried out in a practical manner by showing real children using various concrete and pictorial methods to come to the solution. You can see how using these practical models, such as part-whole and bar models, helps them to find and articulate their answer.

Mastery tips

Mastery Experts give anecdotal advice on where they have used hands-on and real-life elements to inspire their children.

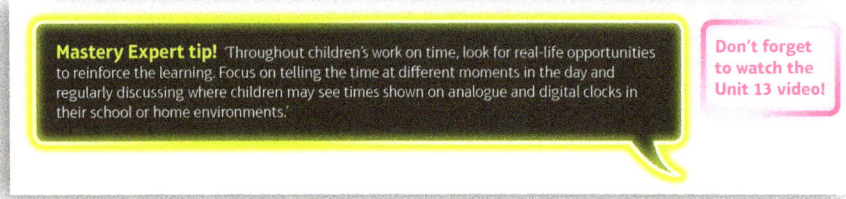

Concrete-Pictorial-Abstract (C-P-A) approach

Each **Share** section uses various methods to explain an answer, helping children to access abstract concepts by using concrete tools, such as counters. Remember, this isn't a linear process, so even children who appear confident using the more abstract method can deepen their knowledge by exploring the concrete representations. Encourage children to use all three methods to really solidify their understanding of a concept.

Pictorial representation – drawing the problem in a logical way that helps children visualise the maths

Concrete representation – using manipulatives to represent the problem. Encourage children to physically use resources to explore the maths.

Abstract representation – using words and calculations to represent the problem.

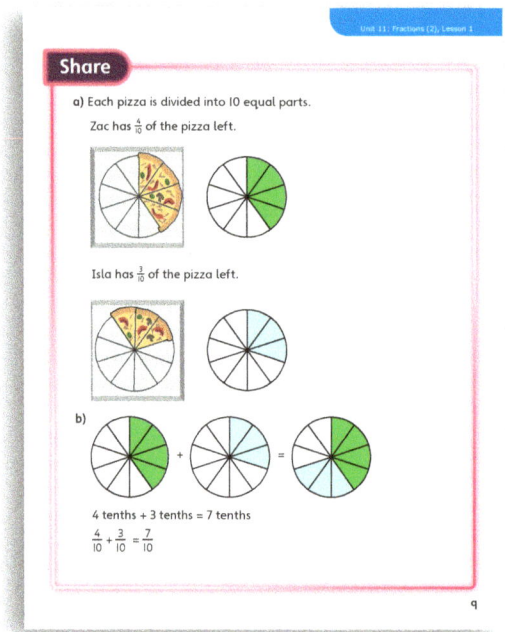

Practical tips

Every lesson suggests how to draw out the practical side of the **Discover** context.

You'll find these in the **Discover** section of the Teacher Guide for each lesson.

> **PRACTICAL TIPS** Ask children to count aloud in 100s when finding the total number of bricks in each part. For support, children could represent the number of bricks using base 10 equipment.

Resources

Every lesson lists the practical resources you will need or might want to use. There is also a summary of all of the resources used throughout the term on page 40 to help you be prepared.

> **RESOURCES**
> **Mandatory:** base 10 equipment
> **Optional:** place value counters, place value grids

Working with children below age-related expectation

This section offers advice on using *Power Maths* with children who are significantly behind age-related expectation. Teacher judgement will be crucial in terms of where and why children are struggling, and in choosing the right approach. The suggestions can of course be adapted for children with special educational needs, depending on the specific details of those needs.

General approaches to support children who are struggling

Keeping the pace manageable
Remember, you have more teaching days than *Power Maths* lessons so you can cover a lesson over more than one day, and revisit key learning, to ensure all children are ready to move on. You can use the + and − buttons to adjust the time for each unit in the online planning. The NCETM's Ready-to-Progress criteria can be used to help determine what should be highest priority.

Same-day intervention
You could go over the Textbook pages or revisit the previous year's work if necessary (see Addressing gaps). Remember that same-day intervention can be within the lesson, as well as afterwards (see page 29). As children start their independent practice, you can work with those who found the first part of the lesson difficult, checking understanding using manipulatives.

Fluency sessions
Fit in as much practice as you can for number bonds and times-tables, etc., at other times of the day. If you can, plan a short 'maths meeting' for this in the afternoon. You might choose to use a Power Up you haven't used already.

Addressing gaps
Use material from the same topic in the previous year to consolidate or address gaps in learning, e.g. Textbook pages and Strengthen activities. The End of unit check will help gauge children's understanding.

Pre-teaching
Find a 5- to 10-minute slot before the lesson to work with the children you feel would benefit. The afternoon before the lesson can work well, because it gives children time to think in between. Recap previous work on the topic (addressing any gaps you're aware of) and do some fluency practice, targeting number facts etc. that will help children access the learning.

Focusing on the key concepts
If children are a long way behind, it can be helpful to take a step back and think about the key concepts for children to engage with, not just the fine detail of the objective for that year group (e.g. addition with a specific number of columns). Bearing that in mind, how could children advance their understanding of the topic?

Providing extra support within the lesson

Support in the Teacher Guide
First of all, use the Strengthen support in the Teacher Guide for guided and independent work in each lesson, and share this with Teaching Assistants, where relevant. As you read through the lesson content and corresponding Teacher Guide pages before the lesson, ask yourself what key idea or nugget of understanding is at the heart of the lesson. If children are struggling, this should help you decide what's essential for all children before they move on.

Annotating pages
You can annotate questions to provide extra scaffolding or hints if you need to, but aim to build up children's ability to access questions independently wherever you can. Children tend to get used to the style of the *Power Maths* questions over time.

Quick recap as lesson starter
The Quick recap for each lesson in the Teacher Guide is an alternative starter activity to the Power Up. You might choose to use this with some or all children if you feel they will need support accessing the main lesson.

Consolidation questions
If you think some children would benefit from additional questions at the same level before moving on, write one or two similar questions on the board. (This shouldn't be at the expense of reasoning and problem-solving opportunities: take longer over the lesson if you need to.)

Hard copy Textbooks
The Textbooks help children focus in more easily on the mathematical representations, read the text more comfortably, and revisit work from a previous lesson that you are building on, as well as giving children ownership of their learning journey. In main lessons, it can work well to use the e-Textbook for **Discover** and give out the books when discussing the methods in the **Share** section.

Reading support
It's important that all children are exposed to problem solving and reasoning questions, which often involve reading. For whole-class work you can read questions together. For independent practice you could consider annotating pages to help children see what the question is asking, and stem sentences to help structure their answer. A general focus on specific mathematical language and vocabulary will help children access the questions. You could consider pairing weaker readers with stronger readers, or read questions as a group if those who need support are on the same table.

Providing extra depth and challenge with *Power Maths*

Just as prescribed in the National Curriculum, the goal of *Power Maths* is never to accelerate through a topic but rather to gain a clear, deep and broad understanding. Here are some suggestions to help ensure all children are appropriately challenged as you work with the resources.

Overall approaches

First of all, remember that the materials are designed to help you keep the class together, allowing all children to master a concept while those who grasp it quickly have time to explore it in more depth. Use the Deepen support in the Teacher Guide (see below) to challenge children who work through the questions quickly. Here are some questions and ideas to encourage breadth and depth during specific parts of the lesson, or at any time (where no part of the lesson sequence is specified):

- **Discover**: 'Can you demonstrate your solution another way?'
- **Share**: Make sure every child is encouraged to give answers and engage with the discussion, not just the most confident.
- **Think together**: 'Can you model your answers using concrete materials? Can you explain your solution to a partner?'
- Practice: Allow all children to work through the full set of questions, so that they benefit from the logical sequence.
- **Reflect**: 'Is there another way of working out the answer? And another way?'
  'Have you found all the solutions?'
  'Is that always true?'
  'What's different between this question and that question? And what's the same?'

Note that the **Challenge** questions are designed so that all children can access and attempt them, if they have worked through the steps leading up to them. There may be some children in a given lesson who don't manage to do the **Challenge**, but it is not supposed to be a distinct task for a subset of the class. When you look through the lesson materials before teaching, think about what each question is specifically asking, and compare this with the key learning point for the lesson. This will help you decide which questions you feel it's essential for all children to answer, before moving on. You can at least aim for all children to try the **Challenge**!

Deepen activities and support

The Teacher Guide provides valuable support for each stage of the lesson. This includes Deepen tips for the guided and independent practice sections, which will help you provide extra stretch and challenge within your lesson, without having to organise additional tasks. If you have a Teaching Assistant, they can also make use of this advice. There are also suggestions for the lesson as a whole in the 'Going Deeper' section on the first page of the Teacher Guide section for that lesson. Every class is different, so you can always go a bit further in the direction indicated, if appropriate, and build on the suggestions given.

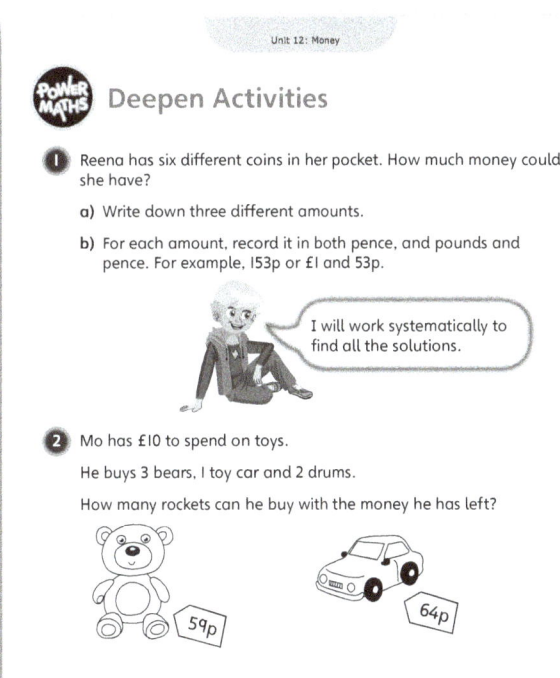

There is a Deepen activity for each unit. These are designed to follow on from the End of unit check, stretching children who have a firm understanding of the key learning from the unit. Children can work on them independently, which makes it easier for the teacher to facilitate the Strengthen activity for children who need extra support. Deepen activities could also be introduced earlier in the unit if the necessary work has been covered. The Deepen activities are on *ActiveLearn* on the Planning page for each unit, and also on the Resources page).

Using the questions flexibly to provide extra challenge

Sometimes you may want to write an extra question on the board or provide this on paper. You can usually do this by tweaking the lesson materials. The questions are designed to form a carefully structured sequence that builds understanding step by step, but, with careful thought about the purpose of each question, you can use the materials flexibly where you need to. Sometimes you might feel that children would benefit from another similar question for consolidation before moving on to the next one, or you might feel that they would benefit from a harder example in the same style. It should be quick and easy to generate 'more of the same' type questions where this is the case.

When you see a question like this one (from Unit 3, Lesson 3), it's easy to make harder examples to do afterwards if you need them. What if the 7 was also blotted out – how many possibilities would there be for the ones digits? For a trickier example, if you set up a similar addition with a tens digit blotted out (and the tens part of the answer given), children would have to factor in the exchanged 10 in order to work out what was missing.

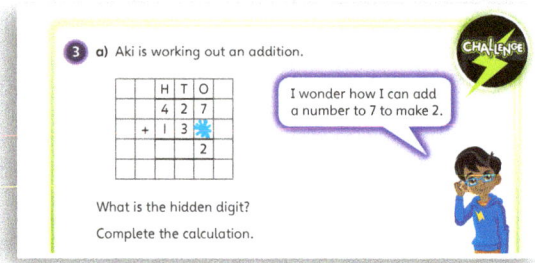

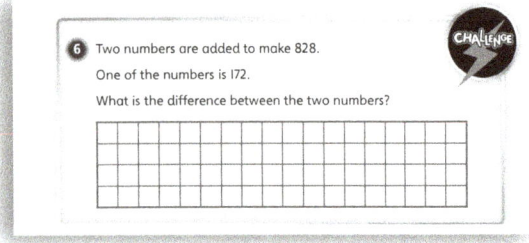

For this example (from Unit 3, Lesson 6), you could ask children to make up their own question(s) for a partner to solve. They could even make up questions to solve themselves! (In fact, for any of these examples you could ask early finishers to create their own question for a partner.)

Here's an example (from Unit 3, Lesson 12) where some of the combinations in the picture feature as questions in the lesson, but others don't. Clearly there are plenty of multi-step problems you could ask using the same information. Children could choose what to buy for their family, or you could tell them about your family and the bikes, helmets and lights you would need. You could also give them a budget to work out the change from buying a list of equipment.

Besides creating additional questions, you should be able to find a question in the lesson that you can adapt into a game or open-ended investigation, if this helps to keep everyone engaged. It could simply be that, instead of answering 5 × 5 etc. on the page, they could build a robot with 5 lots of 5 cubes.

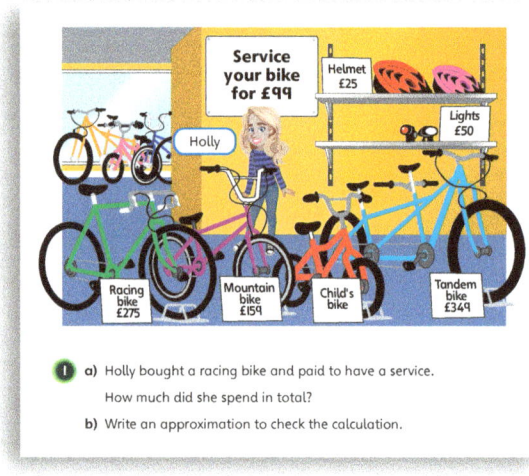

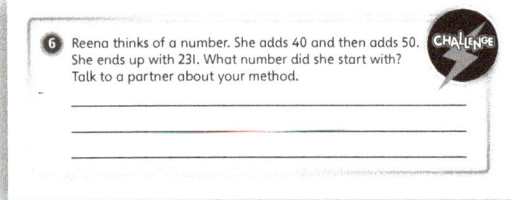

With a question like this (Unit 2, Lesson 9), children could play a game where they say the additions and the end number, and their partner has to work out the start number.

See the bullets above for some general ideas that will help with 'opening out' questions in the books, e.g. 'Can you find all the solutions?' type questions.

Other suggestions

Another way of stretching children is through mixed ability pairs, or via other opportunities for children to explain their understanding in their own way. This is a good way of encouraging children to go deeper into the learning, rather than, for instance, tackling questions that are computationally more challenging but conceptually equivalent in level.

Using *Power Maths* with mixed age classes

Overall approaches

There are many variables between schools that would make it inadvisable to recommend a one-size-fits-all approach to mixed age teaching with *Power Maths*. These include how year groups are merged, availability of Teaching Assistants, experience and preference of teaching staff, range in pupil attainment across years, classroom space and layout, level of flexibility around timetables, and overall organisational structure (whether the school is part of a trust).

Some schools will find it best to timetable separate maths lessons for the different year groups. Others will aim to teach the class together as much as possible using the mixed age planning support on *ActiveLearn* (see the lesson exemplars for ways of organising lessons with strong/medium/weak correlation between year groups). There will also be ways of adapting these general approaches. For example, offset lessons where Year A start their lesson with the teacher, while Year B work independently on the practice from the previous lesson, and then start the next lesson with the teacher while Year A work independently; or teachers may choose to base their provision around the lesson from one year group and tweak the content up/down for the other group.

Key strategies for mixed age teaching

The mixed age teaching webinar on *ActiveLearn* provides advice on all aspects of mixed age teaching, including more detail on the ideas below.

Developing independence over time
Investing time in building up children's independence will pay off in the medium term.

Clear rationale
If someone asked, 'Why did you teach both Unit 3 and 4 in the same lesson/separate lessons?', what would your answer be?

Designing a lesson
1. Identify the core learning for each group
2. Identify any number skills necessary to access the core
3. Consider the flow of concepts and how one core leads to the other

Challenging all children
The questions are designed to build understanding step by step, but with careful thought about the purpose of each question you can tweak them to increase the challenge.

Multiple years combined
With more than two years together, teachers will inevitably need to use the resources flexibly if delivering a single lesson.

Enjoy the positives!

Comparison deepens understanding and there will be lots of opportunities for children, as well as misconceptions to explore. There is also in-built pre-teaching and the chance to build up a concept from its foundations. For teachers there is double the material to draw on! Mixed age teachers require a strong understanding of the progression of ideas across year groups, which is highly valuable for all teachers. Also, it is necessary to engage deeply with the lesson to see how to use the materials flexibly – this is recommended for all teachers and will help you bring your lesson to life!

List of practical resources

Year 3C Mandatory resources

Resource	Lesson
3D shapes (for children to handle – cuboid, cube, prism, pyramid, sphere, cone, cylinder)	**Unit 14** Lesson 8
Analogue clock manipulatives	**Unit 13** Lessons 7, 9, 10
Analogue clock tool	**Unit 13** Lessons 2, 3, 5
Bar models	**Unit 11** Lessons 5, 6
Calendars	**Unit 13** Lesson 6
Clock face with Roman numerals	**Unit 13** Lesson 1
Clock faces (laminated pictures of)	**Unit 13** Lesson 7
Clocks (analogue and digital)	**Unit 13** Lesson 4
Clocks (analogue and digital, showing hours, minutes and seconds)	**Unit 13** Lesson 11
Clocks (digital)	**Unit 13** Lessons 7, 9, 10
Clocks (digital, showing am and pm)	**Unit 13** Lesson 5
Construction materials	**Unit 14** Lesson 9
Counters	**Unit 11** Lessons 5, 6
Dice	**Unit 13** Lesson 4
Flashcards (showing analogue clocks and written times)	**Unit 13** Lesson 2
Fraction strips (or fraction circles)	**Unit 11** Lessons 1, 2
Geoboards (and bands; or square dotted paper to represent geoboards)	**Unit 14** Lesson 3
Key language flashcards ('o'clock', 'half past', 'quarter past', 'quarter to')	**Unit 13** Lesson 2
Multilink cubes	**Unit 14** Lesson 9
Number lines	**Unit 11** Lessons 1, 2
Paper (for folding)	**Unit 14** Lesson 6
Paper (landscape A4, cut into 10 cm strips)	**Unit 14** Lesson 4
Paper (square)	**Unit 14** Lesson 6
Paper (square dotted)	**Unit 14** Lesson 6
Paper (squared)	**Unit 14** Lessons 4, 5
Plastic or paper coins and notes	**Unit 12** Lessons 1, 2, 3, 4, 5
Right-angle measurer (folded paper, ruler, 2D square or rectangle)	**Unit 14** Lesson 2
Rulers	**Unit 14** Lessons 4, 6 **Unit 15** Lessons 3, 4
Scissors	**Unit 14** Lesson 4
Sticks or pencils (of equal length)	**Unit 14** Lesson 7

Year 3C Optional resources

Resource	Lesson
10 cm squares (pre-cut)	**Unit 14** Lesson 4
2D shapes (plastic or wooden)	**Unit 14** Lesson 7
3D shapes (represented as solid shapes and as wireframe models)	**Unit 14** Lesson 8
3D shapes (that open out into nets)	**Unit 14** Lesson 8
Analogue clock manipulatives	**Unit 13** Lesson 7
Bar models	**Unit 11** Lessons 4, 8
Base 10 equipment or rectangular shapes	**Unit 14** Lesson 7
Capacity measuring equipment	**Unit 11** Lesson 8
Cardboard boxes (range of different proportions)	**Unit 14** Lesson 8
Circles (see Practice Book, question 5) with 3 to 10 dots	**Unit 14** Lesson 6
Clock faces (laminated pictures of)	**Unit 13** Lesson 7
Clock faces (with movable hands)	**Unit 14** Lesson 3
Clocks (printed copies of)	**Unit 13** Lesson 1
Coloured rods	**Unit 11** Lessons 3, 4, 8
Connecting cubes	**Unit 15** Lessons 3, 4
Construction materials (snap-together)	**Unit 14** Lesson 9
Counters	**Unit 11** Lessons 4, 7
Cubes	**Unit 15** Lesson 6
Cubes or counters	**Unit 15** Lessons 1, 2
Diagrams of eight-point compass (or chalks so this can be drawn on the playground)	**Unit 14** Lesson 1
Flashcards (showing analogue clock times)	**Unit 13** Lesson 3
Flashcards (showing analogue clocks and written times)	**Unit 13** Lessons 9, 10
Flashcards (showing written times)	**Unit 13** Lesson 5
Hinged rods (pair of)	**Unit 14** Lesson 3
Interlocking cubes	**Unit 11** Lessons 1, 2, 5
Lolly sticks	**Unit 14** Lesson 7
Marbles	**Unit 14** Lesson 5
Matchsticks	**Unit 14** Lesson 7
Mini-whiteboards	**Unit 14** Lesson 7
Mirrors	**Unit 14** Lesson 5
Modelling clay	**Unit 14** Lesson 9
Modern art (examples that contain parallel and perpendicular lines, e.g. Piet Mondrian)	**Unit 14** Lesson 6
Multiplication square	**Unit 11** Lesson 6
Number cards (from 1 to 12)	**Unit 13** Lesson 3
Number lines	**Unit 13** Lesson 8 **Unit 15** Lessons 1, 2, 3, 4, 5
Object (to represent the rover)	**Unit 14** Lesson 1
Paper (large pieces of)	**Unit 13** Lesson 3
Paper (squared)	**Unit 14** Lesson 3 **Unit 15** Lesson 6
Paper circles	**Unit 11** Lesson 2
Paper circles (paper plates)	**Unit 11** Lesson 1
Paper or card (square pieces of)	**Unit 14** Lesson 2
Part-whole models	**Unit 11** Lessons 4, 5, 8
PE equipment	**Unit 14** Lesson 5
Pictures (of a flat, straight horizon)	**Unit 14** Lesson 5
Pictures (of activities for each hour in the day)	**Unit 13** Lesson 7
Pictures (of sunset and sunrise)	**Unit 13** Lesson 7
Pipe cleaners	**Unit 14** Lesson 3
Place value grids	**Unit 12** Lesson 5

Year 3C Optional resources – *continued*

Resource	Lesson
Plastic (or wood) squares and rectangles	**Unit 14** Lesson 4
Plumb line	**Unit 14** Lesson 5
Rulers	**Unit 15** Lessons 5, 7
Rulers (or folded card to make angles: two)	**Unit 14** Lesson 3
Sorting table	**Unit 14** Lesson 7
Sticks and marshmallows	**Unit 14** Lesson 9
Stopwatch	**Unit 13** Lessons 11, 12
Sweets or items to replicate the Discover context	**Unit 11** Lesson 5
Toy figures	**Unit 14** Lesson 1
Toys (soft toys for role play)	**Unit 13** Lesson 7
Wireframe models	**Unit 14** Lesson 9

Getting started with *Power Maths*

As you prepare to put *Power Maths* into action, you might find the tips and advice below helpful.

STEP 1: Train up!

A practical, up-front full day professional development course will give you and your team a brilliant head-start as you begin your *Power Maths* journey. You will learn more about the ethos, how it works and why.

STEP 2: Check out the progression

Take a look at the yearly and termly overviews. Next take a look at the unit overview for the unit you are about to teach in your Teacher Guide, remembering that you can match your lessons and pacing to match your class.

STEP 3: Explore the context

Take a little time to look at the context for this unit: what are the implications for the unit ahead? (Think about key language, common misunderstandings and intervention strategies, for example.) If you have the online subscription, don't forget to watch the corresponding unit video.

STEP 4: Prepare for your first lesson

Familiarise yourself with the objectives, essential questions to ask and the resources you will need. The Teacher Guide offers tips, ideas and guidance on individual lessons to help you anticipate children's misconceptions and challenge those who are ready to think more deeply.

STEP 5: Teach and reflect

Deliver your lesson — and enjoy!

Afterwards, reflect on how it went… Did you cover all five stages? Does the lesson need more time? How could you improve it?

Unit 11
Fractions ②

Mastery Expert tip! 'Children need a rich experience of seeing, touching and creating simple fractions in order to gain a more secure understanding of what a fraction actually is. We created a fraction museum on a whiteboard and each lesson we added examples of our learning with lots of colourful resources and drawings. It really helped my class to embed what they learnt, recall what they already knew and ask questions where they were unsure.'

Don't forget to watch the Unit 11 video!

WHY THIS UNIT IS IMPORTANT

In this unit, children will learn to add and subtract two or more fractions with the same denominator, answering questions in more than one way and comparing the efficiency of each method. They will develop their understanding of solving fraction problems and will learn to solve problems involving fractions of an amount. They will use bar models and other representations to help them to find a unit fraction of an amount and then to find any fraction of an amount. Children will be able to use this knowledge to reason and problem solve – for example, finding the whole if they know a part.

WHERE THIS UNIT FITS

→ Unit 10: Capacity
→ **Unit 11: Fractions (2)**
→ Unit 12: Money

Before they start this unit, it is expected that children:
- understand how to make a whole out of two fractional parts
- understand unit and non-unit fractions
- understand fractions as a number
- can find $\frac{1}{2}$ and $\frac{1}{4}$ of an amount
- can use a bar model to represent problems
- understand the concept of equal parts.

ASSESSING MASTERY

Children who have mastered this unit will be able to add and subtract two or more fractions with the same denominator. They will be able to partition the whole and use their understanding of the whole to solve problems such as $1 - \frac{2}{5} = \square$. Children will be able to find a unit fraction and a non-unit fraction of an amount, using a bar model to support their understanding. They will be able to solve problems such as finding the whole when they are given a part.

COMMON MISCONCEPTIONS	STRENGTHENING UNDERSTANDING	GOING DEEPER
When adding and subtracting fractions, children may add or subtract the denominators as well as the numerators.	Provide opportunities for children to use resources to model fractions, for example, using fraction rods, beads and fraction strips.	Encourage children to make a fraction wall from scratch to deepen their understanding. Challenge children to find different ways to answer the same question. Often, more able children (frequently those who finish quicker) are less inclined to look for alternative options, but emphasise that finding the answer is not as important as the *way* in which they find the answer.
When solving problems such as $\frac{1}{3}$ of $\square$ = 6, children may find $\frac{1}{3}$ of 6 as opposed to multiplying 6 × 3.	Ask children to draw a bar model to support their understanding. They should realise that the bar model is divided into 3 parts. If they know that 1 part is equal to 6, then they know that they need to multiply by 3 to find the whole.	
Children who find it difficult to solve problems may not know where to start and so may guess which method to use.	Offer real-life examples of problem-solving situations. Encourage children to role-play problems and sketch models (such as bar models) to help them make sense of what they need to do.	

Unit 11: Fractions (2)

UNIT STARTER PAGES

Use these pages to introduce the unit focus to children. You can use the characters to explore different ways of working too.

STRUCTURES AND REPRESENTATIONS

Bar model: This representation supports children's reasoning and problem solving with fractions. Bar models will help children understand what calculation they may need to perform and will help them to visualise concepts.

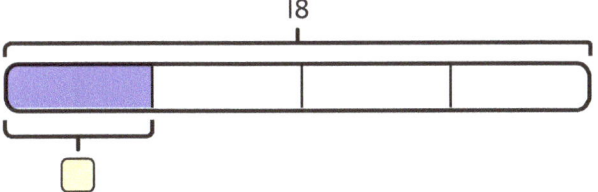

Fraction strip or fraction circle: This is a powerful representation that allows children to organise the information they are given visually, and understand how it should be manipulated in order to find the solution to a problem. It can be used alone, or with a number line to enhance understanding.

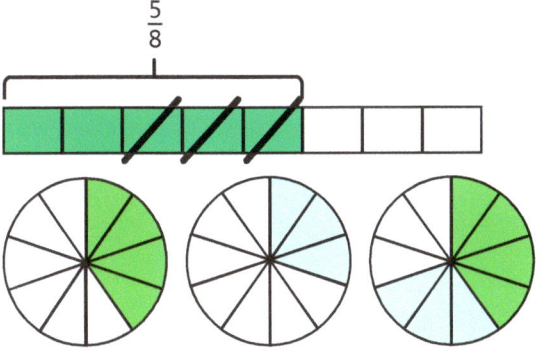

KEY LANGUAGE

There is some key language that children will need to know as part of the learning in this unit.

→ whole, parts, whole parts, equal parts, set of objects, fraction, unit fraction, non-unit fraction, denominator, numerator,

→ partition, split, share, count on, count back, compare, measure, calculate, method

→ whole number, add, subtract, difference, multiply, divide, equal to, greater than (>), less than (<)

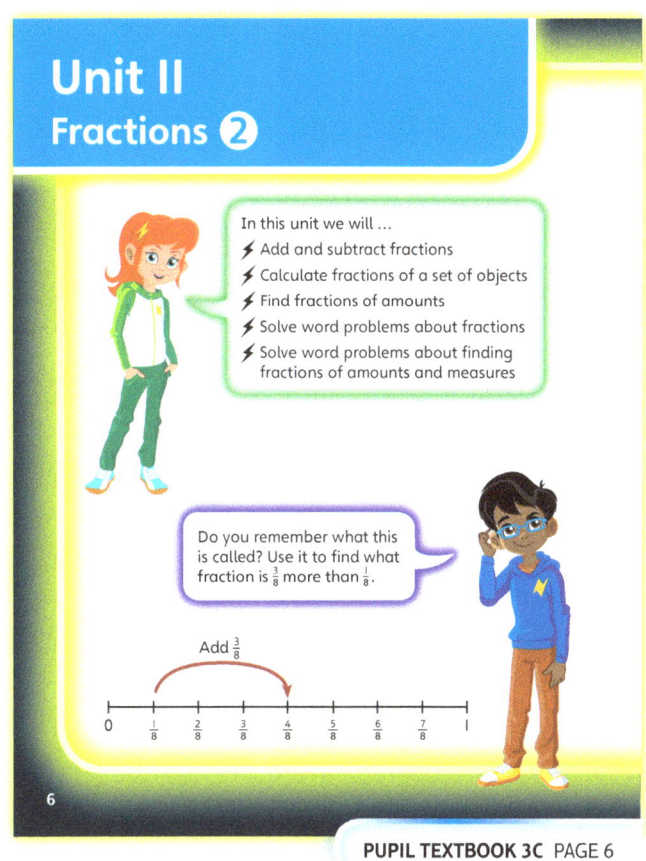

PUPIL TEXTBOOK 3C PAGE 6

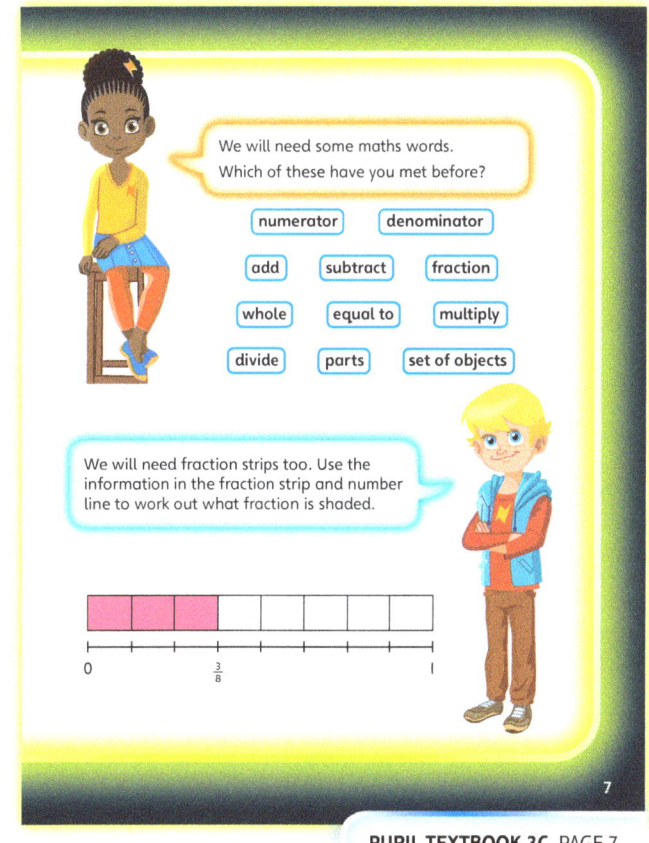

PUPIL TEXTBOOK 3C PAGE 7

Unit 11: Fractions (2), Lesson 1

Add fractions

Learning focus
In this lesson, children will add two or more fractions with the same denominator.

Before you teach
- Are children confident explaining what the numerator and denominator in a fraction show?
- Are children confident describing one whole as a fraction?
- Are children confident counting in different fraction steps?

NATIONAL CURRICULUM LINKS
Year 3 Number – fractions
Add and subtract fractions with the same denominator within one whole [for example, $\frac{5}{7} + \frac{1}{7} = \frac{6}{7}$].

ASSESSING MASTERY
Children will count fraction steps of a constant size, using their understanding of fractions as numbers. This will help them to make sense of adding fractions with the same denominator. They can find pairs of fractions that total one.

COMMON MISCONCEPTIONS
Children may try to add fractions by adding both the numerators and the denominators (for example, $\frac{2}{5} + \frac{1}{5} = \frac{3}{10}$). Ask:
- How many parts has the whole been split into? If we start with $\frac{2}{5}$ and count on another $\frac{1}{5}$, how many do we have now? Has the denominator stayed the same size?

Children may be confused by a whole being made of different fractions. Show them a 'whole', and then split it into fractions. Ask:
- What does the whole look like? How many equal pieces has it been split into? If the whole has been split into 6 equal pieces, it is represented as $\frac{6}{6}$ – what if it is split into 5 pieces instead?

STRENGTHENING UNDERSTANDING
Show fifths on a number line or fraction wall. Practise counting in fifths up to one whole. Provide children with a fraction strip split into 5 equal parts. Ask them to count along it in fifths. Practise addition calculations, such as: 'Start at $\frac{2}{5}$ and count on one more fifth.' Explain that this is written as $\frac{2}{5} + \frac{1}{5} = \frac{3}{5}$.

GOING DEEPER
Ask questions that require children to count fractions within the whole, for example: I started at $\frac{3}{8}$ and counted on $\frac{2}{8}$. Where did I land? I started at $\frac{1}{8}$ and landed at $\frac{6}{8}$. How many eighths have I counted? I counted on $\frac{5}{8}$ and landed at $\frac{6}{8}$. Where did I start? Show how this is adding fractions, by recording as $\frac{3}{8} + \frac{2}{8} = \frac{5}{8}$ or $\frac{1}{8} + \frac{5}{8} = \frac{6}{8}$.

Provide different images of a 'whole'. For instance, when the denominator is 2, a whole can be represented as $\frac{2}{2}$. When the denominator is 5, a whole can be represented as $\frac{5}{5}$.

KEY LANGUAGE
In lesson: fraction, add, fraction strip, number line, calculation

Other language to be used by the teacher: numerator, denominator, unit fraction, non-unit fraction, whole, subtract

STRUCTURES AND REPRESENTATIONS
Bar model, number line, fraction 'pizza'

RESOURCES
Mandatory: fraction strips or fraction circles, number lines
Optional: paper circles (paper plates), interlocking cubes

 In the eTextbook of this lesson, you will find interactive links to a selection of teaching tools.

Quick recap
On the board, draw a circle or bar that is divided into 6 equal parts. Ask children to shade $\frac{1}{6}$. Then ask children to shade in $\frac{5}{6}$ of the circle or bar. Repeat for circles and bars divided into other parts. This will help children understand how to represent simple fractions.

Unit 11: Fractions (2), Lesson 1

Discover

WAYS OF WORKING Pair work

ASK

- Question 1 a): *How many parts has each pizza been cut into? How many parts of each pizza are left?*
- Question 1 a): *What fraction of the pizza is left in the first box? And in the second box?*
- Question 1 b): *How can you add the two fractions together? When you add the fractions together, what is it that you do?*

IN FOCUS This picture shows fractions in a real-life context. Recap the misconception of adding the denominators, by asking children: *How many parts has each pizza been cut into? Does that number change if you eat a slice of the pizza?*

PRACTICAL TIPS On each table, provide a fraction strip or fraction circle split into 10 equal parts. This will help children to visualise one whole made up of 10 parts. It will also provide them with an opportunity to count along the strip in tenths.

Alternatively, offer a practical example. Make 'pizzas' out of paper plates divided into 10 equal pieces. Cut out 4 slices of pizza. Discuss what fraction is being represented. Add 3 more slices of pizza and ask: *How many pieces do we have now? What fraction is this?*

ANSWERS

Question 1 a): Zac has $\frac{4}{10}$ of the pizza left.

Isla has $\frac{3}{10}$ of the pizza left.

Question 1 b): 4 tenths + 3 tenths = 7 tenths

$\frac{4}{10} + \frac{3}{10} = \frac{7}{10}$

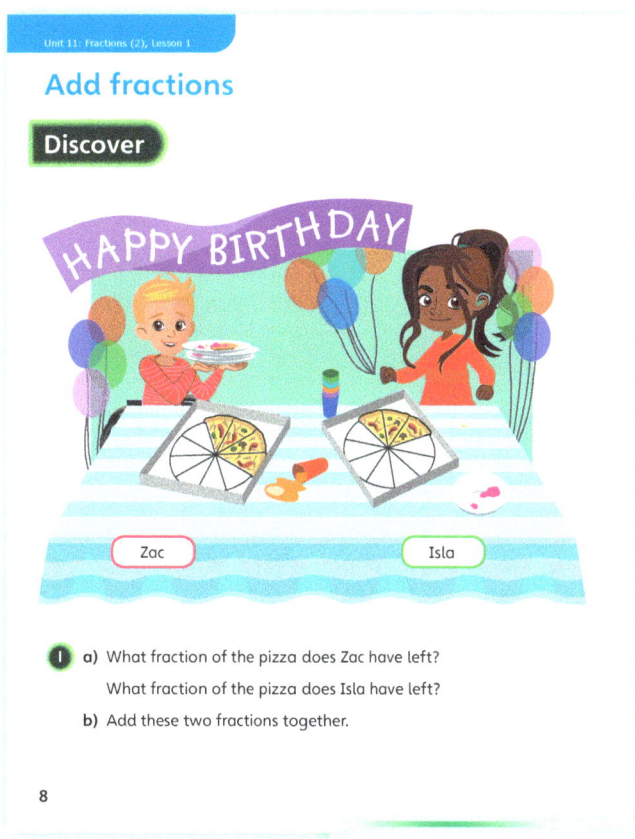

Share

WAYS OF WORKING Whole class teacher led

ASK

- Question 1 a): *What fraction of the pizza is left in each box? How can we represent this as a fraction on the fraction circle?*
- Question 1 b): *How do the diagrams show you how to add these fractions together?*
- Question 1 b): *Why has the numerator changed? Why is the denominator still 10?*

IN FOCUS It is important to give children opportunities to add different fractions with the same numerator. In question 1 b), the diagrams show the pizza that is left in a fraction circle. Children should recognise this from the previous unit. You could also use fraction strips to show counting on, before helping children to transition to using a number line. Ask: *What mistakes might happen when adding fractions?* Question 1 b) shows the two fractions from question 1 a) coming together to show diagrammatically what happens when you add two fractions together.

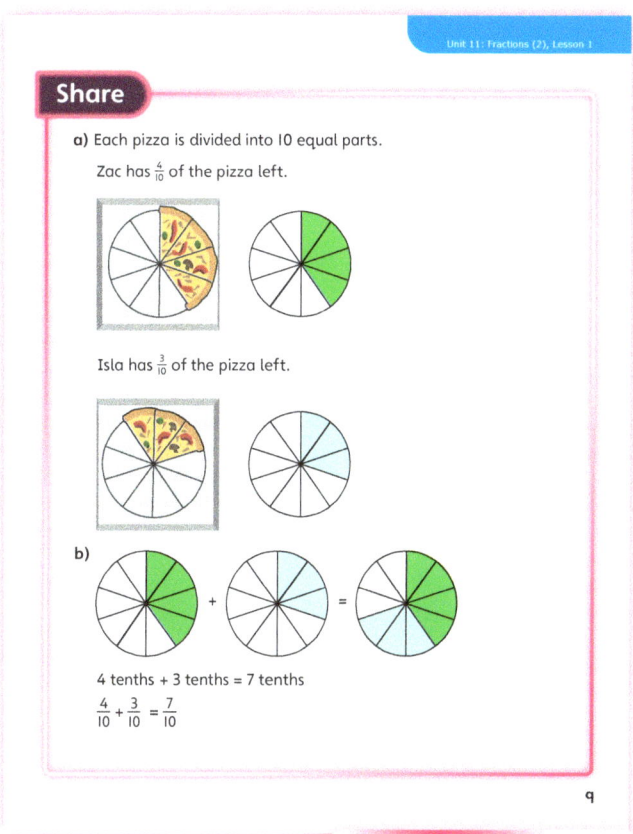

47

Think together

WAYS OF WORKING Whole class teacher led (I do, We do, You do)

ASK
- Question ❶: *What fractions are shown in each diagram?*
- Question ❶: *Will the numerator change? Will the denominator stay the same? How do you know?*
- Question ❷: *What diagrams can you use to check your answers?*

IN FOCUS Questions ❶ and ❷ provide children with practice in adding fractions with the same denominator. Ask them to explain what they are doing as they progress through the questions. Ensure children really understand what happens to the fractions when they are added. Ask: *Why does the denominator not change? What happens if you add the denominators? In your diagram, where are the fractions you are adding? Where is the answer?*

STRENGTHEN In question ❸ b), first help children to understand that because the denominators of all three fractions are 11, children need to find the numerators that will add to make 8. Encourage them to think about their knowledge of number bonds to 8 and to list all the pairs. Ask: *Does the order in which we add the fractions matter?*

DEEPEN In question ❸ b), encourage children to think of as many solutions as they can to $\frac{?}{11} + \frac{?}{11} = \frac{8}{11}$.

ASSESSMENT CHECKPOINT At this point in the lesson, children should be more confident adding fractions with the same denominator. They should understand that fractions are numbers and be confident exploring their properties.

ANSWERS

Question ❶ a): $\frac{4}{8} + \frac{1}{8} = \frac{5}{8}$
Question ❶ b): $\frac{4}{9} + \frac{2}{9} = \frac{6}{9}$
Question ❷ a): $\frac{2}{5} + \frac{1}{5} = \frac{3}{5}$
Question ❷ b): $\frac{2}{11} + \frac{6}{11} = \frac{8}{11}$
Question ❸ a): i) $\frac{3}{5} + \frac{1}{5} = \frac{4}{5}$ iii) $\frac{1}{6} + \frac{3}{6} = \frac{4}{6}$
              ii) $\frac{5}{12} + \frac{1}{12} = \frac{6}{12}$ iv) $\frac{3}{5} + \frac{2}{5} = \frac{5}{5} = 1$
Question ❸ b): There are 7 possible answers:

$\frac{1}{11} + \frac{7}{11} = \frac{8}{11}$
$\frac{2}{11} + \frac{6}{11} = \frac{8}{11}$
$\frac{3}{11} + \frac{5}{11} = \frac{8}{11}$
$\frac{4}{11} + \frac{4}{11} = \frac{8}{11}$
$\frac{5}{11} + \frac{3}{11} = \frac{8}{11}$
$\frac{6}{11} + \frac{2}{11} = \frac{8}{11}$
$\frac{7}{11} + \frac{1}{11} = \frac{8}{11}$

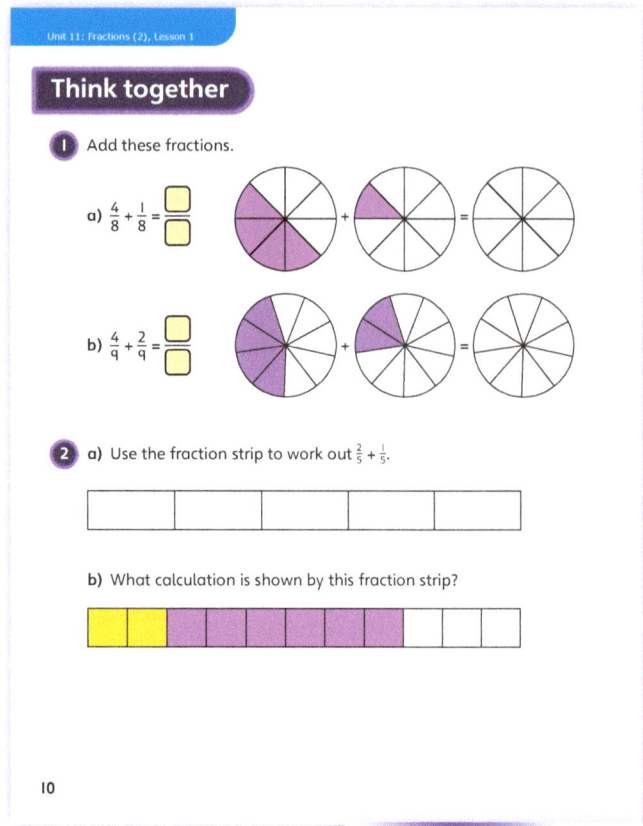

PUPIL TEXTBOOK 3C PAGE 10

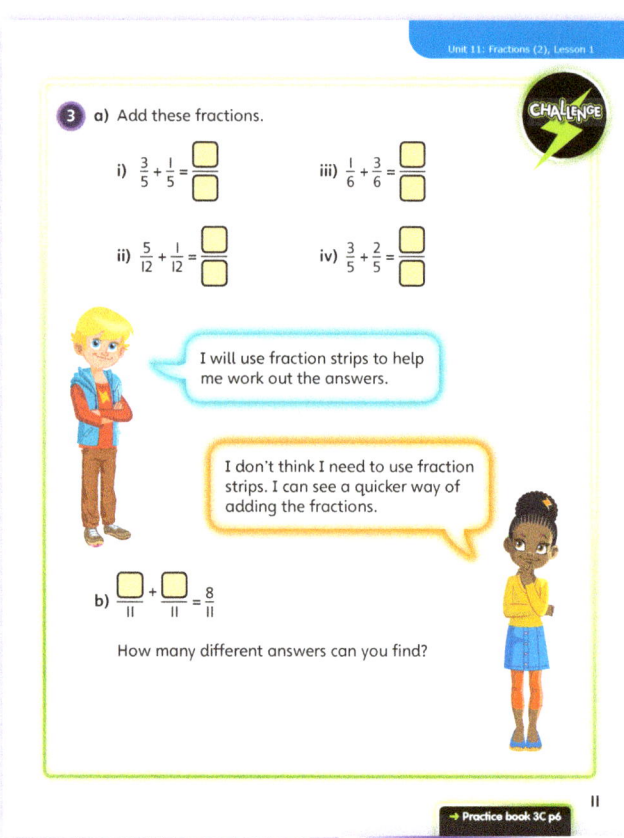

PUPIL TEXTBOOK 3C PAGE 11

Unit 11: Fractions (2), Lesson 1

Practice

WAYS OF WORKING Independent thinking

IN FOCUS Question ❶ provides children with visual images of fractions and requires them to complete the additions. Do children understand how they can shade the shapes to help them find the answers?

Question ❻ requires children to have a secure understanding of what a 'whole' is. Children use number bonds and visual representations to explore pairs of fractions that total 1. If the children know that 5 + 3 = 8, this will help them to recognise that $\frac{5}{8} + \frac{3}{8} = \frac{8}{8}$.

STRENGTHEN In order to support children adding fractions with the same denominator, use fraction circles or fraction strips to show what happens when you add two fractions together. For example, if adding $\frac{3}{10}$ and $\frac{5}{10}$, ask children to shade $\frac{3}{10}$ first and then $\frac{5}{10}$ to show the answer is $\frac{8}{10}$.

DEEPEN You could deepen question ❺ by asking children to prove their ideas using diagrams. Ask: *How many ways can you demonstrate the answer? Explain how you know you have found them all.*

ASSESSMENT CHECKPOINT Children should be confident in understanding how to add fractions. They should understand that they can show calculations with fractions on a number line, in the same way they do with whole numbers.

ANSWERS Answers for the **Practice** part of the lesson can be found in the *Power Maths* online subscription.

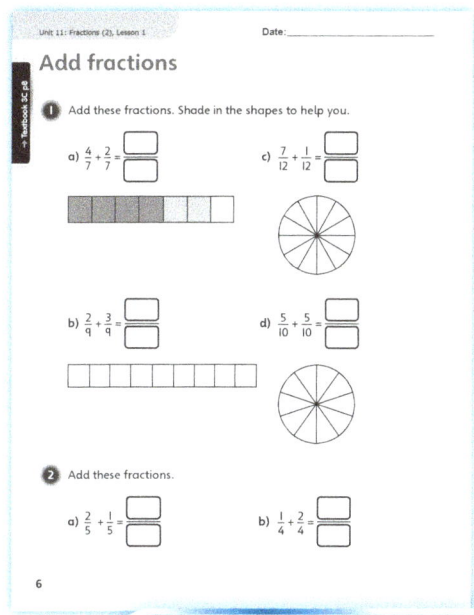

PUPIL PRACTICE BOOK 3C PAGE 6

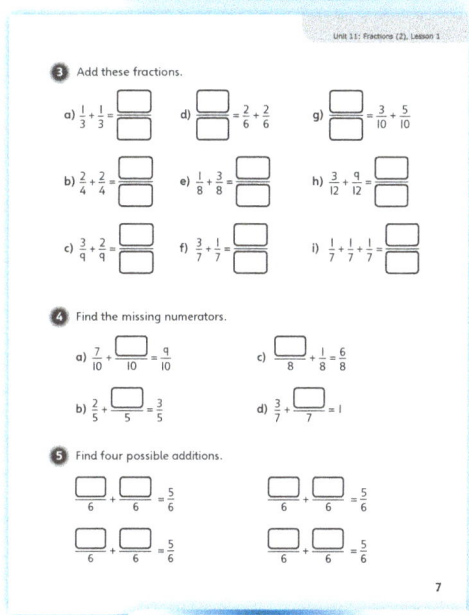

PUPIL PRACTICE BOOK 3C PAGE 7

Reflect

WAYS OF WORKING Independent thinking

IN FOCUS Give children time to explain who they think is correct and why. Once they have recorded their thinking, ask them to share their ideas with a partner. Encourage them to use diagrams to support their thinking. Ask: *Can you explain to Richard how to add fractions with the same denominator?*

ASSESSMENT CHECKPOINT Look for clarity in children's explanations. Rather than learning a rule or shortcut to find the answers, children must know what is happening to the fractions when they are added and why.

ANSWERS Answers for the **Reflect** part of the lesson can be found in the *Power Maths* online subscription.

After the lesson ⏸

- Are children able to confidently explain how to add fractions with the same denominator within one whole?
- Do they fully understand what they are doing and why, or are they relying on a rule to answer questions?
- Can children confidently explain how to use fraction strips or circles to support their answers?
- How will you build in more opportunities to practise adding fractions throughout the school day?

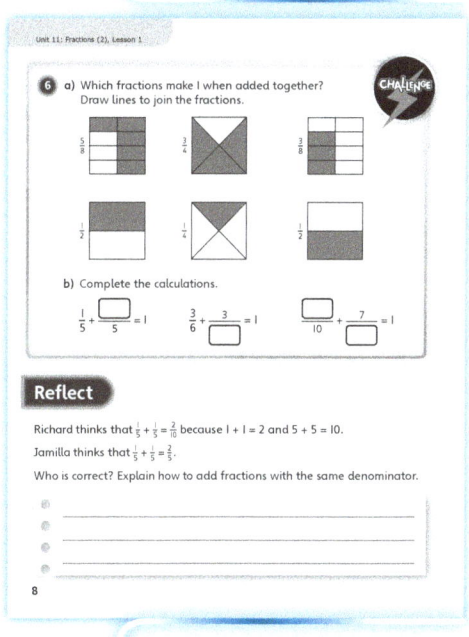

PUPIL PRACTICE BOOK 3C PAGE 8

49

Unit 11: Fractions (2), Lesson 2

Subtract fractions

Learning focus
In this lesson, children will learn to subtract fractions with the same denominator.

Before you teach
- Do children understand what the numerator and denominator show in a fraction?
- Are they confident describing a whole as a fraction?
- Are they comfortable counting back in different fraction steps?

NATIONAL CURRICULUM LINKS

Year 3 Number – fractions

Add and subtract fractions with the same denominator within one whole [for example, $\frac{5}{7} + \frac{1}{7} = \frac{6}{7}$].

ASSESSING MASTERY

Children will draw on their understanding of fractions as numbers and of counting in fraction steps of a constant size. This will help them make sense of subtracting fractions with the same denominator. They can find the difference between two fractions with the same denominator.

COMMON MISCONCEPTIONS

When subtracting fractions, children may subtract the numerators and the denominators separately. For example, they may think that $\frac{3}{5} - \frac{2}{5} = 1$, because 3 – 2 = 1 and 5 – 5 = 0. Ask:

- *If you do $\frac{3}{5} - \frac{2}{5}$, how many parts are in the whole? How many of these parts do you have to start with? How many of them are you subtracting? What do you have left?*

Children may think that whole numbers have the same denominator as the fraction subtracted, for example $1 - \frac{3}{8} = \frac{1}{8} - \frac{3}{8} = \frac{2}{8}$. With this misconception, children will often move the numerators around before subtracting. Ask:

- *If you are working in eighths and you do $1 - \frac{3}{8}$, how many eighths make up the 1 (the whole)? If you calculate $\frac{8}{8} - \frac{3}{8}$ what answer do you get?*

STRENGTHENING UNDERSTANDING

Present children with a real-life problem. Show them an orange and say: *If you have $\frac{3}{5}$ of this orange left and you give a partner 2 of the parts, what fraction of the original orange will you have left? If you had a whole orange and split it into 8 equal parts, what fraction of the orange would be left if you ate 3 of the parts?*

GOING DEEPER

When subtracting with fractions on a number line, children should work in the same way as with whole numbers. Encourage them to use a fraction strip and number line, and to practise counting on and back in different fraction steps. Ask questions that require counting fractions within the whole, for instance: *I started at $\frac{7}{12}$ and counted back $\frac{2}{12}$. Where did I land?* Alternatively, ask: *I started on $\frac{7}{12}$ and landed on $\frac{2}{12}$. How many twelfths have I counted?* Relate the counting to subtracting fractions and record it as $\frac{7}{12} - \frac{2}{12} = \frac{5}{12}$.

KEY LANGUAGE

In lesson: subtract, difference, fraction, whole, calculate

Other language to be used by the teacher: numerator, denominator, unit fraction, non-unit fraction, add, fraction strips

STRUCTURES AND REPRESENTATIONS

Bar model, number line

RESOURCES

Mandatory: fraction strips, number lines

Optional: paper circles, interlocking cubes

 In the eTextbook of this lesson, you will find interactive links to a selection of teaching tools.

Quick recap
Ask children to represent the fraction $\frac{3}{4}$. Share the different methods that children use. Discuss what is the same and what is different about their representations.

50

Discover

WAYS OF WORKING Pair work

ASK

- Question 1 a): *What fraction of the fuel is there in the tank? How do you know?*
- Question 1 a): *What fraction of the fuel will be used on the journey?*
- Question 1 b): *What fraction will be left in the tank?*

IN FOCUS Using a concrete pictorial approach, children will see how a real-life situation can be translated on paper using fraction strips. For question 1 a), ask children to explain how much fuel is in the tank and how much fuel they need to use. In order to calculate the fraction of fuel left, it is important they think of the fractions as numbers. Ask: *What would you do if the question said: 'There are 5 litres of fuel, 3 litres are used. How many litres are left?'*

PRACTICAL TIPS A fraction strip split into 8 equal parts will help children to visualise a whole made up of 8 parts. It will also allow children to count along the strip if needed. Provide a fraction bar for each table.

ANSWERS

Question 1 a): The tank is $\frac{5}{8}$ full.

Question 1 b): 5 eighths – 3 eighths = 2 eighths

$\frac{5}{8} - \frac{3}{8} = \frac{2}{8}$

Share

WAYS OF WORKING Whole class teacher led

ASK

- Question 1 a): *What is the fuel tank split into? How do you know? What fraction of the tank is full? How is this represented on a fraction strip?*
- Question 1 b): *How much fuel will be used on the journey? Why do we need to subtract? How is subtraction shown? How much fuel is left?*

IN FOCUS It is important to give children opportunities to subtract different fractions with the same denominator. In question 1 a), it is important to represent the fuel on a fraction strip. Discuss how the fraction strip represents the fuel left. In question 1 b), use the crossing out method to find how much fuel is left. Some children may try to simplify $\frac{2}{8}$ to $\frac{1}{4}$. Avoid deliberately simplifying at this stage, as equivalent fractions should be encountered (with the support of a representation such as a fraction wall).

STRENGTHEN In question 1 b), relate counting back in eighths to the subtraction of fractions – for example, help children to see that counting back '$\frac{4}{8}$, $\frac{3}{8}$, $\frac{2}{8}$ takes 3 jumps of $\frac{1}{8}$' is equivalent to to the subtraction $\frac{5}{8} - \frac{3}{8} = \frac{2}{8}$. Ask children to practise $\frac{3}{8} - \frac{2}{8}$, $\frac{7}{8} - \frac{2}{8}$ and so on. Clarify any misconceptions. To strengthen the idea of 'whole', ask children to calculate what fraction of the whole tank is empty.

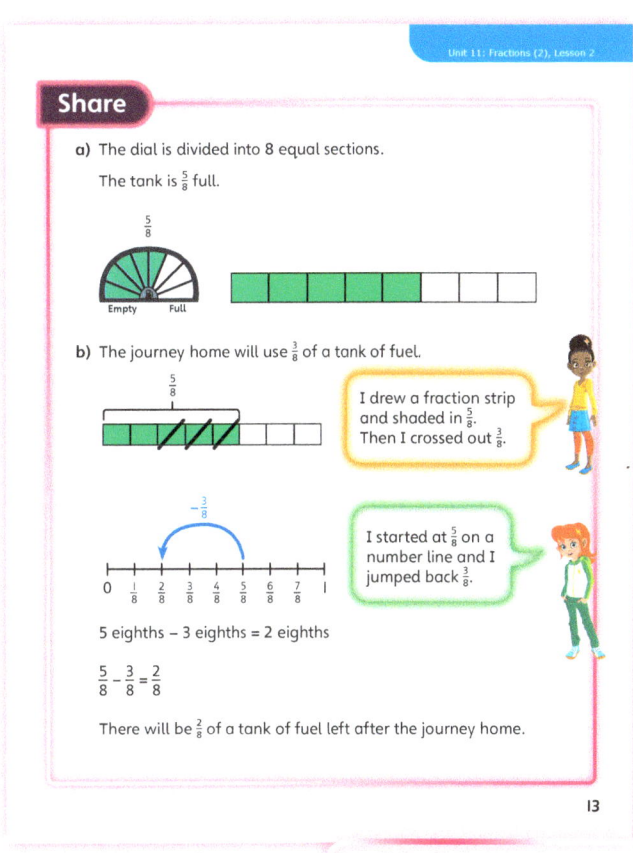

51

Think together

WAYS OF WORKING Whole class teacher led (I do, We do, You do)

ASK

- Question ❶: *What fractions are shown in each diagram?*
- Question ❶: *How can you show each subtraction? What is the subtraction?*
- Question ❷: *Will the numerators change? Will the denominators stay the same? How do you know? How can you show the subtraction using a diagram?*
- Question ❸ a): *How many different answers can you find?*
- Question ❸ b): *What does the difference mean? How can you use the number line to help you?*

IN FOCUS Question ❶ provides children with different visual representations of subtracting fractions with the same denominator. Question ❷ removes the visual support. Ask children to explain their working to ensure they really understand what happens to the fractions when they are subtracted. Ask: *Why is the denominator not changing? What happens if you subtract the denominators? Would the answer be different?*

STRENGTHEN For question ❸, strengthen the link between fractions being numbers and subtracting fractions with the same denominator. Use simple calculations, such as 7 − 4 = 3, to show that finding the difference is another model of subtraction.

DEEPEN For question ❸ b), draw two fraction strips or number lines, one above the other. Ask: *If the difference between the fractions is $\frac{3}{10}$, what will the fractions be? How can you show a difference of $\frac{3}{10}$ in the diagram? How will you find all the answers? How can you record your answer?*

ASSESSMENT CHECKPOINT At this point in the lesson, children should be confident in subtracting fractions with the same denominator. They should be able to use a number line to count back.

ANSWERS

Question ❶ a): $\frac{7}{8} - \frac{5}{8} = \frac{2}{8}$

Question ❶ b): $\frac{9}{10} - \frac{2}{10} = \frac{7}{10}$

Question ❷ a): $\frac{4}{5} - \frac{1}{5} = \frac{3}{5}$

Question ❷ b): $\frac{5}{7} - \frac{1}{7} = \frac{4}{7}$

Question ❷ c): $\frac{10}{11} - \frac{6}{11} = \frac{5}{11}$

Question ❷ d): $\frac{7}{8} - \frac{7}{8} = \frac{0}{0} = 0$

Question ❸ a): There are 5 possible answers:

$\frac{7}{7} - \frac{5}{7} = \frac{2}{7}$ $\frac{4}{7} - \frac{2}{7} = \frac{2}{7}$

$\frac{6}{7} - \frac{4}{7} = \frac{2}{7}$ $\frac{3}{7} - \frac{1}{7} = \frac{2}{7}$

$\frac{5}{7} - \frac{3}{7} = \frac{2}{7}$

Question ❸ b): $\frac{10}{10} - \frac{7}{10} = \frac{3}{10}$ $\frac{6}{10} - \frac{3}{10} = \frac{3}{10}$

$\frac{9}{10} - \frac{6}{10} = \frac{3}{10}$ $\frac{5}{10} - \frac{2}{10} = \frac{3}{10}$

$\frac{8}{10} - \frac{5}{10} = \frac{3}{10}$ $\frac{4}{10} - \frac{1}{10} = \frac{3}{10}$

$\frac{7}{10} - \frac{4}{10} = \frac{3}{10}$

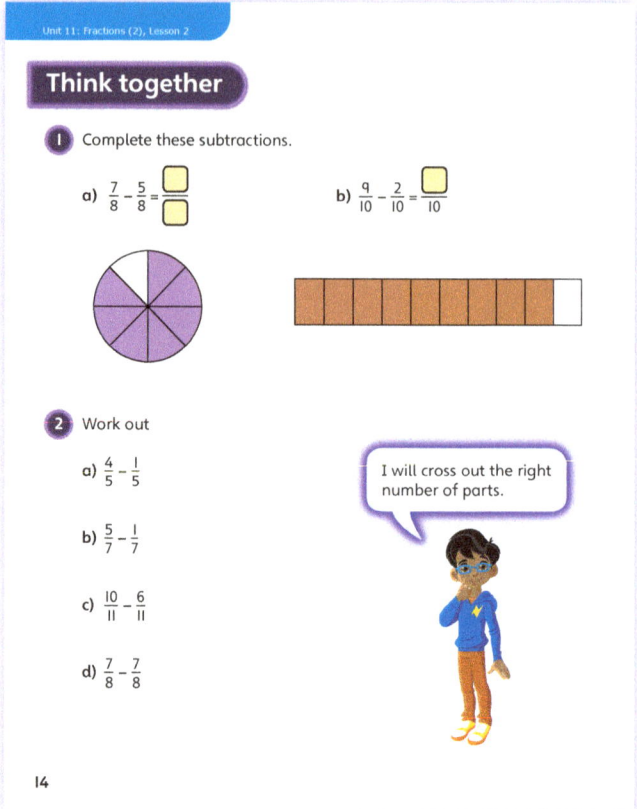

PUPIL TEXTBOOK 3C PAGE 14

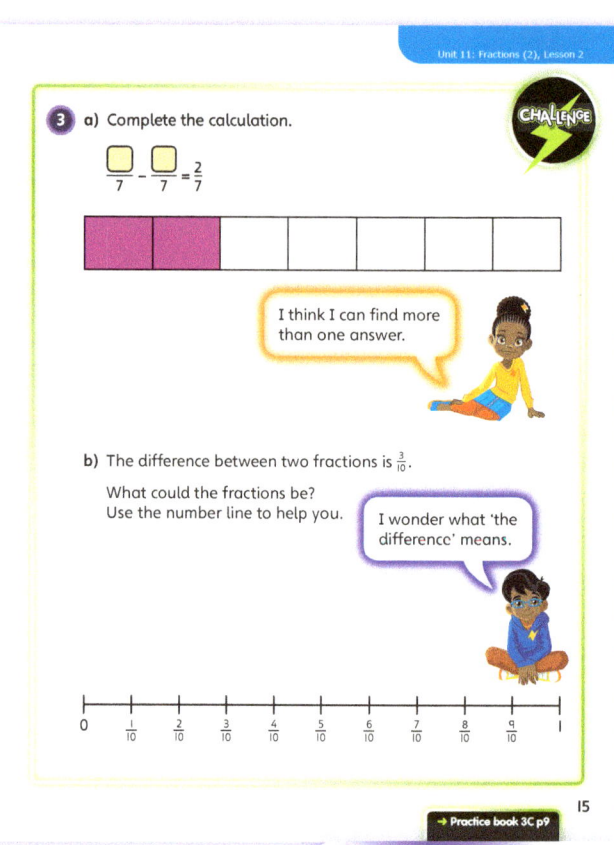

PUPIL TEXTBOOK 3C PAGE 15

Unit 11: Fractions (2), Lesson 2

Practice

WAYS OF WORKING Independent thinking

IN FOCUS Questions ❶ and ❷ provide children with visual images of fractions and require them to complete the subtractions. They should shade in the relevant parts and cross out what they are subtracting. Children should begin to notice that when the denominators are the same, they can just subtract the numerators. Question ❸ provides plenty of practice of the subtracting fractions concept.

STRENGTHEN Question ❸ requires children to subtract fractions with the same denominator, without a visual representation. Encourage children to explain their answers. Ask: *How can you be sure you are correct? What can you do to check your answer?* Address any misconceptions around working with whole numbers. Discuss how number lines might be useful.

DEEPEN You could extend question ❹ by providing a different answer and asking children to find as many questions as they can that will give the answer. In question ❺, explore what happens when adding and subtracting fractions at the same time. Ask: *Would the answer be different if you changed the order of the calculations?*

ASSESSMENT CHECKPOINT Children should have a good understanding of how to subtract fractions. They should be able to show calculations with fractions on a number line in the same way that they do with whole numbers.

ANSWERS Answers for the **Practice** part of the lesson can be found in the *Power Maths* online subscription.

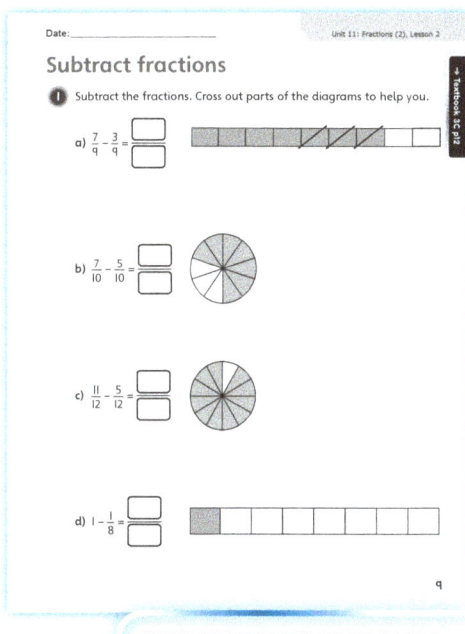

PUPIL PRACTICE BOOK 3C PAGE 9

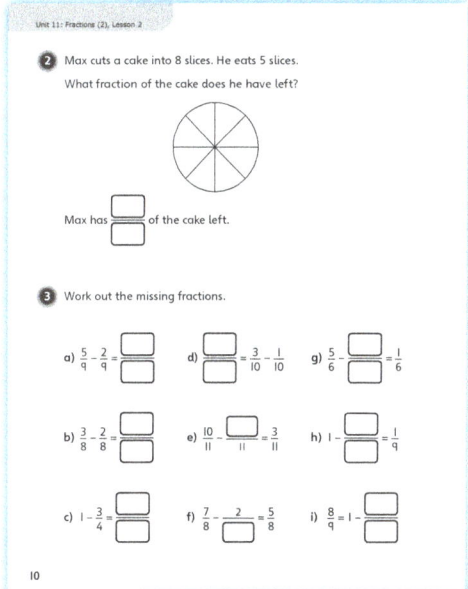

PUPIL PRACTICE BOOK 3C PAGE 10

Reflect

WAYS OF WORKING Pair work

IN FOCUS Once they have recorded their thinking, give children an opportunity to share their ideas with a partner. Encourage them to use diagrams to support their thinking. Can children explain how Reena could find the difference between the fractions?

ASSESSMENT CHECKPOINT Look for clarity in children's explanations. Children should know that when finding the difference between two fractions they need to subtract them, as they would with two whole numbers. They can use fraction strips, fraction circles or number lines to support their answers.

ANSWERS Answers for the **Reflect** part of the lesson can be found in the *Power Maths* online subscription.

After the lesson ⏸

- Can children confidently explain how to subtract fractions with the same denominator within one whole?
- Can they explain how to use number lines and fraction strips or circles to support their answers?
- Are children confident that they need to subtract to find the difference between two fractions?

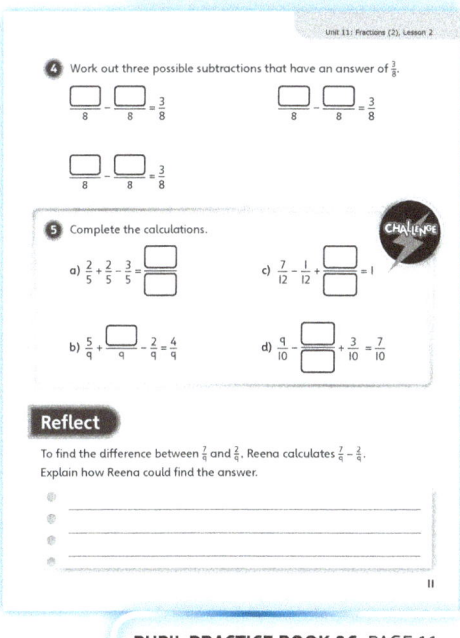

PUPIL PRACTICE BOOK 3C PAGE 11

53

Unit 11: Fractions (2), Lesson 3

Partition the whole

Learning focus
In this lesson, children partition a whole into two (or more) fractions that have the same denominator.

Before you teach
- Check children can represent a fraction using a bar model or fraction circle.
- Do children know how to add two simple fractions with the same denominator?
- Ensure that children know that a whole can be represented by a fraction where the numerator and denominator are equal (for example, $\frac{6}{6}$).

NATIONAL CURRICULUM LINKS

Year 3 Number – fractions

Add and subtract fractions with the same denominator, within one whole [for example, $\frac{5}{7} + \frac{1}{7} = \frac{6}{7}$].

ASSESSING MASTERY

Children can work out what they need to add to a given fraction to make the whole (for example, $\frac{2}{5} + \square = 1$). Children realise that to find the missing number they have to find two numerators that add up to make the denominator. Children can also use their knowledge to solve problems such as $1 - \frac{1}{6} = \square$.

COMMON MISCONCEPTIONS

Children often struggle with the concept of the whole written as a fraction. Explain that the whole can be written in many different ways and present children with examples, such as $\frac{4}{4}$ or $\frac{5}{5}$ or $\frac{6}{6}$. Show bar models or fractions strips to represent these fractions and help children to see that a whole is where the numerator and denominator are equal. When tackling problems such as $\frac{4}{9} + \frac{?}{9} = 1$, children need to remember that the whole in this case is $\frac{9}{9}$.

STRENGTHENING UNDERSTANDING

To support children solving problems such as $\frac{3}{5} + \square = 1$, ask children to draw the fraction $\frac{3}{5}$. They may represent this as a bar model with $\frac{3}{5}$ shaded or as a fraction circle. Ask: *What other fraction do you need to shade in to make 1 (or the whole)?* Children should be able to see that they need to shade in $\frac{2}{5}$ to make the whole.

GOING DEEPER

Encourage children to partition the whole in different ways. For example, how many answers can they find for $\frac{?}{10} + \frac{?}{10} = 1$. Encourage children to take a systematic approach to working out all the possible answers.

KEY LANGUAGE

In lesson: whole, numerator, denominator, partition

Other language to be used by the teacher: add, subtract

STRUCTURES AND REPRESENTATIONS

Part-whole models, bar models

RESOURCES

Optional: coloured rods

 In the eTextbook of this lesson, you will find interactive links to a selection of teaching tools.

Quick recap
Show children a 0–1 number line divided into 8 equal parts. As a class, label the eighths, $\frac{1}{8}, \frac{2}{8}$ and so on. Use the number line to practise simple addition and subtractions: for example, $\frac{5}{8} - \frac{2}{8} = ?$ or $\frac{4}{8} + \frac{4}{8} = ?$.

Discover

WAYS OF WORKING Pair work

ASK

- Question 1 a): *How many parts does the fraction strip have? What fraction is red? What fraction is yellow? How can this help you complete the part-whole model?*
- Question 1 b): *How can you complete the fraction strip in another way? How many different ways can you find?*

IN FOCUS Questions 1 a) and b) focus on eighths. Children use the fraction strip to help them to partition the whole. Ensure children understand that the whole is represented by the fraction strip, as this will help them to partition the whole. In question 1 a), help children, if necessary, to see the $\frac{3}{8}$ and $\frac{5}{8}$. Some children may just put 3 and 5, in which case make sure they see this as a fraction.

For question 1 b), children draw their own fraction strip (with 8 parts) and shade it, partitioning the whole in a different way. Once children have done this, ask them if they can find another way. Ask: *What do you notice about the numerators and denominators?*

PRACTICAL TIPS Have printed fraction strips ready for children to shade in.

ANSWERS

Question 1 a):

Question 1 b):

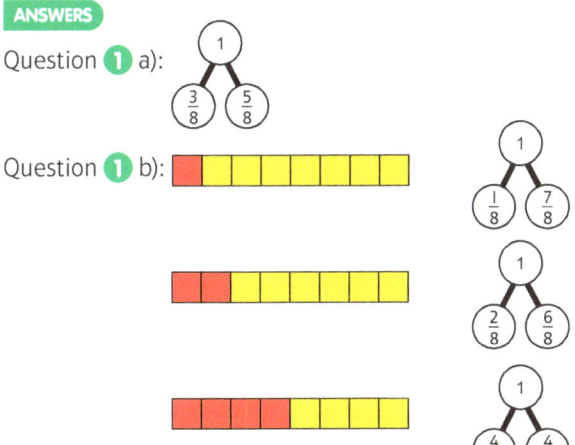

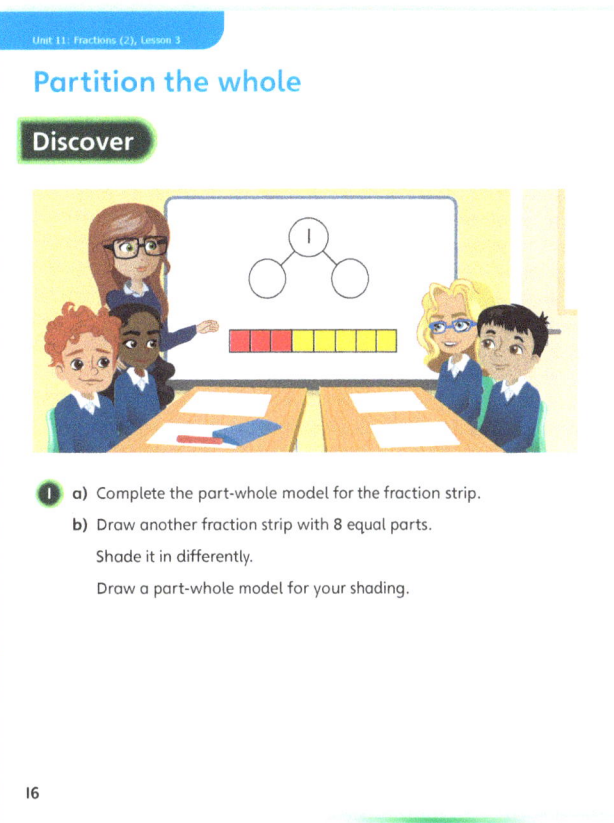

PUPIL TEXTBOOK 3C PAGE 16

Share

WAYS OF WORKING Whole class teacher led

ASK

- Question 1 a): *Can you see why $\frac{3}{8}$ and $\frac{5}{8}$ make the whole?*
- Question 1 b): *Did you find any other ways to partition the whole? Are they the same? Did you find them all?*

IN FOCUS For question 1 a), show children how the whole can be partitioned into $\frac{3}{8}$ and $\frac{5}{8}$ using shading. In question 1 b), take different answers from the children to try to find all the possible solutions. Some children may have shaded their strips in a different way (for example, 5 red and 3 yellow) but with the same partitioning. Explain that this is the same as for question 1 a), the parts are just in different circles.

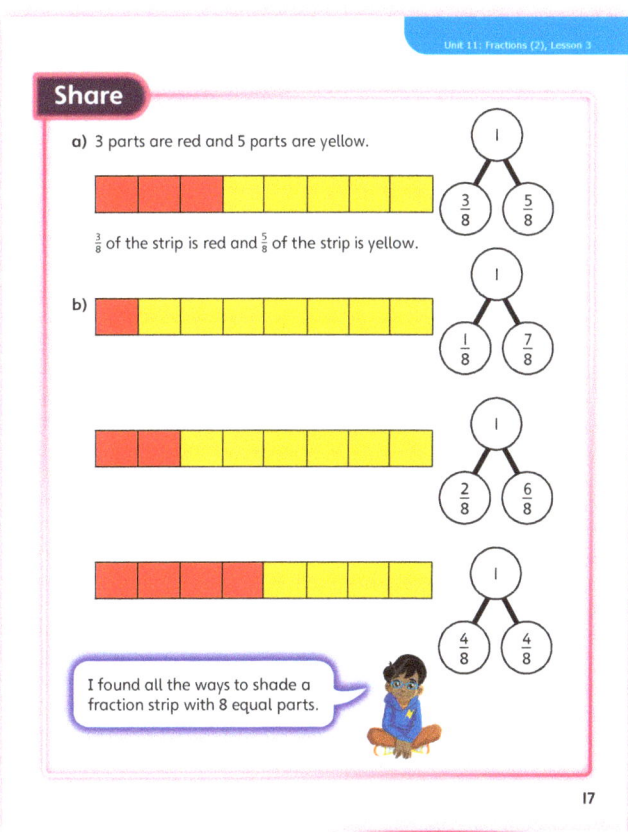

PUPIL TEXTBOOK 3C PAGE 17

Unit 11: Fractions (2), Lesson 3

Think together

WAYS OF WORKING Whole class teacher led (I do, We do, You do)

ASK
- Question ❶: *What fraction of the circle is red? What fraction is yellow? How can you use this to complete the part-whole model?*
- Question ❷: *What are the missing fractions? What is the whole each time? What method did you use to find the missing part?*
- Question ❸: *How can you use what you learnt earlier to find the answer?*

IN FOCUS In question ❶ children are given a fraction circle shaded $\frac{1}{6}$ red and $\frac{5}{6}$ yellow to help them see that $\frac{1}{6} + \frac{5}{6} = 1$. Discuss with children why these two fractions make a whole. In question ❷ children may want to draw a fraction strip or circle to support their understanding initially. It is useful if children start to see that the numerators add together to make the denominator. Try to draw this generalisation out. Provide children with some further examples if necessary. In question ❸ children apply their knowledge of parts that make a whole to work out subtractions.

STRENGTHEN To support children solving problems such as those in question ❷, ask them to draw the fraction that is already known. They may represent this as a bar model or a fraction circle, with the total number of parts given by the denominator of the known part, and the shaded number of parts given by the numerator of the known part. Ask: *What other fraction do you need to shade in to make 1 (or the whole)?* Explain that this provides the missing part. Ask if they notice anything about the sum of the two numerators.

DEEPEN Encourage children to partition the whole in different ways. For example, how many answers can they find for the following: $\frac{?}{10} + \frac{?}{10} = 1$. Encourage children to take a systematic approach to working out all the possible answers.

ASSESSMENT CHECKPOINT Questions ❶ and ❷ will help you determine whether children are confident with partitioning the whole.

ANSWERS

Question ❶: $\frac{1}{6}$ and $\frac{5}{6}$

Question ❷ a): $\frac{1}{4}$

Question ❷ b): $\frac{2}{9}$

Question ❷ c): $\frac{3}{10}$

Question ❷ d): $\frac{13}{20}$

Question ❸ a): Taking one part away from the whole, leaves the other part.
$1 - \frac{3}{5} = \frac{2}{5}$

Question ❸ b): $1 - \frac{2}{3} = \frac{1}{3}$
$1 - \frac{5}{9} = \frac{4}{9}$
$1 - \frac{1}{12} = \frac{11}{12}$

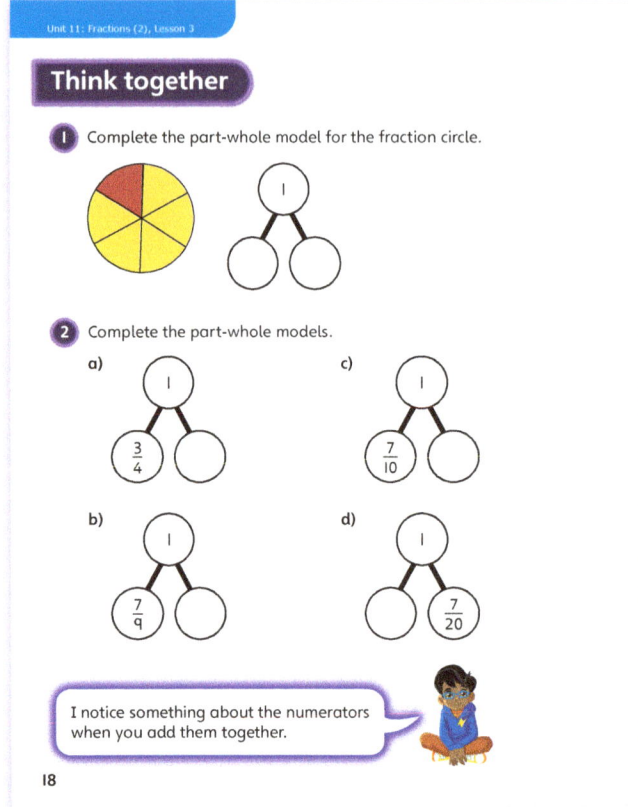

PUPIL TEXTBOOK 3C PAGE 18

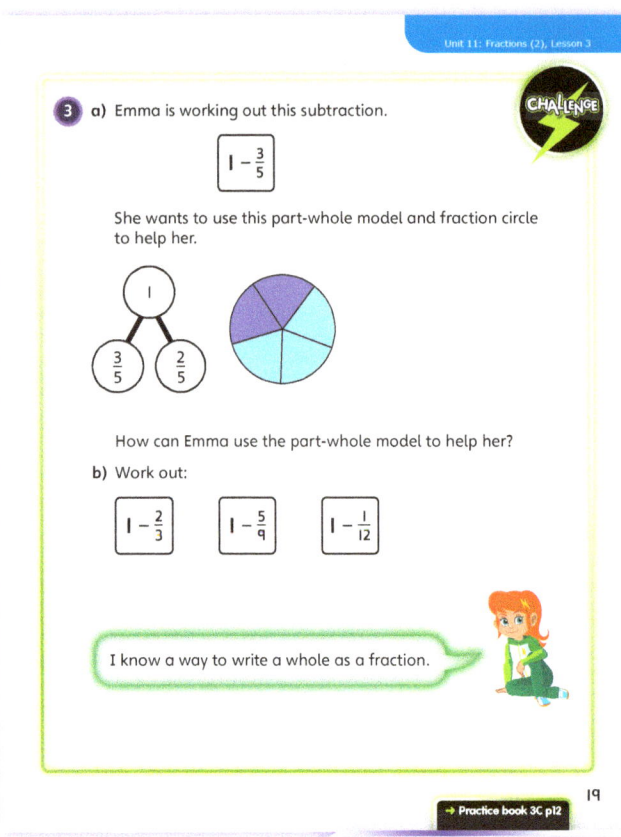

PUPIL TEXTBOOK 3C PAGE 19

Unit 11: Fractions (2), Lesson 3

Practice

WAYS OF WORKING Independent thinking

IN FOCUS In questions ❶ and ❷ children are provided with structured support to partition the whole. In question ❸ children may want to draw diagrams to support their understanding of why the sums are equal to 1. Children may write or explain the generalisation that the numerators sum to the denominator. In question ❹ children are required to use their generalisation to find the missing parts. Question ❻ challenges children to apply their understanding of partitioning the whole to complete subtractions.

STRENGTHEN To support children solving problems such as those in questions ❸ and ❹, ask them to draw the fraction part already known. They may represent this as a bar model or a fraction circle, with the total number of parts given by the denominator of the known part, and the shaded number of parts given by the numerator of the known part. Ask: *What other fraction do you need to shade in to make 1 (or the whole)?* Explain that this provides the missing part. Ask if they notice anything about the sum of the two numerators.

DEEPEN Ask children to partition the whole in different ways. For example, how many answers can they find to the following: $\frac{?}{10} + \frac{?}{10} = 1$. Encourage children to take a systematic approach to working out all the possible answers.

THINK DIFFERENTLY In question ❺ children can partition the whole in whatever way they would like. Encourage them to use different denominators to show the breadth of their understanding. The whole can be presented in many different ways and this makes it a great thing to discuss.

ASSESSMENT CHECKPOINT Use questions ❶ to ❹ to check that children are fluent and confident in partitioning the whole.

ANSWERS Answers for the **Practice** part of the lesson can be found in the *Power Maths* online subscription.

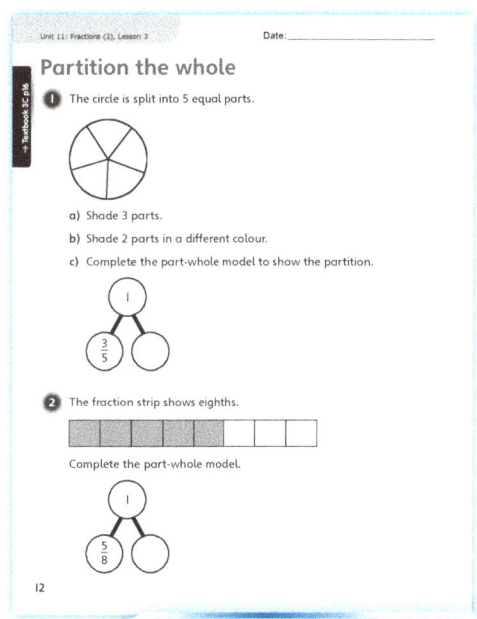

PUPIL PRACTICE BOOK 3C PAGE 12

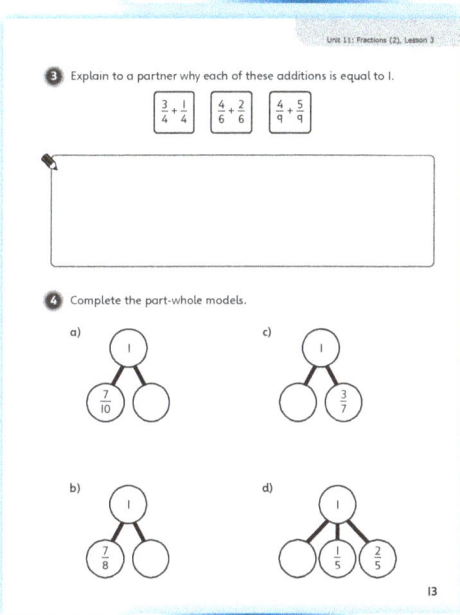

PUPIL PRACTICE BOOK 3C PAGE 13

Reflect

WAYS OF WORKING Pair work

IN FOCUS Children should discuss the question as a pair and rehearse their answer before discussing as a class. Children may want to write down some examples from the lesson of fractions that make a whole.

ASSESSMENT CHECKPOINT Children should be able to reason that the numerators should add up to make the same number as the denominator. This shows that the fraction makes a whole.

ANSWERS Answers for the **Reflect** part of the lesson can be found in the *Power Maths* online subscription.

After the lesson

- Can children partition the whole in different ways, finding missing parts?
- Do children know that if the numerators of two fractions add to give the denominator then the fractions make a whole?

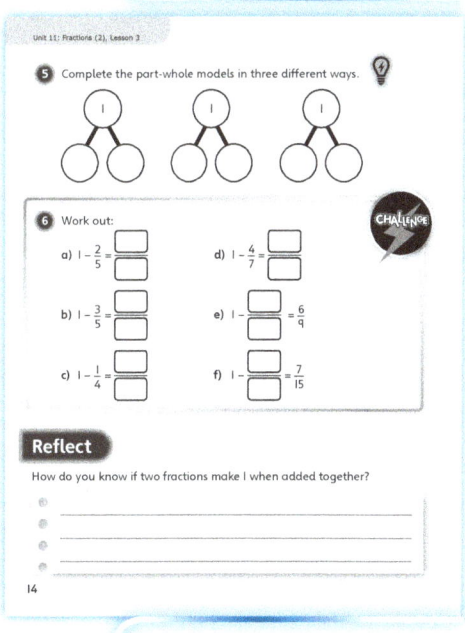

PUPIL PRACTICE BOOK 3C PAGE 14

Unit 11: Fractions (2), Lesson 4

Problem solving – add and subtract fractions

Learning focus
In this lesson, children will learn to reason mathematically and solve problems by adding and subtracting fractions.

Before you teach
- Which resources will children find the most useful to represent the given information?
- How will you ensure that children are able to interpret the different types of question they will encounter?

NATIONAL CURRICULUM LINKS

Year 3 Number – fractions

Solve problems that involve addition and subtraction of fractions.

Add and subtract fractions with the same denominator, within one whole [for example, $\frac{5}{7} + \frac{1}{7} = \frac{6}{7}$].

ASSESSING MASTERY

Children can recognise the operation needed to answer a word problem. They can write the necessary calculation to answer the problem and successfully find and write the correct answer.

COMMON MISCONCEPTIONS

In previous units, children have added and subtracted fractions, using images such as a number line, following explicit instructions. In this unit, children will decide which operation to use. However, they may find it difficult to identify exactly what the question is asking. Instead of applying their knowledge of fractions, this uncertainty can lead them to guess at the answer. Ask:
- *Can you draw a picture or a diagram that describes the problem?*

STRENGTHENING UNDERSTANDING

Resources such as fraction strips should be readily available for children. Encourage children to make sense of any problem before attempting to solve it, and show them how to organise their thinking. Ask: *What is it about? Can you describe it?* If they are unsure, ask them to read the question again. Ask: *Have you used all the information given?*

GOING DEEPER

Challenge children to find different ways to find the answer to the same question. Sometimes, children who work out the answer more quickly are less inclined to look for different ways to answer the question, moving on as soon as they have found an answer. Remind them that the route to finding the answer can be more important than the answer itself.

KEY LANGUAGE

In lesson: fraction, more, add, subtract

Other language to be used by the teacher: calculate, word problem, less, greatest, addition, subtraction

STRUCTURES AND REPRESENTATIONS

Bar model, number line

RESOURCES

Optional: bar models, counters, coloured rods, part-whole models

 In the eTextbook of this lesson, you will find interactive links to a selection of teaching tools.

Quick recap
Practise adding and subtracting fractions with the same denominator. Ask children questions such as $\frac{3}{8} + \frac{2}{8}$ and $\frac{7}{9} - \frac{2}{9}$ and so on. Discuss with children how to find the answers and how to use diagrams to support them as necessary.

Unit 11: Fractions (2), Lesson 4

Discover

WAYS OF WORKING Pair work

ASK
- Question 1 a): *What fraction of the food did they eat on Monday? And on Tuesday? How can you work out how much food they ate in total?*
- Question 1 b): *If $\frac{4}{10}$ of the food has been eaten, how much is left? Has more been used or is there more left?*

IN FOCUS Use these pictures as a springboard to discuss the food that children use at home. Ask: *If you know how much food is used each day of the week, how can you work out how much food is used in the whole week?*

PRACTICAL TIPS Use fraction strips with 10 equal parts to help children visualise the question. It might be useful to revisit how the number line can be used to add and subtract fractions within one whole.

ANSWERS

Question 1 a): $\frac{1}{10} + \frac{3}{10} = \frac{4}{10}$

They ate $\frac{4}{10}$ of the food in total on Monday and Tuesday.

Question 1 b): $1 - \frac{4}{10} = \frac{6}{10}$ of the food is left in the box.

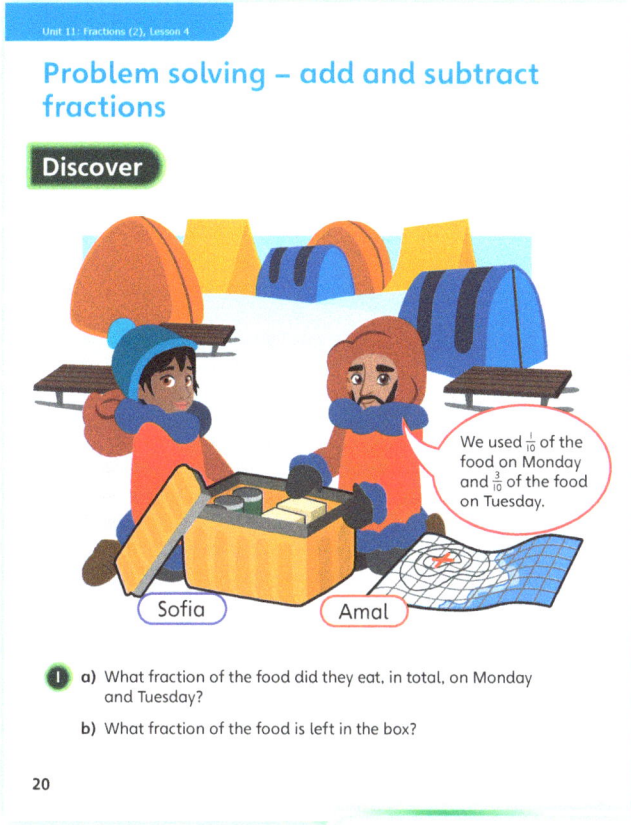

PUPIL TEXTBOOK 3C PAGE 20

Share

WAYS OF WORKING Whole class teacher led

ASK
- Question 1 a): *To work out how much food they ate on Monday and Tuesday, what information do you need to know? How can a bar model help you?*
- Question 1 b): *How can you work out how much food is left? What fraction of food was there to start with? What fraction of food has been used in total on Monday and Tuesday?*

IN FOCUS In this section, children must interpret the questions accurately in order to solve the problems. They also need to be confident in adding and subtracting fractions within a whole.

STRENGTHEN Set up two fraction strips at the front of the class or on the whiteboard. Next to the first fraction strip write: Monday $1 - \frac{1}{10} = \frac{9}{10}$. Next to the second fraction strip write: Tuesday $\frac{9}{10} - \frac{3}{10} = \frac{6}{10}$. Ask: *Is this answer correct? How can you explain it? What does each step represent?*

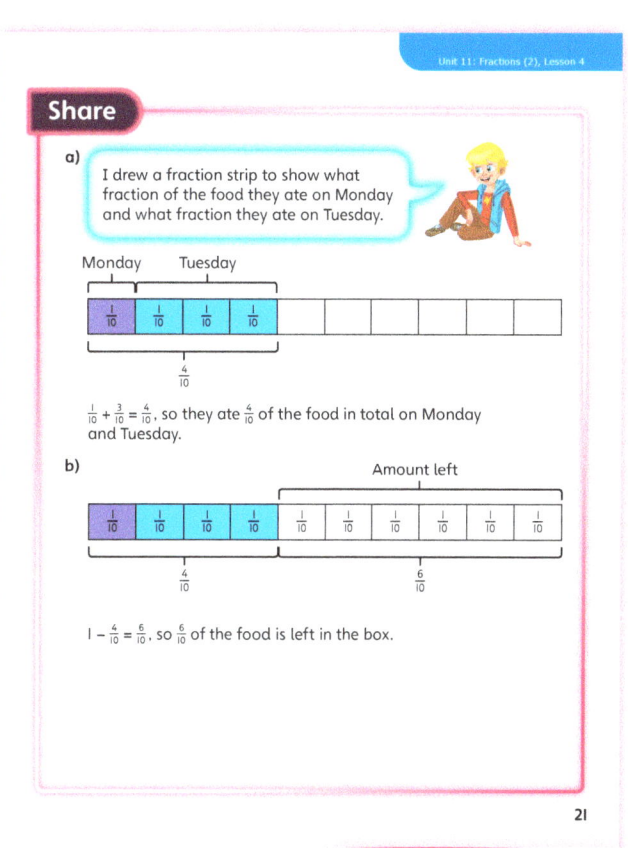

PUPIL TEXTBOOK 3C PAGE 21

Unit 11: Fractions (2), Lesson 4

Think together

WAYS OF WORKING Whole class teacher led (I do, We do, You do)

ASK
- Question ❶: *What fraction of the journey did Sofia and Amal walk? What fraction of the journey did they ski?*
- Question ❷: *What fraction of the tents are blue? And red? How can you find the fraction that are yellow?*
- Question ❸: *How can you use a fraction strip to help you?*

IN FOCUS Questions ❶ and ❷ are two-step problems. First, children need to find the total amount; then they need to subtract the answer from 1. Question ❸ progresses from comparing fractions, to finding the difference between two fractions.

STRENGTHEN Some children may find it difficult to answer question ❸. Discuss Astrid's suggestion that children find the fraction that Max ate first. Some children may misinterpret this information and assume that Max ate $\frac{1}{5}$ of the packet. Ask them to read the question carefully. Discuss how they can check if their answer is correct.

DEEPEN In question ❸, use a fraction strip split into 5 equal parts. The fraction strips should be accessible for all children. Concrete visual resources are a powerful tool to help deepen children's understanding of fractions. Challenge children to think of fraction word problems, swap them with a partner and draw or use representations to model the answers.

ASSESSMENT CHECKPOINT At this point in the lesson, children should be able to understand and interpret the questions and identify the operations required to solve them. They should also be able to explain any manipulatives they have chosen to use.

ANSWERS

Question ❶ a): They have travelled $\frac{6}{8}$ of the journey so far.
Question ❶ b): $\frac{2}{8}$ of the journey is left.
Question ❷: $\frac{5}{9}$ of the tents are yellow.
Question ❸ a): They ate $\frac{3}{5}$ of the packet altogether.
Question ❸ b): $\frac{2}{5}$ of the packet is left.

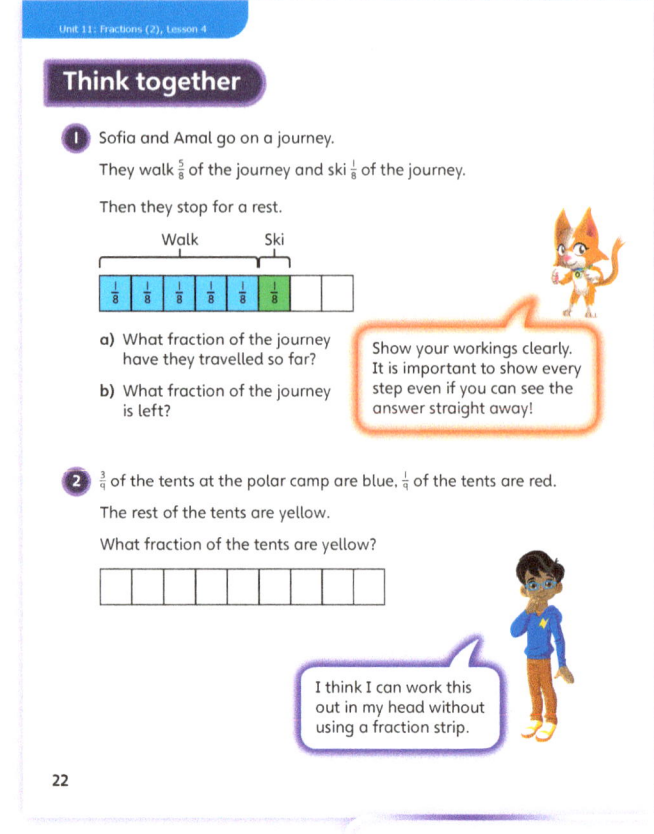

PUPIL TEXTBOOK 3C PAGE 22

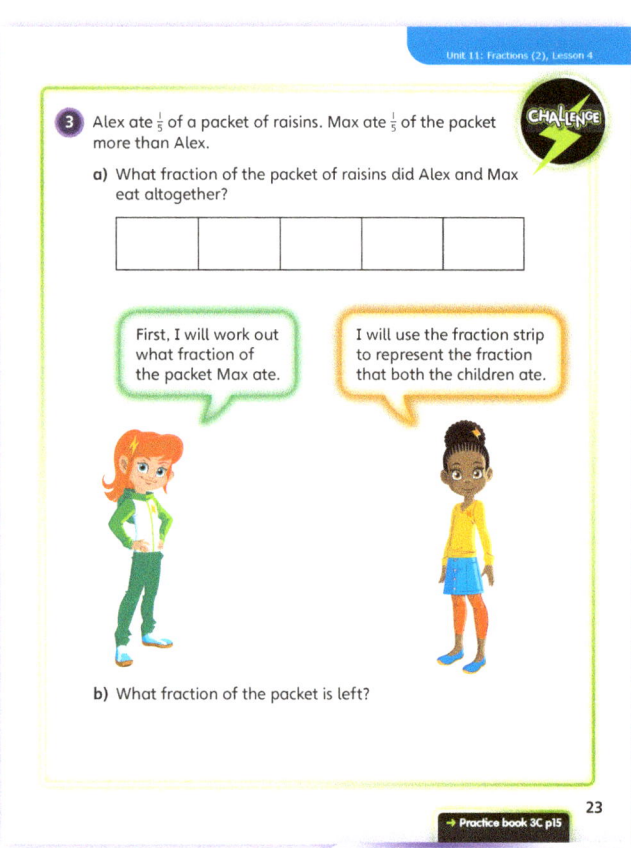

PUPIL TEXTBOOK 3C PAGE 23

Unit 11: Fractions (2), Lesson 4

Practice

WAYS OF WORKING Independent thinking

IN FOCUS Question ③ challenges children to create six different questions that all have the same answer. Children need to use a systematic approach when solving this problem in order to organise their thinking and results.

STRENGTHEN Encourage any children who are finding question ④ difficult to use a fraction strip or number line, or both. Remind them to read the question slowly and record their findings step-by-step. Encourage children to label their findings for each step, so they can share their thinking through the stages and check their answers at the end.

DEEPEN Ask children to work in pairs. They should each set their partner a given number of questions like those in question ③, but for a different fraction (for example, $\frac{7}{9}$). Once they have each completed their partner's questions, they should check one another's work. This will help develop their fluency in calculating with fractions that have the same denominator.

ASSESSMENT CHECKPOINT By this point in the lesson, children should be showing confidence in solving problems. They should be able to use representations to demonstrate a question visually, and use a systematic approach to solve it.

ANSWERS Answers for the **Practice** part of the lesson can be found in the *Power Maths* online subscription.

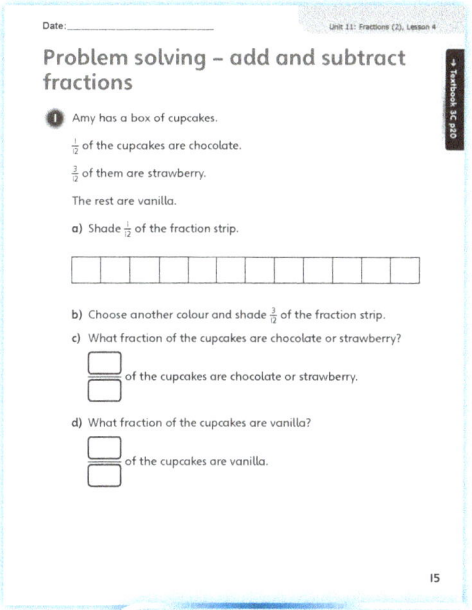

Reflect

WAYS OF WORKING Independent thinking

IN FOCUS Give children an opportunity to develop their own fraction problems independently. Can they answer the questions themselves? What method do children prefer to use to answer their own questions? Ask them to swap problems with a partner. They should compare the methods they used. Ask: *Whose method is more efficient?*

ASSESSMENT CHECKPOINT Children should be using different representations fluently to solve fraction problems. Assess their written explanations of what they think they should pay attention to when solving fraction problems. Look and listen for children's reasoning and be ready to clarify any misconceptions.

ANSWERS Answers for the **Reflect** part of the lesson can be found in the *Power Maths* online subscription.

After the lesson ⏸

- Can children confidently work in different ways to solve a problem?
- Are they confident in choosing different types of representation to solve problems independently?
- What opportunities will you provide for them to practise solving multi-step problems outside the lesson?

Unit 11: Fractions (2), Lesson 5

Unit fractions of a set of objects

Learning focus
In this lesson, children will find a unit fraction of a set of objects. Using a bar model or strips of paper to fold, children will find these fractional amounts and begin to link finding fractions of amounts to dividing by the denominator.

Before you teach
- Will children's factual fluency be a barrier to understanding the concept of the lesson?
- What additional resources can be provided to help children for whom this barrier exists?

NATIONAL CURRICULUM LINKS

Year 3 Number – fractions

Recognise, find and write fractions of a discrete set of objects: unit fractions and non-unit fractions with small denominators.

ASSESSING MASTERY

Children can find a unit fraction of a set of objects. They can describe how to use resources to represent the stages within calculations. They can interpret what each number within a calculation represents and how this links to the question context.

COMMON MISCONCEPTIONS

Children may confuse the number of parts and the number of items in each part. For example, if they are finding $\frac{1}{4}$ of 20, they may think this means that the 20 objects must be split into groups of 4 as opposed to being shared between the 4 groups. Ask:
- *What does the denominator represent?*

STRENGTHENING UNDERSTANDING

Children should be allowed to use counters and bar models for as long as necessary to help them to understand the concept of finding a fraction of an amount. Once this concept is understood, the resources should be removed to allow children to practise their number facts.

GOING DEEPER

Challenge children to find unit fractions of different amounts to explore when it is and when it is not possible to split an amount into equal parts and how the size of these parts changes as the denominator changes.

KEY LANGUAGE

In lesson: set of objects, fraction, denominator, divide, represent, calculate, share, whole

Other language to be used by the teacher: group, numerator, unit fraction

STRUCTURES AND REPRESENTATIONS

Bar model, part-whole model

RESOURCES

Mandatory: counters, bar models

Optional: sweets or items to replicate the **Discover** context, part-whole models

 In the eTextbook of this lesson, you will find interactive links to a selection of teaching tools.

Quick recap

Play a game of division bingo. Children write down 6 numbers from 1 to 12 on a 3 × 2 grid. Ask division questions that have 1 to 12 as their answers. Focus on multiplication facts from times-tables that children should know by this stage.

Unit 11: Fractions (2), Lesson 5

Discover

WAYS OF WORKING Pair work

ASK
- Question 1 a): *What does it mean to find a fraction of an amount?*
- Question 1 a): *What representation could be used to find a fraction of an amount?*
- Question 1 a): *How many cakes are there in the picture? How much of each topping do they need for one cake?*
- Question 1 b): *Are you told how many chocolate swirls there are in a bag? What fraction of a bag of chocolate swirls does the recipe say is needed for the cake?*

IN FOCUS Throughout this section of the lesson, children are required to find a fraction of an amount where it is possible to see each part that makes the whole. Encourage children to make links with the denominator and the number of parts the whole must be split into.

PRACTICAL TIPS Getting children to replicate the job of the factory workers would be a fun and practical way for them to connect with the context presented to them in the questions. Provide them with quantities of items that can be shared equally into the required number of parts.

ANSWERS

Question 1 a): 10 marshmallows are needed for 1 cake.

Question 1 b): 9 chocolate swirls are needed for 1 cake.

PUPIL TEXTBOOK 3C PAGE 24

Share

WAYS OF WORKING Whole class teacher led

ASK
- Question 1 a): *Why has the whole been split into 3 parts?*
- Question 1 b): *How many parts has the bar model been split into? Why are we splitting the whole into that number of parts?*

IN FOCUS In this part of the lesson, children are shown step-by-step how the bar model can be used to calculate a fraction of a quantity. Children need to become comfortable with each step of this process and understand why the representation and resources are manipulated in different ways at different points of the calculation.

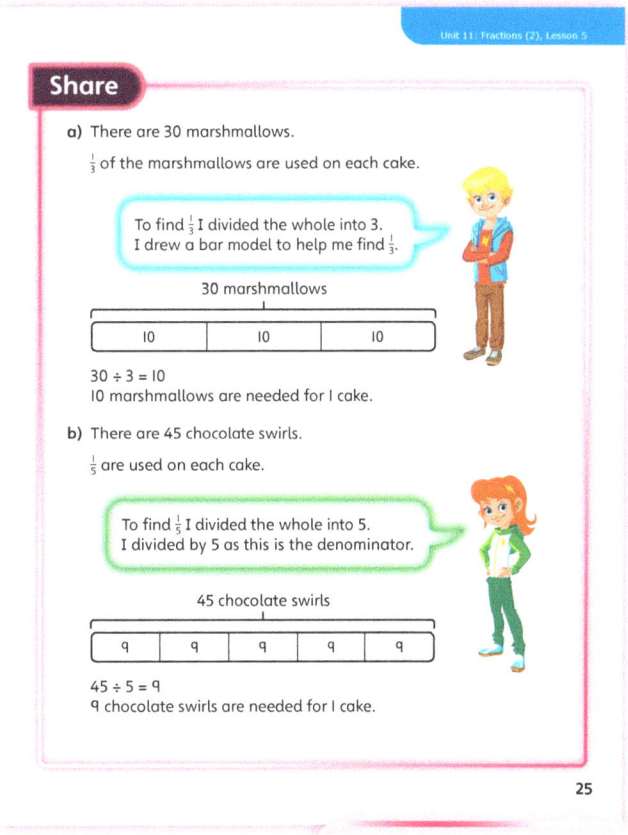

PUPIL TEXTBOOK 3C PAGE 25

Unit 11: Fractions (2), Lesson 5

Think together

WAYS OF WORKING Whole class teacher led (I do, We do, You do)

ASK
- Question ❶: *How many apples are on the tree? How many parts has the bar model been split into? Why? What is each part worth?*
- Question ❷: *How many equal groups will the footballs be split into? How many footballs will there be in each group?*
- Question ❸ a): *How can you use the array to help you see the answer? Can you draw a bar model for each one?*

IN FOCUS In question ❶, children are using a bar model to support them in finding a unit fraction of an amount. Discuss with children why the bar model has been divided into 6 parts (because of the denominator).

In question ❷, children work out how many footballs each child will get. Discuss throughout how this relates to the calculation $\frac{1}{3}$ of 18.

In question ❸ a), children are presented with an array of coins and they have to work out the relevant fraction of an amount. They should use the bar model and array to support them. Encourage them to point or circle the different fractions. (For example $\frac{1}{4}$ of 24 can be seen by pointing to the 4 rows of 6 counters, and so on.)

STRENGTHEN Children should continue to work with resources until they are secure with the concept of finding a fraction of an amount. If they find it difficult to identify the number of parts the whole must be split into, provide them with ready-made bar models to simplify the process of sharing the set of objects between the parts.

DEEPEN Children could be asked to compare unit fractions of the same amount without having to calculate them. For example: which is bigger, $\frac{1}{2}$ of 24 or $\frac{1}{4}$ of 24? Ask: *Is there any way you can find this out without working out the actual values?* In question ❸, children could be asked to order the size of the answers based on the unit fractions.

ASSESSMENT CHECKPOINT Use the questions to determine whether children are confident with finding a simple unit fraction of an amount. If they are not, support them by providing some more fluency-based examples.

ANSWERS

Question ❶: $\frac{1}{6}$ of 30 apples = 5 apples

Question ❷ a): They will get 6 balls each.

Question ❸ a): $\frac{1}{2}$ of £24 = £12 $\frac{1}{3}$ of £24 = £8
$\frac{1}{4}$ of £24 = £6 $\frac{1}{12}$ of £24 = £2

Question ❸ b): $\frac{1}{8}$ of 24 = 3

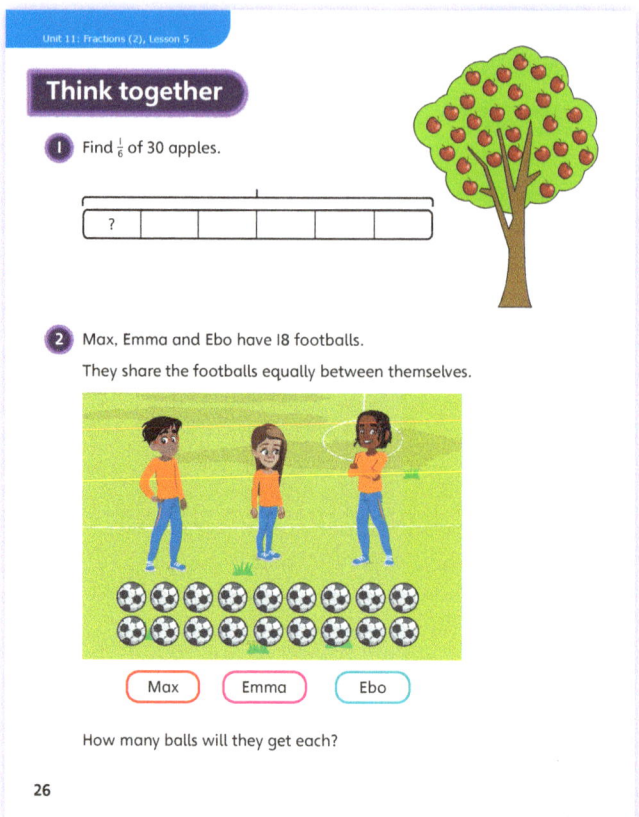

PUPIL TEXTBOOK 3C PAGE 26

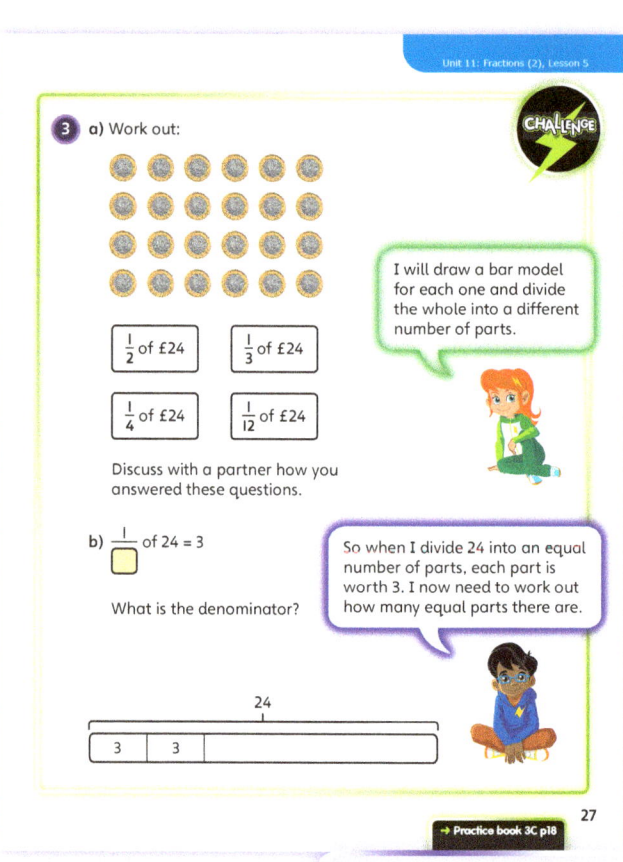

PUPIL TEXTBOOK 3C PAGE 27

Unit 11: Fractions (2), Lesson 5

Practice

WAYS OF WORKING Independent thinking

IN FOCUS During this section of the lesson, children are initially provided with bar models and sentence scaffolds to replicate the representations they used in the **Think together** section. These scaffolds are gradually removed throughout the lesson but children should be encouraged to continue to draw their own models to show the steps as they work through **Practice**.

In question ③ children should be encouraged to try to answer the questions without the need for a bar model. They should consider the denominator and how it links to the division that they need to do.

STRENGTHEN If children find producing their own bar models too difficult, they could be given ready-made bar models. If children understand the concept but are making calculation errors, they could be provided with relevant facts to use, rather than relying on resources to share the whole into parts.

DEEPEN Give children the chance to explore the link between the number of parts and the number of objects in each part and to describe the pattern that they find. For example, ask: *Calculate $\frac{1}{5}$ of 20 and $\frac{1}{4}$ of 20. What do you notice? Do you think this pattern always occurs?*

ASSESSMENT CHECKPOINT Children should be able to explain the steps of their workings and make links between the pictorial and abstract representations of each problem. If children rely on counters throughout the lesson, this is an indication that they do not have the appropriate factual fluency or that they do not understand the concept in sufficient depth. It is unlikely that they will have made the link between the value of the denominator and the number to divide the whole by.

ANSWERS Answers for the **Practice** part of the lesson can be found in the *Power Maths* online subscription.

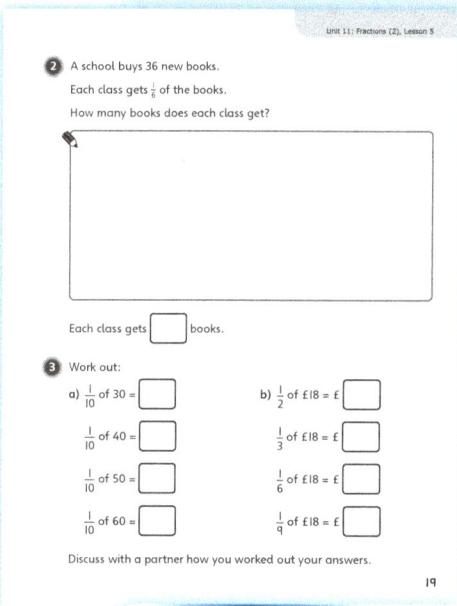

PUPIL PRACTICE BOOK 3C PAGE 18

PUPIL PRACTICE BOOK 3C PAGE 19

Reflect

WAYS OF WORKING Pair work

IN FOCUS This question allows children to summarise how to calculate a unit fraction of a set of objects. Encourage children to use the appropriate mathematical language throughout their explanations. Ask them to write down their response before they tell a partner. They can use diagrams to aid their explanations.

ASSESSMENT CHECKPOINT The strength of children's written response can be used to assess how secure their understanding is of finding a fraction of a set of objects.

ANSWERS Answers for the **Reflect** part of the lesson can be found in the *Power Maths* online subscription.

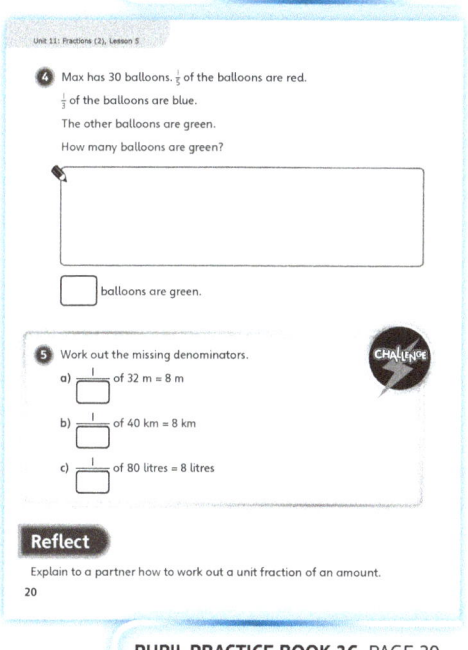

PUPIL PRACTICE BOOK 3C PAGE 20

After the lesson

- Are children confident finding a unit fraction of a set of objects?
- Are children ready to calculate a non-unit fraction of a set of objects?

Unit 11: Fractions (2), Lesson 6

Non-unit fractions of a set of objects

Learning focus
In this lesson, children will find non-unit fractions of a set of objects. They will link dividing by the denominator and multiplying by the numerator in order to find the solution.

Before you teach
- Are children confident finding a unit fraction of a set of objects?

NATIONAL CURRICULUM LINKS

Year 3 Number – fractions

Recognise, find and write fractions of a discrete set of objects: unit fractions and non-unit fractions with small denominators.

ASSESSING MASTERY

Children can find a non-unit fraction of a set of objects and describe how resources can be used to represent the stages within a calculation. They should be able to interpret what each number within a calculation represents and understand how this links to the context they are provided with.

COMMON MISCONCEPTIONS

Some children may continue the method they learnt in the previous lesson and simply calculate a unit fraction of a set of objects, disregarding the different numerator. Ask:
- *What does the denominator represent? What does the numerator represent?*

STRENGTHENING UNDERSTANDING

Continue to allow children to use counters and bar models for as long as necessary for them to understand the concept of finding a non-unit fraction of an amount. A multiplication square or similar resource may also be helpful.

GOING DEEPER

To deepen their understanding, provide children with calculations that have missing information in different places. For example, $\frac{?}{4}$ of 20 = 15 or $\frac{3}{?}$ of 25 = 15.

KEY LANGUAGE

In lesson: set of objects, denominator, numerator, unit fraction, non-unit fraction, whole, part, calculate, fraction, share, group

STRUCTURES AND REPRESENTATIONS

Bar model, part-whole model

RESOURCES

Mandatory: counters, bar models

Optional: multiplication square

 In the eTextbook of this lesson, you will find interactive links to a selection of teaching tools.

Quick recap

Ask children to find a unit fraction of an amount. Give examples, such as $\frac{1}{5}$ of 35. Ask children to work with a partner and create a diagram that shows that the answer is 7. Ask other similar questions to ensure that all children understand how to find a unit fraction of an amount.

Unit 11: Fractions (2), Lesson 6

Discover

WAYS OF WORKING Pair work

ASK

- Question 1 a): *How do we find $\frac{1}{5}$ of an amount? What diagram can you draw to show this?*
- Question 1 b): *What fraction of a bowl of raspberries is needed for the cake? How do you think you can find $\frac{2}{5}$? How do you think you can change your diagram?*

IN FOCUS It is important for children to make links to the previous lesson, and the similar context will help them to do this. They must identify that the initial strategy to complete the problem is the same as the one used in the previous lesson. This is the purpose of question 1 a). The **Quick recap** activity will also help. In question 1 b) children have to think about how they will find $\frac{2}{5}$, given that they know $\frac{1}{5}$. Throughout, encourage children to draw a bar model to support them.

PRACTICAL TIPS Children should continue to replicate the context of the lesson and share a set of objects between a different number of bowls. You could then ask them to find a non-unit fraction of the objects in one bowl.

ANSWERS

Question 1 a): $\frac{1}{5}$ of 35 = 7

Question 1 b): 14 raspberries are needed for 1 cake.

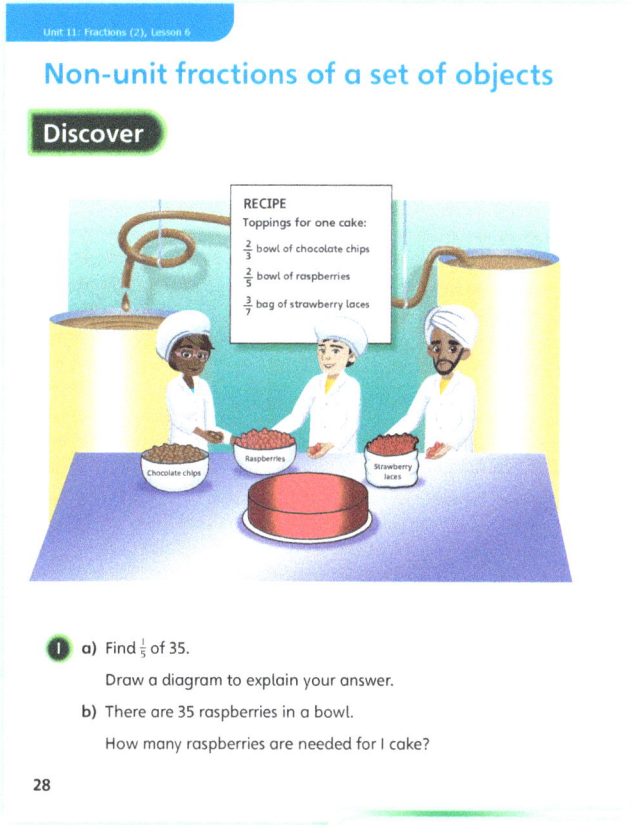

PUPIL TEXTBOOK 3C PAGE 28

Share

WAYS OF WORKING Whole class teacher led

ASK

- Question 1 a): *What is the same as and what is different from the method used in the previous lesson?*
- Question 1 b): *Why is there an additional multiplication step in this method?*
- Question 1 a): *How many parts are shaded on the bar model? What does each shaded part on the bar model show?*

IN FOCUS When children re-enact the scenario given in **Discover**, they should think about what bar model they will use to find the non-unit fraction of the amount you have specified. Children should link each step that they complete to the appropriate abstract equation in order to understand why they either divide or multiply at different points. Question 1 a) recaps the learning from the previous lesson. In question 1 b), children change their bar model to show what they need to do in order to find two-fifths.

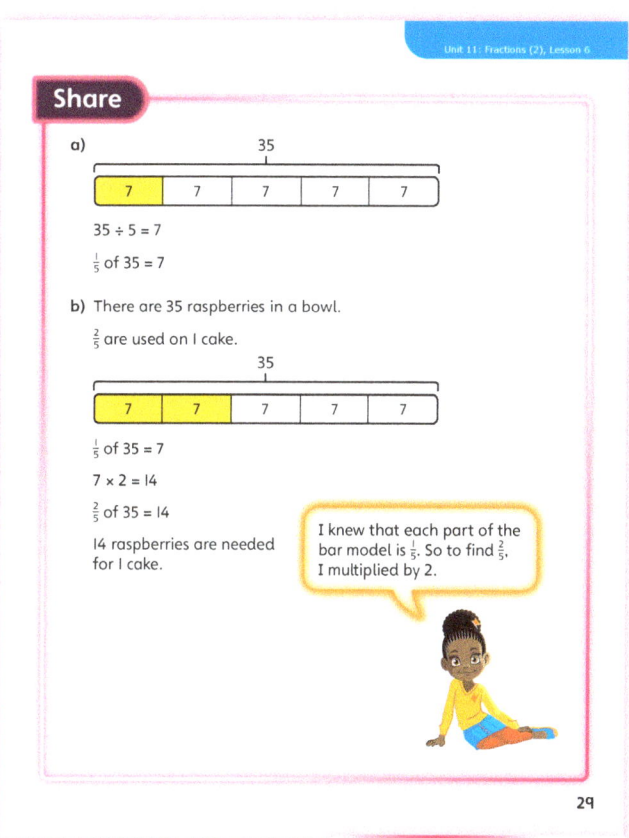

PUPIL TEXTBOOK 3C PAGE 29

67

Unit 11: Fractions (2), Lesson 6

Think together

WAYS OF WORKING Whole class teacher led (I do, We do, You do)

ASK
- Question 1 a): *Why do we calculate the unit fraction of an amount initially?*
- Question 1 b): *How is this unit fraction of the amount used to calculate a non-unit fraction of the same amount?*

IN FOCUS In question 1, children initially find the unit fraction of an amount and then use this quantity to calculate non-unit fractions of the same amount. Children should begin to understand that when they initially divide by the denominator they have found the unit fraction of the set of objects, and that when they then multiply by different numerators they are finding different non-unit fractions.

STRENGTHEN Children should continue to use resources throughout this section of the lesson until they fully understand the concept. It may help them to use a scaffold, similar to those provided in the **Share** section, to link each calculation to the manipulation of the resources at different points.

DEEPEN Children should describe the patterns they see when finding progressive fractions, for example $\frac{1}{5}, \frac{2}{5}, \frac{3}{5}, \frac{4}{5}$ and $\frac{5}{5}$ of the same amount. They should link the difference between each progressive amount to the initial unit fraction that was calculated.

ASSESSMENT CHECKPOINT If children can successfully complete question 2 and explain what they have done at each step in order to answer the question, it should offer a good indication of their understanding at this point in the lesson.

ANSWERS

Question 1 a): $\frac{1}{6}$ of 18 = 3

Question 1 b): $\frac{4}{6}$ of 18 = 12
$\quad\quad\quad\quad\quad$ 4 × 3 = 12

Question 1 c): $\frac{5}{6}$ of 18 = 15
$\quad\quad\quad\quad\quad$ 5 × 3 = 15.

Question 2 a): 8

Question 2 b): 12

Question 3: Children should show:
$\frac{3}{5}$ of £15 = $\frac{3}{4}$ of £12 = £9.

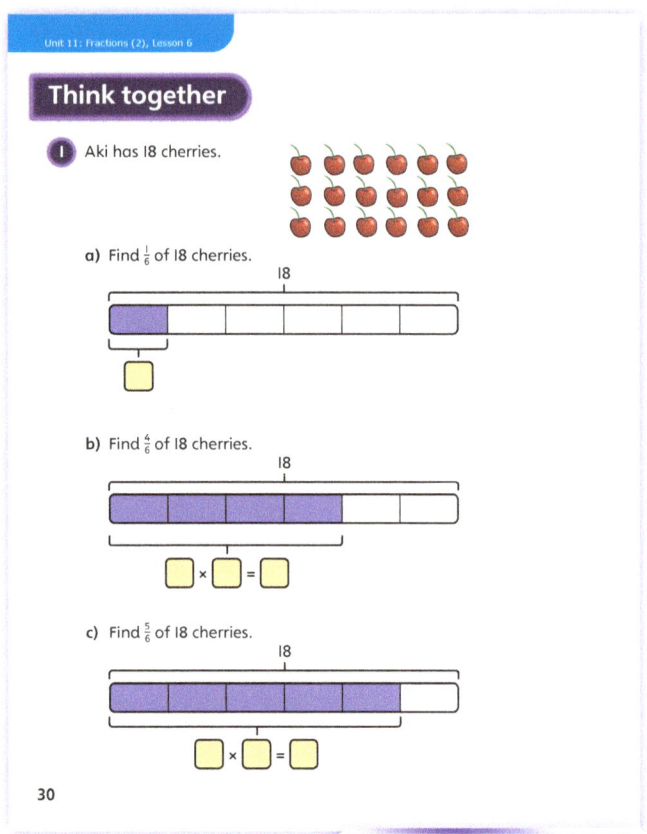

PUPIL TEXTBOOK 3C PAGE 30

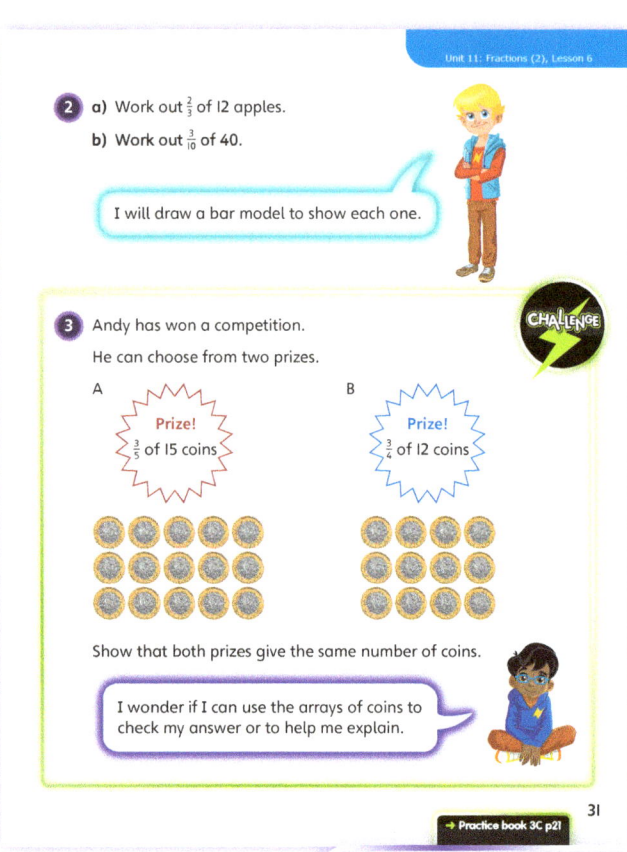

PUPIL TEXTBOOK 3C PAGE 31

Unit 11: Fractions (2), Lesson 6

Practice

WAYS OF WORKING Independent thinking

IN FOCUS In question ❶, children are provided with question scaffolding to illustrate that their first step should be to calculate the unit fraction of a set of objects, and that this can then be used to calculate the non-unit fraction of the same amount. This scaffolding is then removed for later questions in this section. In question ❷, children have to find $\frac{2}{5}$ of a different amount. Children may use bar models to support their understanding. They should notice that the method remains the same, and that just the starting amount changes. In question ❸, bar models are provided to support understanding.

STRENGTHEN Children may require a scaffold, similar to that provided for the initial questions within the lesson, for all questions. This may just be a prompt to first find the unit fraction and then multiply by the numerator to find the non-unit fraction, or it may be more question-specific, requiring appropriate visual images.

DEEPEN Ask children to devise their own problems, similar to **Challenge** question ❺, and swap with a partner to solve. Provide children with calculations that have missing information in different places. This will help deepen their understanding of the concept – for example, $\frac{?}{4}$ of 20 = 15 or $\frac{3}{?}$ of 25 = 15.

ASSESSMENT CHECKPOINT Use questions ❶, ❷ and ❸ to check that children are fluent and confident in finding a non-unit fraction of an amount.

ANSWERS Answers for the **Practice** part of the lesson can be found in the *Power Maths* online subscription.

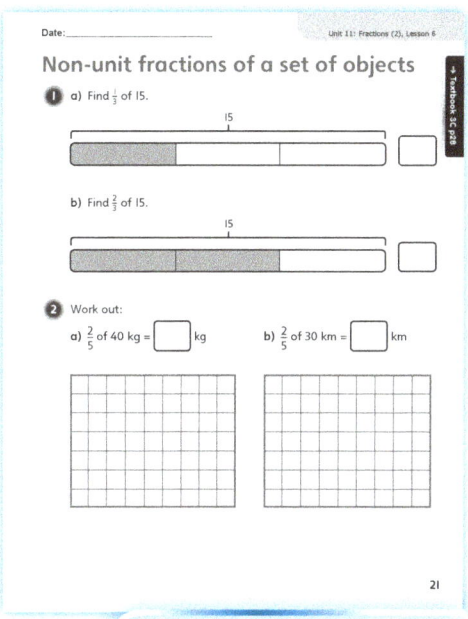

PUPIL PRACTICE BOOK 3C PAGE 21

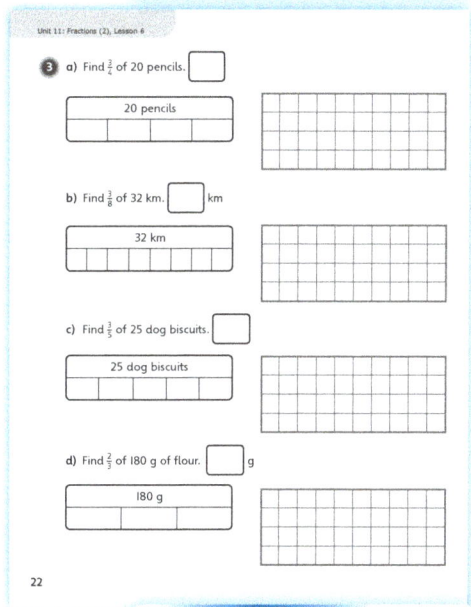

PUPIL PRACTICE BOOK 3C PAGE 22

Reflect

WAYS OF WORKING Independent thinking, pair work

IN FOCUS This question helps children to formalise the method of finding a non-unit fraction of an amount by using a bar model. Children should be encouraged to use the appropriate mathematical language to explain the different features of their bar model and how it allows them to calculate $\frac{2}{5}$ of £35. Ask children to work independently initially and to then share their response with either a partner or the whole class. They should explain the steps they took to draw the bar model and to work out the answer.

ASSESSMENT CHECKPOINT The strength of the children's verbal explanation and accuracy of their bar model can be used to assess how secure their understanding is of the main objectives of the lesson.

ANSWERS Answers for the **Reflect** part of the lesson can be found in the *Power Maths* online subscription.

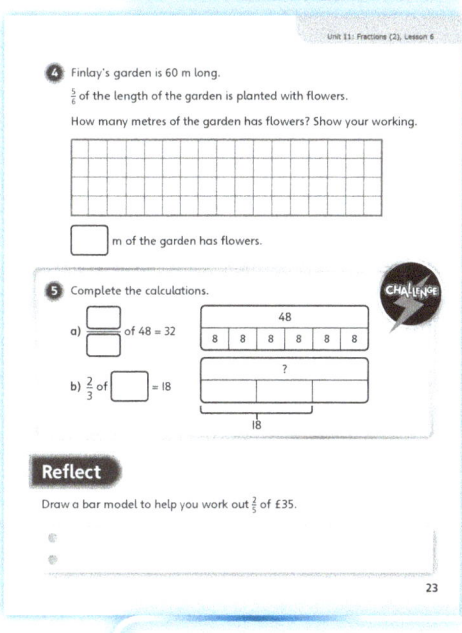

PUPIL PRACTICE BOOK 3C PAGE 23

After the lesson ⏸

- Are children ready to move on to finding increasingly difficult non-unit fractions of an amount in the next lesson?
- Do children require additional practice in finding a non-unit fraction of a countable set of objects first?

69

Unit 11: Fractions (2), Lesson 7

Reason with fractions of an amount

Learning focus
In this lesson, children will find non-unit fractions of a set of objects not shown visually and use given fractional amounts to calculate an unknown whole.

Before you teach
- Are children secure finding a unit and a non-unit fraction of an amount?
- Can children explain how a bar model can be used to find a fraction of an amount?

NATIONAL CURRICULUM LINKS

Year 3 Number – fractions

Recognise, find and write fractions of a discrete set of objects: unit fractions and non-unit fractions with small denominators.

ASSESSING MASTERY

Children can reason with fractions of an amount and solve problems where they are given a fraction and they have to find the whole or find a different fraction of the amount. In order to be successful with problems like this, it is important that children understand the concepts from the previous lesson and can use a bar model to explain how to find a fraction of a given amount.

COMMON MISCONCEPTIONS

When solving problems such as $\frac{1}{5}$ of ☐ = 10, children will often say that the answer is 2 as they have found $\frac{1}{5}$ of 10. Ask children to pause and take a step back and read the question. Ask:
- *How does this question differ from ones that you worked out in previous lessons? What is the same? What is different? What information have you been told?*

Ask them to represent the problem using a bar model. They should see how this time the whole bar model is unknown and they know one-fifth. By representing as a bar model, they should be able to see that each part is worth 10 and that to find the whole they need to multiply 10 by 5.

STRENGTHENING UNDERSTANDING

Children should continue to use bar model diagrams to help them conceptualise what they are being asked to work out and the steps that are needed to find a solution. Ask children what each number in a calculation tells them and where this information can go on the bar model.

GOING DEEPER

Ask children to create their own problems where they have to find the whole. Encourage children to check their answers. This will allow children to demonstrate their deep conceptual understanding and ability to select appropriate values that work with different fractional quantities.

KEY LANGUAGE

In lesson: set of objects, fraction, numerator, denominator, non-unit fraction, calculate, represent, divide, multiply, share, whole

Other language to be used by the teacher: original amount

STRUCTURES AND REPRESENTATIONS

Bar model

RESOURCES

Optional: counters

 In the eTextbook of this lesson, you will find interactive links to a selection of teaching tools.

Quick recap

Ask children to find the following: $\frac{1}{5}$ of 20, $\frac{2}{5}$ of 20, $\frac{3}{5}$ of 20, $\frac{4}{5}$ of 20 and $\frac{5}{5}$ of 20. What is the same about the methods they use? What is different? Encourage children to show their understanding by drawing a bar model to help their explanations.

Unit 11: Fractions (2), Lesson 7

Discover

WAYS OF WORKING Pair work

ASK

- Question 1 a): *What information do you know? How can you represent this information on a bar model? How many parts is the bar model split into? What is one part equal to? What is the whole equal to?*
- Question 1 b): *What is one tenth of the number? How can you work out nine tenths? Can you do this in one step?*

IN FOCUS In question 1 a), children realise (from the bar model) that they have been given one of the parts and that they have to work out the whole. Encourage children to explain why they put each number where they did on the bar model. Children may put 6 as the whole. If so, get them to think about this carefully. In question 1 b), children are required to work out $\frac{9}{10}$, given that they know $\frac{1}{10}$. Encourage children to find different ways to work out the answer.

PRACTICAL TIPS Encourage children to draw or represent the information that they are given within a question. This will help them to conceptualise what they need to do to solve the question. As the quantities within the question increase, children may have to be creative when using resources to represent each question.

ANSWERS

Question 1 a): Mr Lopez's number is 60.

Question 1 b): $\frac{9}{10}$ of Mr Lopez's number is 54.

PUPIL TEXTBOOK 3C PAGE 32

Share

WAYS OF WORKING Whole class teacher led

ASK

- Question 1 a): *How does the bar model represent the information?*
- Question 1 a): *Do we know a whole or a part? How do you know?*
- Question 1 b): *Can you explain each method to find $\frac{9}{10}$? How do the methods differ?*

IN FOCUS Throughout **Share**, use the bar model representations to help children understand how to find the whole. Children may think the 6 is the whole but explain that this time the 6 represents one of the parts. Use the bar model to help explain that to find the whole, children need to multiply by 10. In question 1 b), use the bar models to explain two different methods they can use to work out $\frac{9}{10}$. This will help children to reason with finding a fraction of an amount.

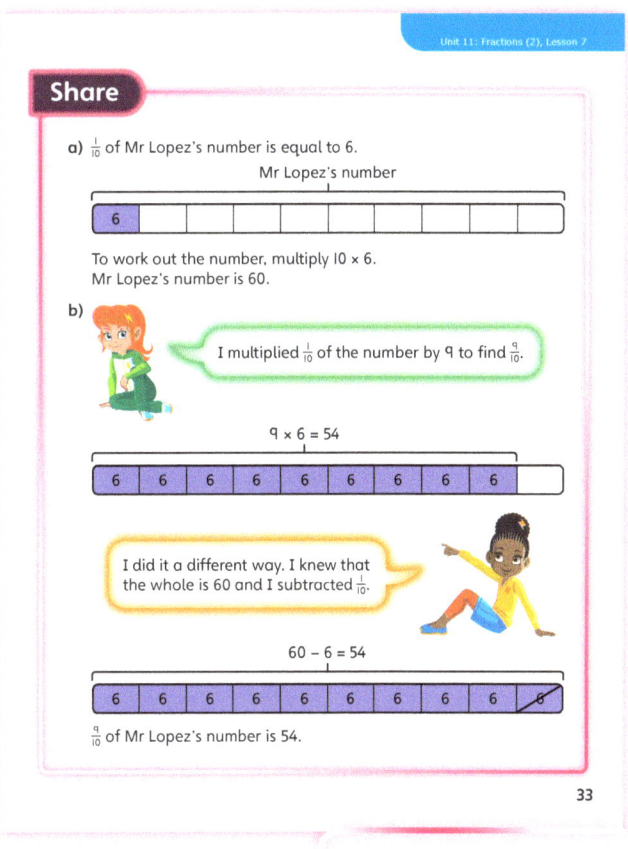

PUPIL TEXTBOOK 3C PAGE 33

Unit 11: Fractions (2), Lesson 7

Think together

WAYS OF WORKING Whole class teacher led (I do, We do, You do)

ASK
- Question 1: *How does the bar model represent Reena's comment? What is each part worth? How can you find the whole?*
- Question 2: *What does the bar model show? How can you use it to find $\frac{1}{5}$? How can you use it to find the whole?*
- Question 3: *What are Reena's and Richard's numbers? How can you find $\frac{1}{6}$ of the total of their numbers? Is there a quicker way?*

IN FOCUS In question 1, talk children through how and why the bar model represents Reena's statement. Ask children to explain why one of the parts is equal to 3. Discuss how they can use multiplication to find Reena's number. In question 2, children reason further by dividing to find $\frac{1}{5}$ when they know $\frac{2}{5}$. The bar model representation will help them see why they need to divide by 2. The remaining questions provide different examples of problems, each one asking something a bit different. Discuss different methods that children may use to get to the answers. In question 3, children work in pairs to work out $\frac{1}{6}$ of the total of Richard and Reena's number. Discuss different methods children may use and compare them. Ask: *Which method is most efficient?*

STRENGTHEN Children may find it useful to use coloured rods to physically make a bar model and manipulate its parts. Alternatively, some children may require bar models that match each question so they can record the values that they are given and organise their thinking.

DEEPEN Children should be encouraged to work backwards from their answers to check that they are correct. This will deepen their understanding of working in both directions from and to the whole set of objects.

ASSESSMENT CHECKPOINT Children should be able to justify the steps of their calculations based on the context of the question and the bar model that has either been given or drawn.

ANSWERS

Question 1: Reena's number is 24.

Question 2 a): $\frac{1}{5}$ of Richard's number is 6.

Question 2 b): Richard's number is 30.

Question 2 c): $\frac{4}{5}$ of Richard's number is 24.

Question 2 d): $\frac{9}{10}$ of Richard's number is 27.

Question 3: Children should describe the two methods to show that the answers are the same.
$\frac{1}{6}$ of 24 + $\frac{1}{6}$ of 30 = 4 + 5 = 9
$\frac{1}{6}$ of (24 + 30) = $\frac{1}{6}$ of 54 = 9

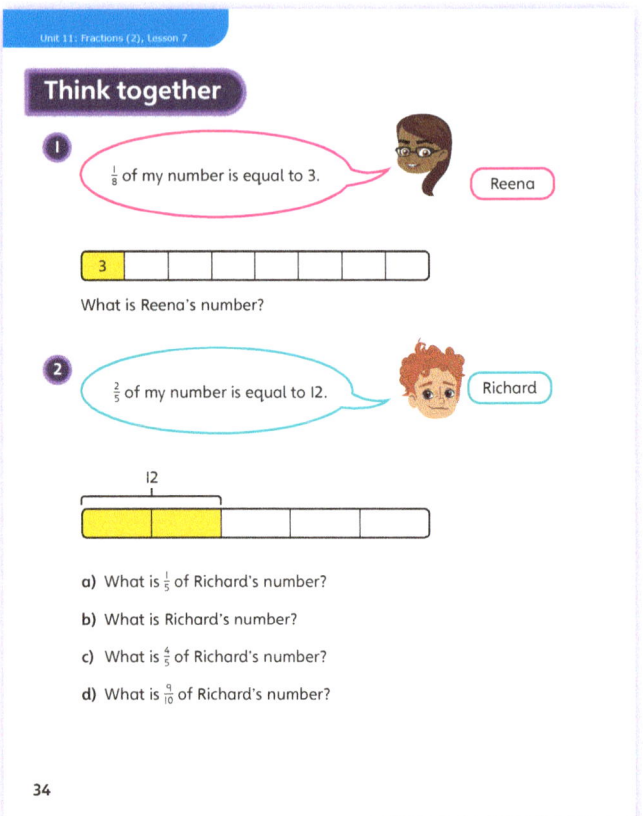

PUPIL TEXTBOOK 3C PAGE 34

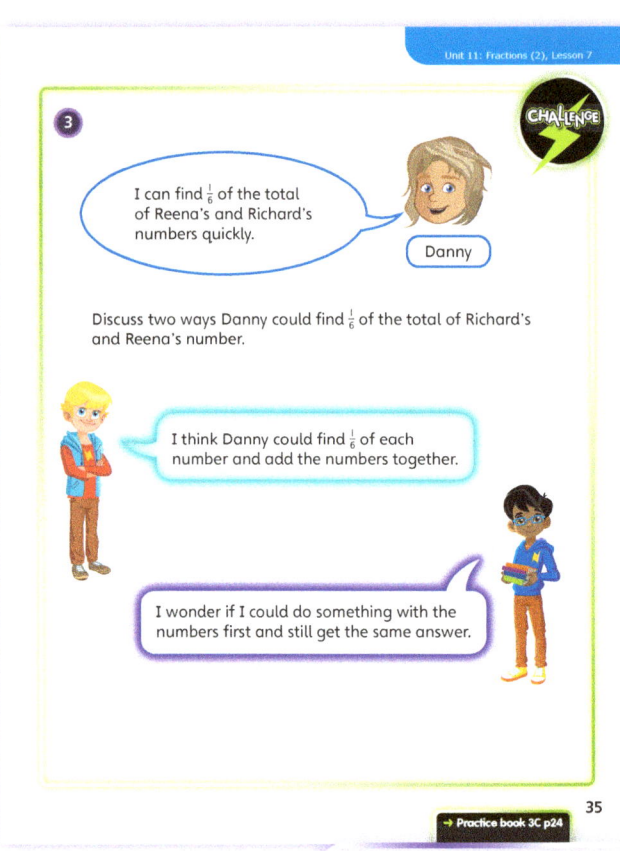

PUPIL TEXTBOOK 3C PAGE 35

Unit 11: Fractions (2), Lesson 7

Practice

WAYS OF WORKING Independent thinking

IN FOCUS Initially children are given scaffolds to show the steps of their calculations. Encourage children to show these steps in questions where this scaffold is not given. Where a bar model is not given, they should also be encouraged to draw their own.

For question ②, children may realise they just need to multiply the denominator by the answer. They should be encouraged to check that their answer is correct by finding the fraction of the amount.

In question ③, children are given $\frac{1}{8}$ and they have to use reasoning to work out other fractions of the whole, without necessarily working out the whole.

Question ④ requires children to first divide to work out $\frac{1}{4}$.

STRENGTHEN Children should continue to work with resources to help them interpret the information given in the question and to draw bar models accordingly. If they still find this difficult, children could be given bar models that represent each question and that require them to label the given information accurately.

DEEPEN Children could be challenged to create word problems from existing questions that are not contextualised. Or they can create their own problems from scratch – this would require them to select appropriate values that work with different fractional quantities.

THINK DIFFERENTLY Question ⑤ requires children to find the most efficient way to solve the problem. Some children may find $\frac{1}{5}$ of each number and add. Others may realise they can add first and then find $\frac{1}{5}$ of the total. Discuss with children which way they prefer.

ASSESSMENT CHECKPOINT Questions ① to ④ will help you understand how confident children are with reasoning with fractions of amounts. Can they find the whole fluently? Can they find other fractions of an amount, when they are given one fraction already?

ANSWERS Answers for the **Practice** part of the lesson can be found in the *Power Maths* online subscription.

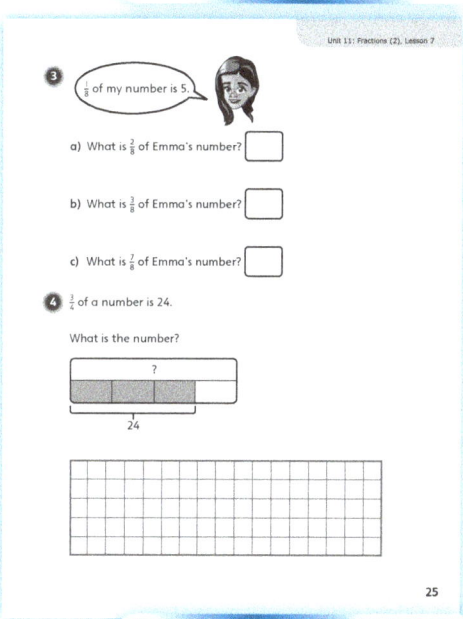

PUPIL PRACTICE BOOK 3C PAGE 24

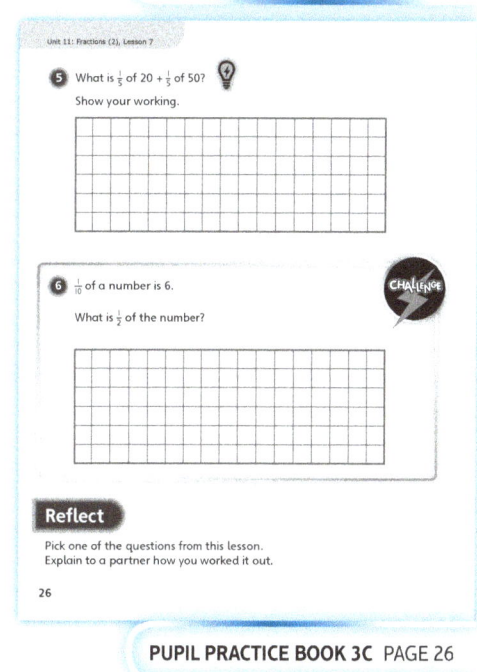

PUPIL PRACTICE BOOK 3C PAGE 25

Reflect

WAYS OF WORKING Pair work

IN FOCUS Children pick one of the questions from this lesson and explain to a partner how they worked out the answer. Their partner should listen and ask them to explain why the answer and method are correct.

ASSESSMENT CHECKPOINT The strength and accuracy of children's explanations will allow you to gauge how secure children are with their understanding of the lesson objective.

ANSWERS Answers for the **Reflect** part of the lesson can be found in the *Power Maths* online subscription.

PUPIL PRACTICE BOOK 3C PAGE 26

After the lesson

- Can children find the whole given a fraction of the whole?
- Can children find other fractions of an amount, given information about a part (for example, find $\frac{4}{5}$ if they are told $\frac{1}{5}$)?

Unit 11: Fractions (2), Lesson 8

Problem solving – fractions of measures

Learning focus
In this lesson, children will learn to reason mathematically, and solve problems involving fractions and money by adding and subtracting fractions.

Before you teach
- What resources will you be using in the lesson? Which resources will children find most useful?
- Can children use division to find fractions of an amount and then consider what is left?
- What previous opportunities have children had to find measurements?

NATIONAL CURRICULUM LINKS

Year 3 Number – fractions

Recognise, find and write fractions of a discrete set of objects: unit fractions and non-unit fractions with small denominators.

Recognise and use fractions as numbers: unit fractions and non-unit fractions with small denominators.

ASSESSING MASTERY

Children can recognise the operation needed to answer a word problem. They can write the necessary calculation to answer the problem and find the correct solution.

COMMON MISCONCEPTIONS

Some children will generalise and assume that, for example 'all quarters' are the same. They do not understand that the size of the whole determines the size of the fractional part. Ask:
- Which is bigger, $\frac{1}{4}$ of a 2 litre bottle or $\frac{1}{4}$ of a 500 ml bottle?

Some children misunderstand the idea that 'the bigger the denominator the smaller the part' and ignore the numerators when comparing fractions. For instance, they may think that $\frac{1}{4} > \frac{3}{5}$ because quarters are bigger than fifths. Ask:
- Which is greater, $\frac{1}{4}$ or $\frac{3}{5}$? How do you know? What information have you looked at to decide?

STRENGTHENING UNDERSTANDING

Use bar models to support children's understanding of the problem. Ask children what each line in the question tells them about the bar model. Ask: *How many parts should the bar model be? How do you know? Do you know the whole or do you need to work it out?*

GOING DEEPER

The more hands-on experience children gain, the deeper their understanding will be. Provide real-life examples of fractions of measurements, for instance a cake recipe that includes fractions of different amounts. Give children time to discuss and compare the different amounts and encourage them to refer to a fraction wall to check their answers.

KEY LANGUAGE

In lesson: fraction, amount, subtract, method

Other language to be used by the teacher: word problem, mass, capacity, scales, litre (l), add, calculate, measure, greater

STRUCTURES AND REPRESENTATIONS

Bar model, fraction wall

RESOURCES

Optional: bar models, capacity measuring equipment, coloured rods, part-whole models

 In the eTextbook of this lesson, you will find interactive links to a selection of teaching tools.

Quick recap
In this lesson children need to apply fractions of amounts to word problems. Check that children can find any fraction of an amount. Encourage children to show their understanding by drawing an appropriate bar model.

Unit 11: Fractions (2), Lesson 8

Discover

WAYS OF WORKING Pair work

ASK

- Question 1 a): *What information is important? Can you draw a bar model to help you?*
- Question 1 b): *How can you work out how much fish is left in the bucket? Is there more than one way to find the answer?*

IN FOCUS Questions 1 a) and b) apply earlier understanding to the context of measure and word problems. In question 1 a), children need to use the information to find out how much fish the penguins need. Children should be encouraged to draw a bar model for support. In question 1 b), children explore different ways to find out how much fish is left.

PRACTICAL TIPS Recreate the scenario in **Discover** with something accessible to the class (for example, act out a problem with a bucket of water). Make sure you have equipment on hand to deal with spillages!

ANSWERS

Question 1 a): The zookeeper feeds the penguins 8 kg of fish.

Question 1 b): There are 12 kg of fish left.

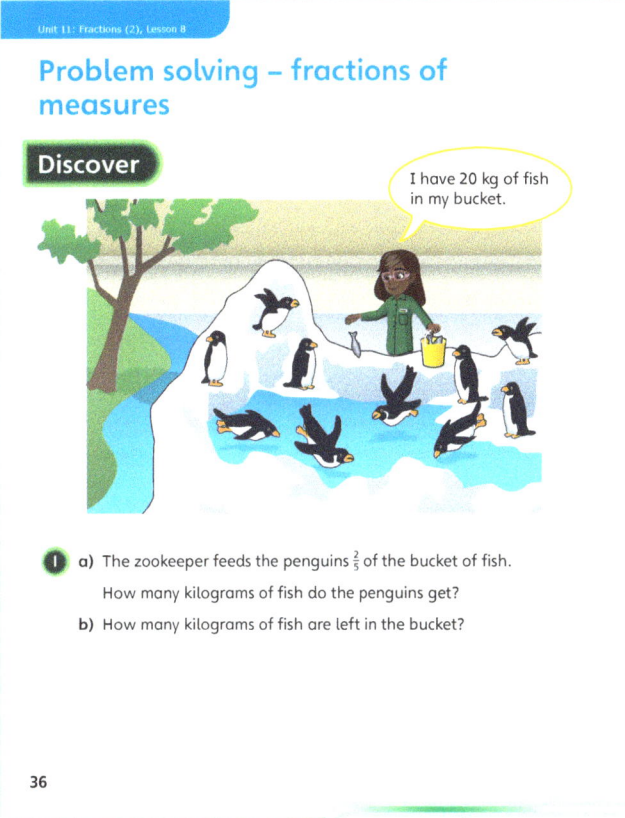

PUPIL TEXTBOOK 3C PAGE 36

Share

WAYS OF WORKING Whole class teacher led

ASK

- Question 1 a): *How does the bar model represent the problem? Why is each part worth 4 kg? How do you know that 8 kg of fish are eaten?*
- Question 1 b): *What method did Flo use first to find what is left? Why does finding $\frac{3}{5}$ of the bucket give the same answer? Which method do you prefer?*

IN FOCUS For question 1 a), encourage children to draw the bar model, putting on all the relevant information. Remind children why the bar model is divided into 5 equal parts and how to find the value of each part. Then explain why you need to multiply by 2 to find $\frac{2}{5}$. In question 1 b), discuss the two methods that children could use to find how much fish is left in the bucket. Ask children which method they prefer. The second method is conceptually more difficult and so children may at this stage just do a subtraction.

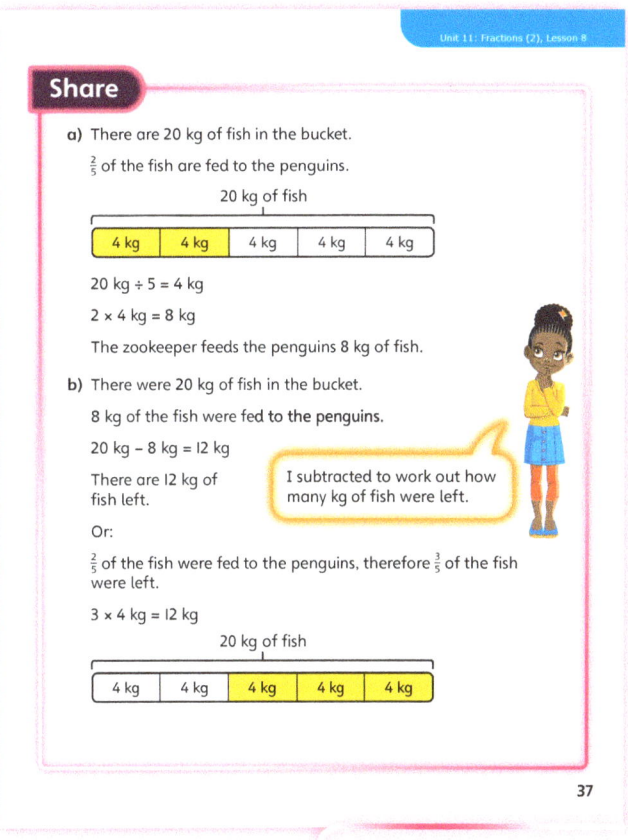

PUPIL TEXTBOOK 3C PAGE 37

Think together

WAYS OF WORKING Whole class teacher led (I do, We do, You do)

ASK

- Question ❶: *How does the bar model represent the situation? How can you find the number of each coloured frog?*
- Question ❷: *What would the bar model look like for this question? How many parts would you divide this into? How much is each part worth? How can you find out how much is left? Is there more than one way to do this?*
- Question ❸: *How much feed is left in each bag? Is there enough feed for the zebras? Do you need to work out all the information before you know?*

IN FOCUS These questions practise finding fractions of an amount in worded contexts. In question ❶, children work out the number of orange frogs and the number of green frogs. To work out the number of remaining frogs they may either do a subtraction or find a fraction of the total. In question ❷, children should be encouraged to draw a bar model – though some children may be able to find the answer without doing this. In question ❸, children can approach solving the problems in different ways. For example, some children may realise they don't need to work out the amount of feed left in the second bag, as there cannot be enough feed anyway. Discuss Flo's suggestion for finding a quarter of a large amount. Most of these problems have multiple steps that must be completed to find the answer. Encourage children to show all the steps in their working out.

STRENGTHEN Use bar models to support children's understanding of the problem. Ask children what each line in the question tells them about the bar model. Ask: *How many parts should the bar model be? How do you know? Do you know the whole or do you need to work it out?*

DEEPEN Discuss how you can find fractions of amounts without going into decimals – for example, find $\frac{2}{5}$ of 1 kg or $\frac{1}{2}$ of $\frac{1}{2}$ kg. Children should use their knowledge of converting units.

ASSESSMENT CHECKPOINT Children should understand the questions and explain what is required. They should be able to explain the operations used to solve the questions and demonstrate using manipulatives or drawing diagrams to support their understanding and reasoning.

ANSWERS

Question ❶ a): There are 18 orange and green frogs altogether.

Question ❶ b): There are 6 yellow frogs.

Question ❷: Max has 8 treats left.

Question ❸: There is not enough feed.
$\frac{1}{4}$ of 800 g + $\frac{3}{4}$ of 200 g = 200 g + 150 g = 350 g
350 g < 450 g

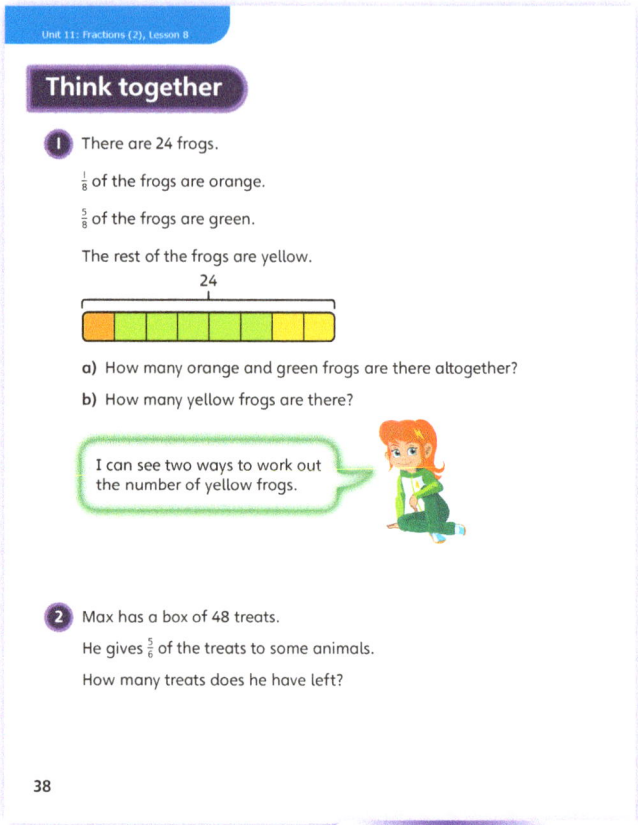

PUPIL TEXTBOOK 3C PAGE 38

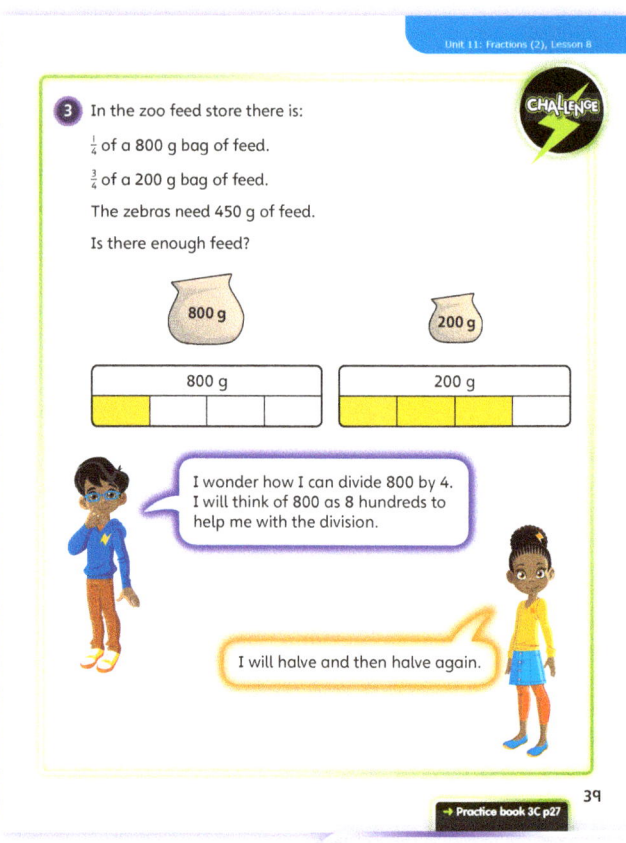

PUPIL TEXTBOOK 3C PAGE 39

Unit 11: Fractions (2), Lesson 8

Practice

WAYS OF WORKING Independent thinking

IN FOCUS This **Practice** provides an opportunity for children to solve fraction of amount problems set in different contexts, including many multi-step problems. Children should be encouraged to draw bar models to help them determine the calculations that they need to do. Question ❶ involves finding unit fractions of amounts and then working out the total. In question ❷, look for the methods that children use. Do they find $\frac{3}{5}$ and subtract or do they find $\frac{2}{5}$? Encourage children to discuss both methods. In question ❹, children first have to find the right-hand side and then find the whole by multiplying.

STRENGTHEN For all questions, encourage children to use a fraction strip, number line, or both, to work through the steps of the problem. Remind children to read the questions carefully and record their findings. Encourage children to label their bar models, so they can follow their work and check their answers at the end.

DEEPEN Question ❺ challenges children's ability to solve problems. They will need to be systematic, and organise their thinking and results. Deepen their understanding by asking them to describe the features of the bar model they drew and explain how it helped them to answer the question.

ASSESSMENT CHECKPOINT Use the questions to determine whether children can solve problems involving finding fractions of amount. Can they draw bar models to help them work out the steps they need to take to solve the problems?

ANSWERS Answers for the **Practice** part of the lesson can be found in the *Power Maths* online subscription.

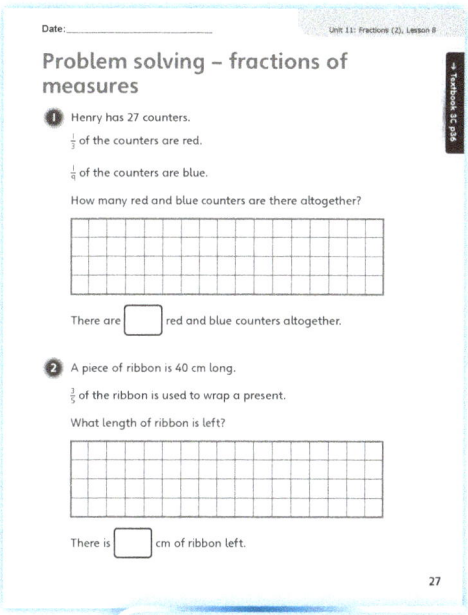

PUPIL PRACTICE BOOK 3C PAGE 27

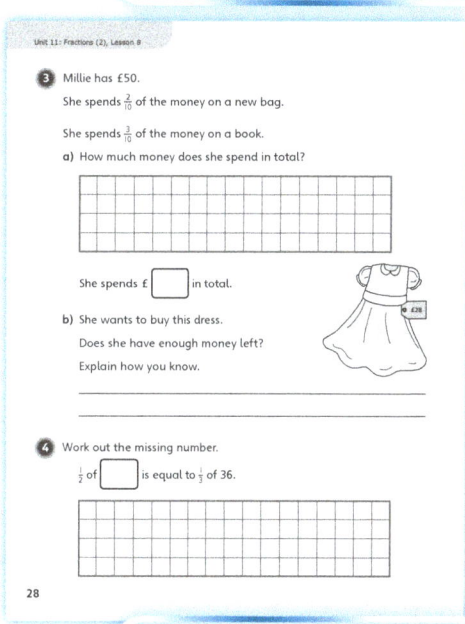

PUPIL PRACTICE BOOK 3C PAGE 28

Reflect

WAYS OF WORKING Pair work

IN FOCUS This activity gives children a chance to reflect on the method they used to solve question ❷. Share the solutions as a class. Discuss the two possible ways that children may solve this problem and discuss why both methods work. Does the class prefer one way to the other? Does it matter?

ASSESSMENT CHECKPOINT Ensure that children have at least one method to solve question ❷.

ANSWERS Answers for the **Reflect** part of the lesson can be found in the *Power Maths* online subscription.

After the lesson ⏸

- What opportunities will children have to compare fractions of measures outside this lesson?
- What opportunities will children have to problem-solve independently?
- Are children confident in choosing different types of representation to solve problems?

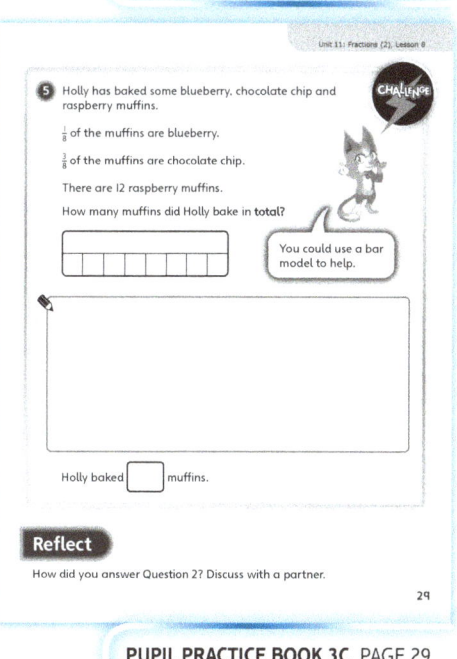

PUPIL PRACTICE BOOK 3C PAGE 29

77

Unit 11: Fractions (2)

End of unit check

Don't forget the unit assessment grid in your *Power Maths* online subscription.

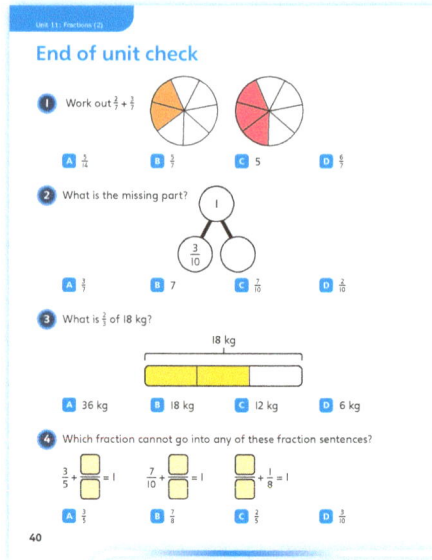

PUPIL TEXTBOOK 3C PAGE 40

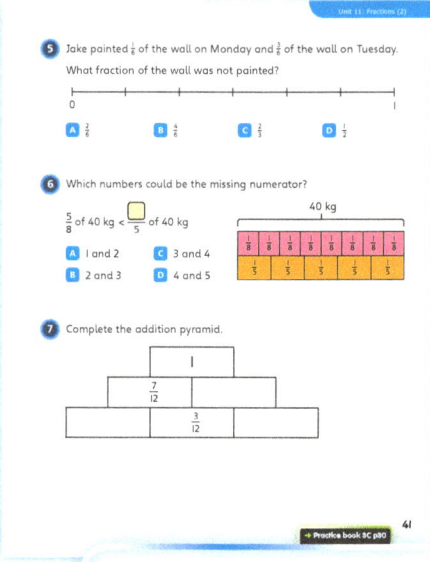

PUPIL TEXTBOOK 3C PAGE 41

WAYS OF WORKING Group work adult led

IN FOCUS
- Question ❶ checks children's understanding of adding two fractions with the same denominator. Visual representations are provided for support.
- Question ❷ assesses children's ability to partition the whole. Ask: *What do the numerators need to add up to?*
- Question ❸ assesses children's ability to find a non-unit fraction of an amount. Ask: *What do you do first? Then what? How does the bar model help you solve the problem?*
- Question ❹ assesses children's ability to add and subtract fractions to make 1. Ask: *For each of these calculations, how can 1 be represented as a fraction? Why?*
- Question ❺ assesses children's ability to solve a problem by adding and subtracting fractions. Ask: *Can you solve the question in two ways?*
- Question ❻ assesses children's ability to find fractions of measures.
- Question ❼ provides a fraction pyramid that checks whether children can add and subtract fractions confidently. Each pair of fractions should add to the number directly above them.

ANSWERS AND COMMENTARY

Children will demonstrate mastery by finding equivalent fractions and comparing fractions. They can use bar models and number lines to support their answers. Children can add and subtract fractions with confidence and can solve fraction problems including problems involving fractions of measures.

Q	A	WRONG ANSWERS AND MISCONCEPTIONS	STRENGTHENING UNDERSTANDING
1	B	Choosing A implies children do not understand the concept of adding fractions. They have added the denominators.	To help children gain fluency in their understanding of fractions: • Make sure children have access to a fraction wall throughout the day. • Ensure all the representations of number lines are labelled clearly. • Give children access to fraction rods throughout the lesson. • Ensure children have access to unit fractions made up of different numbers of equal parts.
2	C	Choosing A, B or D may indicate a lack of understanding of partitioning the whole.	
3	C	Choosing D implies that children have found $\frac{1}{3}$ as opposed to $\frac{2}{3}$.	
4	A	Choosing B, C or D may indicate that children lack experience in adding and subtracting fractions to make one whole.	
5	A	Choosing B may indicate children misread the question. Choosing C or D may indicate a lack of understanding of how to subtract fractions and identify equivalent fractions.	
6	D	Choosing A, B or C may indicate that children lack an understanding of comparing fractions of measures.	
7	Row 2: $\frac{5}{12}$ Row 3: two fractions that add to $\frac{9}{12}$	Check that the fractions in each row add up to 1.	

Unit 11: Fractions (2)

My journal

WAYS OF WORKING Independent thinking

ANSWERS AND COMMENTARY

Children should notice that there are 18 counters in the whole (in the top bar) and these have been divided into 3 equal parts. There are 6 counters in each part.

Children may record things such as:
- 18 ÷ 3 = 6, which represents 18 shared into 3 equal groups, giving 6 in each group.
- 18 ÷ 6 = 3, which represents 18 grouped into groups of 6s, means that there are 3 groups.
- They may write a fact family such as 3 × 6 = 18 or 6 × 3 = 18.

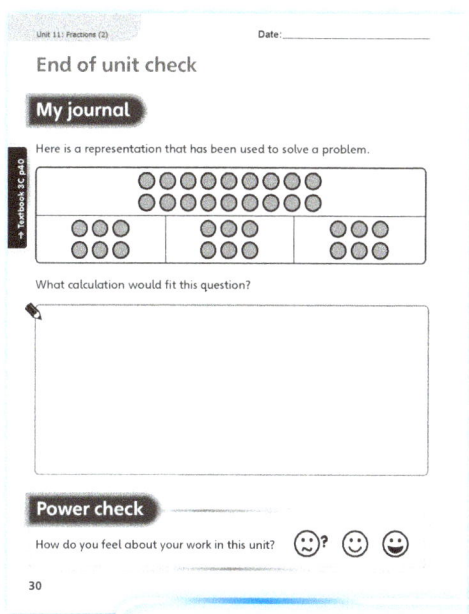

PUPIL PRACTICE BOOK 3C PAGE 30

Power check

WAYS OF WORKING Independent thinking

ASK
- *What did you know about adding and subtracting fractions before you started this unit? What new things have you learnt?*
- *How confident do you feel about adding and subtracting fractions? How does this differ from adding and subtracting with whole numbers?*
- *What do you feel you could improve on in this unit? What do you need more help or practice with?*

Power play

WAYS OF WORKING Pair work

IN FOCUS This game will assess children's ability to add and subtract fractions with the same denominator. Children can use a bar model or number line to support their answers and show their calculations clearly.

ANSWERS AND COMMENTARY When using three cards, some children are not sure whether to add or subtract first. Provide more opportunities to practise counting on and back on a number line using fraction steps. Encourage children to pay attention to the result. Does it change depending on which calculation they do first?

PUPIL PRACTICE BOOK 3C PAGE 31

After the unit

- Is your classroom a 'fraction-rich' environment? How many opportunities are there for children to engage with fractions around the school environment? Are the fraction representatives visible and available for all children to use?

Strengthen and **Deepen** activities for this unit can be found in the *Power Maths* online subscription.

Unit 12
Money

Mastery Expert tip! 'Money is a great opportunity to bring real-life experience into the classroom. Send some ideas home about what parents and children could discuss when out shopping, for example. Children could add up two amounts in their head or find the change. Showing that maths is everywhere really helps reinforce the relevance of the subject.'

Don't forget to watch the Unit 12 video!

WHY THIS UNIT IS IMPORTANT

In this unit, children convert between pounds and pence for the first time. Although children know from Year 2 that there are 100 pence in one pound, they have not converted between them. Children will move on to converting amounts such as 720p into pounds and pence and vice versa. Notation with the decimal point is not used until Year 4, so children will continue to use the structure of *x* pounds and *y* pence or £*x* and *y*p. This will help them unitise 100 pence as £1. Next, children solve addition and subtraction problems relating to money. They add the units separately and convert for amounts crossing a pound. Working with money in context is an important skill and questions are asked in money contexts throughout *Power Maths*. This unit will give children the confidence to deal with problems that arise in future units, and it is vital that they understand the key concepts, particularly around conversion between pounds and pence.

WHERE THIS UNIT FITS

→ Unit 11: Fractions
→ **Unit 12: Money**
→ Unit 13: Time

In this unit, children will apply their knowledge of addition and subtraction to solve number problems involving money, including across a whole pound. In the subsequent units, contexts involving money feature regularly.

Before they start this unit, it is expected that children:
- know the symbols for pounds (£) and pence (p)
- can work out the value of coins and notes by counting the pounds and pence separately
- understand what change is and how to work it out in simple cases
- can select coins and notes that make a particular amount and recognise different ways of doing this.

ASSESSING MASTERY

By the end of the unit, children will be able to convert amounts from pounds and pence to just pence and vice versa. They should also be able to solve addition and subtraction problems (including finding change).

COMMON MISCONCEPTIONS	STRENGTHENING UNDERSTANDING	GOING DEEPER
Children may write amounts such as £7 and 6 pence as 76 pence. Discussing the place value of each digit should help them overcome this.	Encourage children to use coins and number lines throughout to support understanding. For example, when adding two amounts, children may find it useful to make the amounts from coins and add the coins together.	Solve problems such as: *Altogether, a toy train and toy car cost £2. The train costs 30 pence more than the car. How much does each item cost?* *Amelia has £5. She buys 3 cupcakes which cost 85p each. How much change does she get?*
Children may make mistakes when working out amounts of change. Use a number line to help support children.	Representing an addition and subtraction problem as a bar model may help children understand the structure of a problem.	

Unit 12: Money

UNIT STARTER PAGES

Use these pages to introduce the unit focus to children. You can use the characters to explore different ways of working too.

STRUCTURES AND REPRESENTATIONS

Coins: Using coins to physically model the problems will help children visualise and manipulate different amounts, particularly when crossing a pound.

Bar model: This model will help children visualise and understand the structure of the problems.

£12 + ☐ = £20

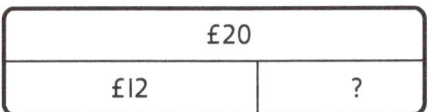

Number line: This model helps children visualise the order of numbers. It can help them count on or back to find a total amount and can be more efficient than using concrete resources.

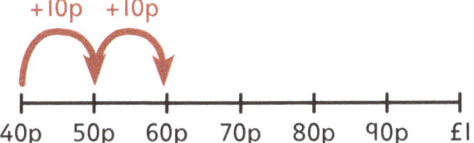

KEY LANGUAGE

There is some key language that children will need to know as part of the learning in this unit.

→ pounds (£) and pence (p)
→ convert
→ total
→ difference
→ change

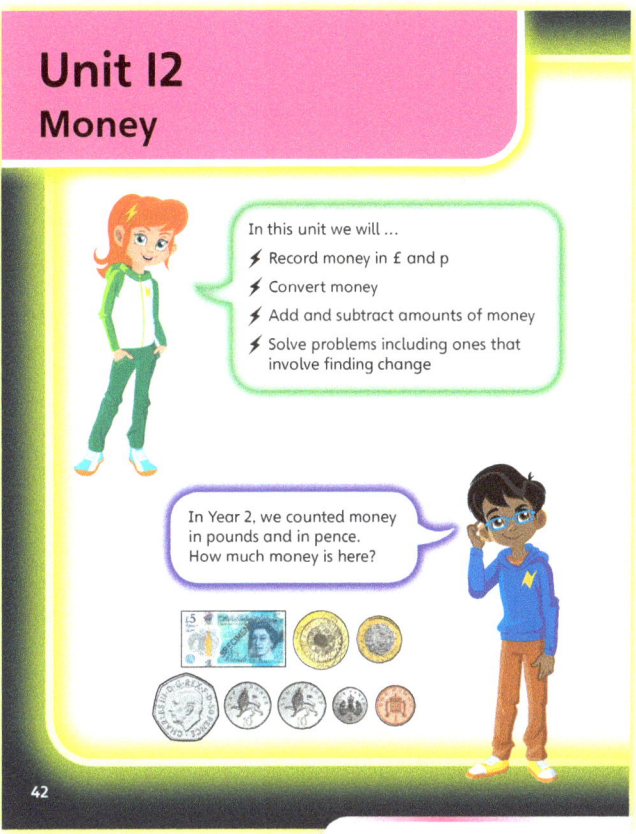

PUPIL TEXTBOOK 3C PAGE 42

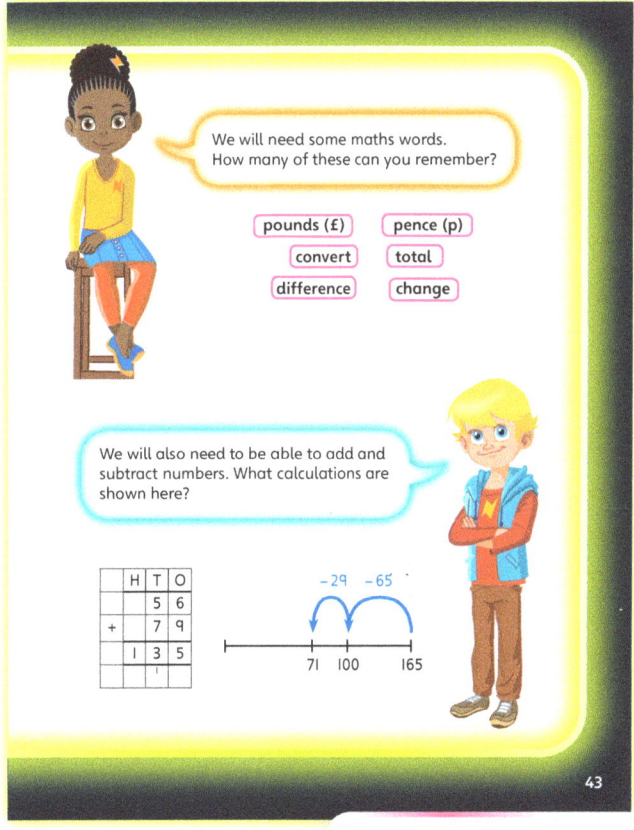

PUPIL TEXTBOOK 3C PAGE 43

Unit 12: Money, Lesson 1

Pounds and pence

Learning focus

In this lesson, children will learn the value of each coin and note and understand what these values represent. Children will answer questions that involve finding a total amount and express these amounts using the word 'and'.

Before you teach

- Can children count in multiples of coin denominations (1s, 2s, 5s, 10s, 20s, 50s)?
- Can children add simple 1- and 2-digit numbers?
- Do children understand the concept of place value and that there are ten 1s in 10 and ten 10s in 100?

NATIONAL CURRICULUM LINKS

Year 3 Measurement – money
Add and subtract amounts of money to give change, using both £ and p in practical contexts.

ASSESSING MASTERY

Children can recognise and use all coins and notes to make different totals. They can use a variety of different coins and notes to make the same total and reason about different amounts using vocabulary such as greater than, less than and equal to.

COMMON MISCONCEPTIONS

Children may not understand the value of each coin. They may think that £1 and 1p have the same value or that 20p has a greater value than £1. Ask:
- *How many pence is this coin worth? Can you show it on a number line?*

Children may also think that if they have more coins they have a greater total. Ask:
- *How many pence do you have all together?*

STRENGTHENING UNDERSTANDING

Role play can be a good way to help children understand the value of coins. Children can exchange ten 1p coins for a 10p coin or five 10p coins for a 50p coin, and so on. Create an activity where children have to find other children in the room with the same amount of money as them. Link this to place value counters where children have exchanged ten 1s for a 10 counter or ten 10s for a 100 counter.

GOING DEEPER

Ask children questions that require them to make an amount using the fewest/greatest number of coins and/or notes possible. Children could also create amounts using an exact number of coins/notes, for example: *Make 78p using exactly five coins. Make £15 and 25 pence using exactly three notes and seven coins.*

KEY LANGUAGE

In lesson: pounds, pence, total

Other language to be used by the teacher: amount

STRUCTURES AND REPRESENTATIONS

Number lines

RESOURCES

Mandatory: plastic or paper coins and notes

 In the eTextbook of this lesson, you will find interactive links to a selection of teaching tools.

Quick recap

Children need to know the different coins and notes. Show a coin or note on the whiteboard and ask children to write down the value. You may want to extend this by displaying multiple copies of the same coin and asking children to count how much in total. This practises simple counting of amounts using coins.

Unit 12: Money, Lesson 1

Discover

WAYS OF WORKING Pair work

ASK
- Question 1 a): *How has the money been laid out to help you count it? Does the order matter?*
- Question 1 a): *Can you show the amount on a number line?*
- Question 1 b): *What coins does Sofia have in her purse?*

IN FOCUS Review the names and values of the coins with children and use question 1 a) to check whether children can count amounts of money together. Pay attention to how they count. Do they separate the pounds and pence? Do they group the same amounts together?

PRACTICAL TIPS Role-play the opening scene with children and represent Lee and Sofia's money using plastic coins and notes or by drawing pictures.

ANSWERS

Question 1 a): Lee has 12 pounds and 88 pence, so £12 and 88p.

Question 1 b): Sofia gave Lee a £2 coin and 20p, 5p, 2p and 1p coins.

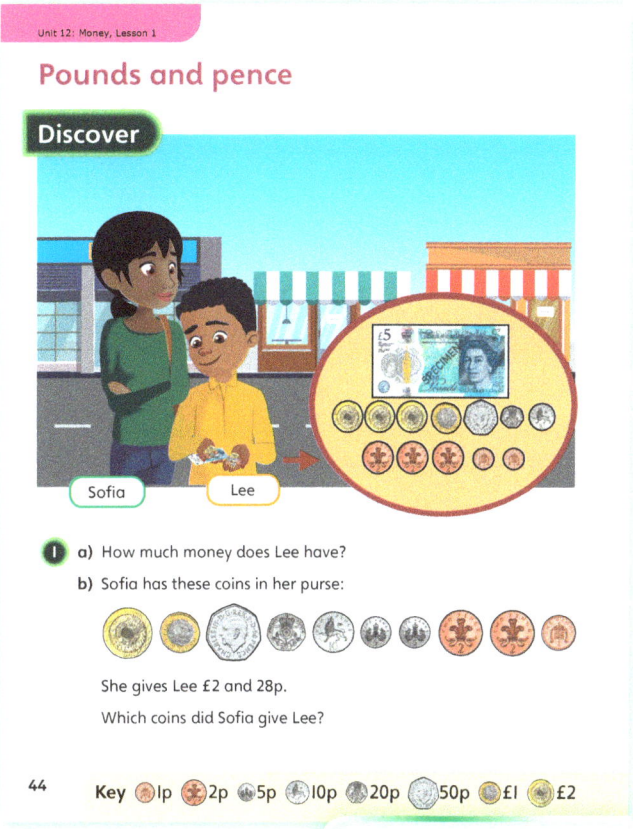

PUPIL TEXTBOOK 3C PAGE 44

Share

WAYS OF WORKING Whole class teacher led

ASK
- Question 1 a): *Which are the pounds and which are the pence? Which coins or notes have the biggest value? How does separating the pounds and pence make it easier to count? Does putting the money in size order help? How could you make £10?*
- Question 1 b): *Is there only one way Sofia can make £2 and 28p?*

IN FOCUS Children should see the value in separating pounds and pence in order to count them and also look for familiar number bonds to make 10. When counting on in money, children should recognise that they should first use their knowledge of number bonds to 100 to make a full £1 and then continue counting in pounds.

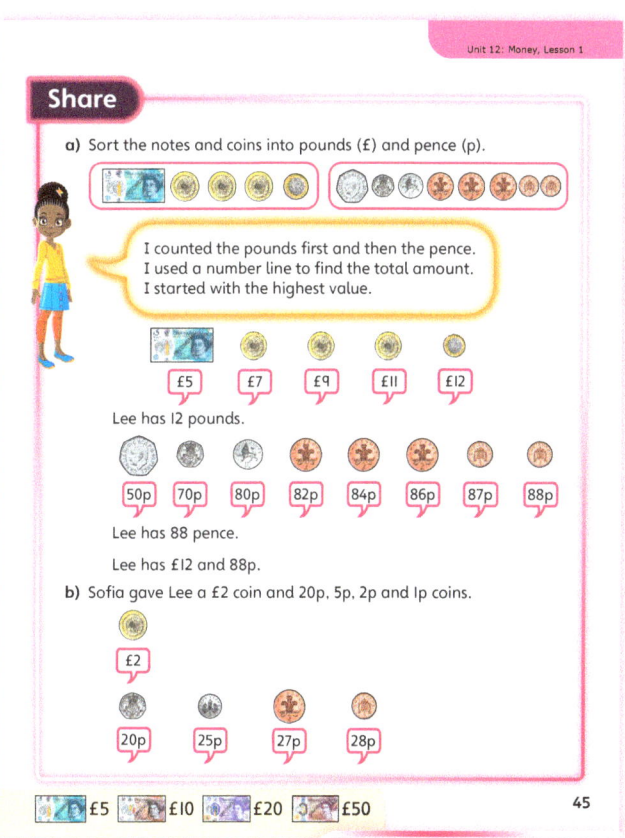

PUPIL TEXTBOOK 3C PAGE 45

Think together

WAYS OF WORKING Whole class teacher led (I do, We do, You do)

ASK

- Question ❶: *How can you arrange the money to make it easier to count? Which amounts add to make £10?*
- Question ❷: *Is there more than one way to make £25 and 37p?*
- Question ❸: *Will everyone's table be completed in exactly the same way?*

IN FOCUS Question ❷ gives children the opportunity to begin to reason about finding different ways to make the same amount. Encourage children to use their knowledge of number bonds to count efficiently. Question ❸ challenges children to work within specific criteria. At first glance they will think that there is only one possible answer for each part of the table but through discussion they will find that this is not always the case. For example, five coins that make £1 could be five 20p coins or 50p, 20p, 10p, 10p, 10p.

STRENGTHEN To support understanding, represent amounts using plastic coins or real money. Encourage children to count aloud and record their answer each time they make 50p, £1 or £10.

DEEPEN In questions ❷ and ❸, encourage children to explore all the possible ways of making an amount rather than just one way. Can they work systematically to check that they have found all possible combinations of notes and coins to total a particular amount? In question ❸, can they record their results clearly in a table? Are there always more possible combinations for a greater number of coins?

ASSESSMENT CHECKPOINT Can children use an effective method to count money, such as separating pounds and pence? Do children recognise that there is more than one way to make an amount using coins and notes?

ANSWERS

Question ❶: Sofia has 31 pounds and 66 pence.
Sofia has £31 and 66p.

Question ❷: Answers will vary depending on which notes and coins children choose. Check their chosen money totals £25 and 37p.
For example: Lee could take two 10 pound notes, one 5 pound note, one 20 pence coin, one 10 pence coin, one 5 pence coin and one 2 pence coin.

Question ❸: There is more than one answer.
One possible solution is:
4 coins: 50p, 20p, 20p, 10p
5 coins: 20p, 20p, 20p, 20p, 20p
6 coins: 50p, 10p, 10p, 10p, 10p, 10p
7 coins: 20p, 20p, 20p, 10p, 10p, 10p, 10p
8 coins: 50p, 10p, 10p, 10p, 5p, 5p, 5p, 5p
9 coins: 20p, 10p, 10p, 10p, 10p, 10p, 10p, 10p, 10p.

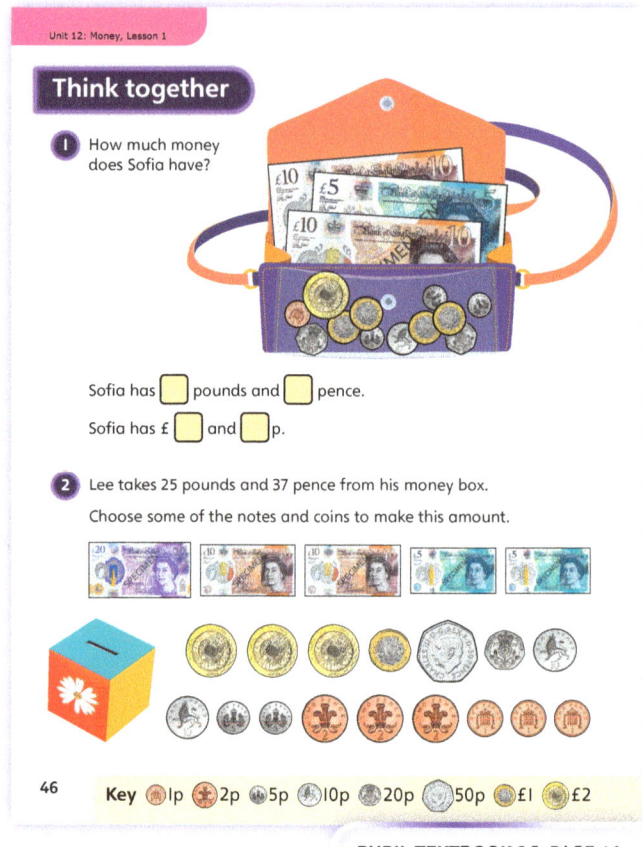

PUPIL TEXTBOOK 3C PAGE 46

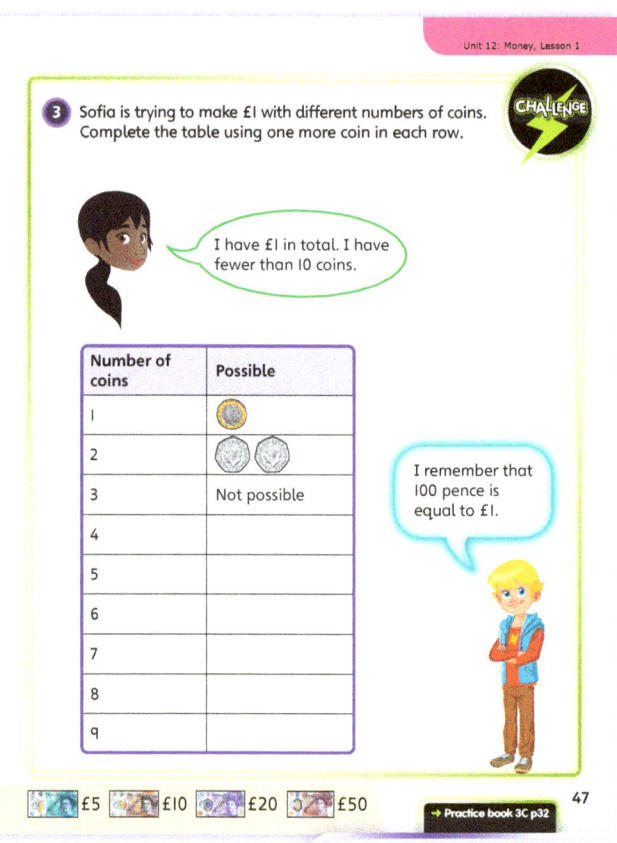

PUPIL TEXTBOOK 3C PAGE 47

Unit 12: Money, Lesson 1

Practice

WAYS OF WORKING Independent thinking

IN FOCUS Questions ❶ and ❷ aim to consolidate children's understanding of counting money and children should be able to represent their answers as pounds and pence or £ and p. Question ❸ compares two ways to make the same amount using a specific number of coins. This encourages children to compare values of coins and swap them for different coins. In question ❹, children need to think about the values of coins to decide the fewest needed to make an amount. Question ❺ provides a more open-ended challenge with multiple possible answers.

STRENGTHEN Throughout the exercise children can use plastic or paper coins to help them. They could place the coins in piles or draw a box around them each time they make £1 or 50p to help them count.

DEEPEN Ask children to work systematically to find all the possible solutions to question ❺. Can they come up with their own problem that has exactly 10 different solutions?

ASSESSMENT CHECKPOINT In questions ❶ and ❷, check children's strategy. Do they start with the pounds then move on to the pence or is their approach more random? In question ❸, check whether children are able to understand that different combinations of coins and notes can have the same value.

ANSWERS Answers for the **Practice** part of the lesson can be found in the *Power Maths* online subscription.

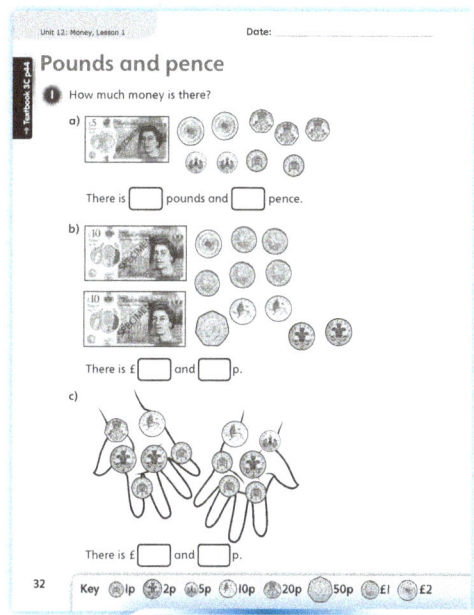

PUPIL PRACTICE BOOK 3C PAGE 32

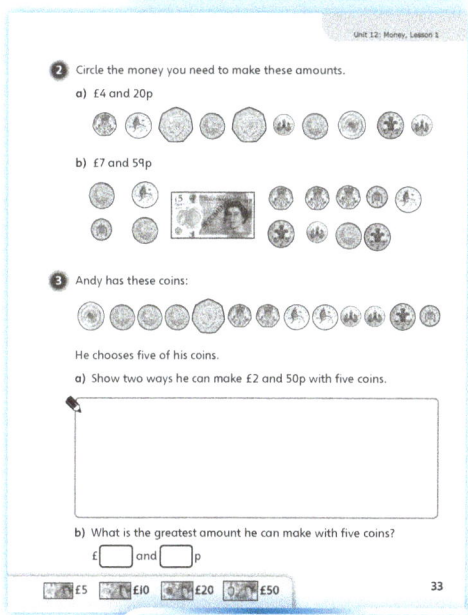

PUPIL PRACTICE BOOK 3C PAGE 33

Reflect

WAYS OF WORKING Independent thinking

IN FOCUS This question checks children's understanding of money, including counting money and the differences in value between coins.

ASSESSMENT CHECKPOINT Children are here required to explain the mistake that has been made. Do they realise that the £1 and 1p hold a different value and that £1 is greater than 1p?

ANSWERS Answers for the **Reflect** part of the lesson can be found in the *Power Maths* online subscription.

After the lesson ⏸

- Can children use an effective and efficient strategy to count money?
- Can children make the same value using different combinations of coins and/or notes?
- Do children understand that different coins hold different values? In particular, that £1 is greater than 1p, 2p, 10p, 20p and 50p?

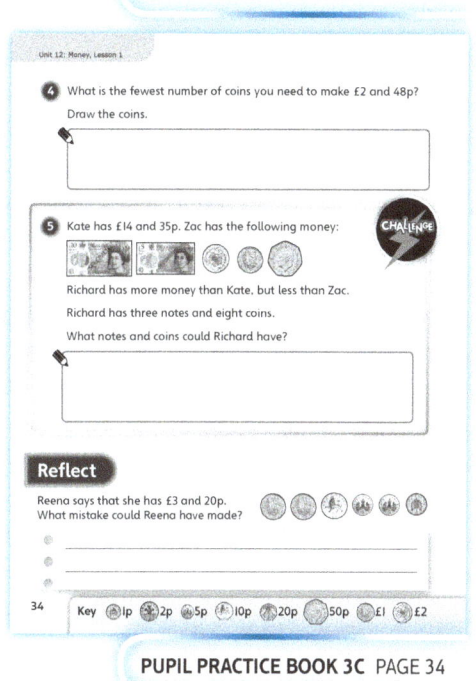

PUPIL PRACTICE BOOK 3C PAGE 34

85

Unit 12: Money, Lesson 2

Convert pounds and pence

Learning focus
In this lesson, children will convert between pounds and pence. They understand that 100 pence make one pound and can write an amount in pounds or pence from representations of coins.

Before you teach
- Do children know the value of all the coins?
- Can children multiply and divide by 100?

NATIONAL CURRICULUM LINKS

Year 3 Measurement – money

Add and subtract amounts of money to give change, using both £ and p in practical contexts.

ASSESSING MASTERY

Children can convert between pounds and pence using images of coins.

COMMON MISCONCEPTIONS

Children may not understand that 100 pence is equivalent to £1 and may simply add the numbers that are on the coins. For example, they may think a £1 coin and a 20 pence coin is 21 pence in total because they have seen the '1' on the £1 coin and the '20' written on the 20 pence coin and have added them together. Ask:
- How many pence are in £1? How many pence are in 20p?

STRENGTHENING UNDERSTANDING

Children who are struggling with converting between pence and pounds should use plastic or paper coins to help them see things in a concrete way. Encourage children to say the units of each coin.

GOING DEEPER

Ask children to look at money notes as well as coins. Can they say how many pence there are in, for example, £20? Challenge children by giving them an amount in pounds and asking how many different ways they can make it using coins. For example, £2 could be made from £1, 50p, 20p, 10p, 10p, 5p, 2p, 2p, 1p. This could be represented on a part-whole diagram.

KEY LANGUAGE

In lesson: pounds, pence, amount

Other language to be used by the teacher: convert

STRUCTURES AND REPRESENTATIONS

Part-whole models

RESOURCES

Mandatory: plastic or paper coins and notes

 In the eTextbook of this lesson, you will find interactive links to a selection of teaching tools.

Quick recap

Do children know the value of all the coins? Can they multiply and divide by 100? Ask: *How many 50 pence coins make £1? How can you check? How many 20p coins make £1? What about other coins?* This will help children understand that 100p is equal to £1.

Unit 12: Money, Lesson 2

Discover

WAYS OF WORKING Pair work

ASK
- Question 1 a): *How many pence are in £1?*
- Question 1 a): *How can you make £1 from the coins Sofia puts in the machine?*
- Question 1 b): *What silver coins do you know?*

IN FOCUS Question 1 a) requires children to convert a number of coins into pounds. Children need to know 100 pence makes one pound. Encourage children to think about how they can make £1 from the coins that Sofia puts in the machine. Question 1 b) is more open-ended and asks children to explore how to make £1 from any amount of the same silver coin. Encourage children to find all solutions to this problem.

PRACTICAL TIPS To help children in question 1 b), remove the monetary context for a while. You could revisit counting in 5s, 10s, 20s or 50s. For example, by counting in 5s to 100, children can see that there are twenty 5s in 100. Putting this back into the monetary context shows that Lee could put in twenty 5p coins to make £1. Provide children with plastic money so they can model Sofia and Lee's coins.

ANSWERS

Question 1 a): Sofia put £2 and 61p into the machine.

Question 1 b): Lee could have put in these coins:
two 50p coins, ten 10p coins, five 20p coins, twenty 5p coins.

Share

WAYS OF WORKING Whole class teacher led

ASK
- Question 1 a): *How many pence are in £1? Can you group the coins together to make £1? Is there more than one way of doing this?*
- Question 1 a): *How many pounds can you make? What coins are left? How much is this? What is the total amount of money Sofia put in the machine?*
- Question 1 b): *Which silver coins could Lee put into the machine? What do you need to think about when making £1? How many of these coins would make £1? Could Lee use a different coin to make £1?*

IN FOCUS For question 1 a), show children pictures of the groups of coins that make £1. Can children explain why these groups make £1? Emphasise the 100 pence and £1 equivalence. Encourage children to consider different groups of coins that make £1. Show children the coins that are left. Can children explain why the total is £2 and 61p? In question 1 b), discuss with children which silver coins exist. Show children pictures of each silver coin and go through each coin separately. Can children explain why each of these makes £1? Again, emphasise that these coins all make 100 pence.

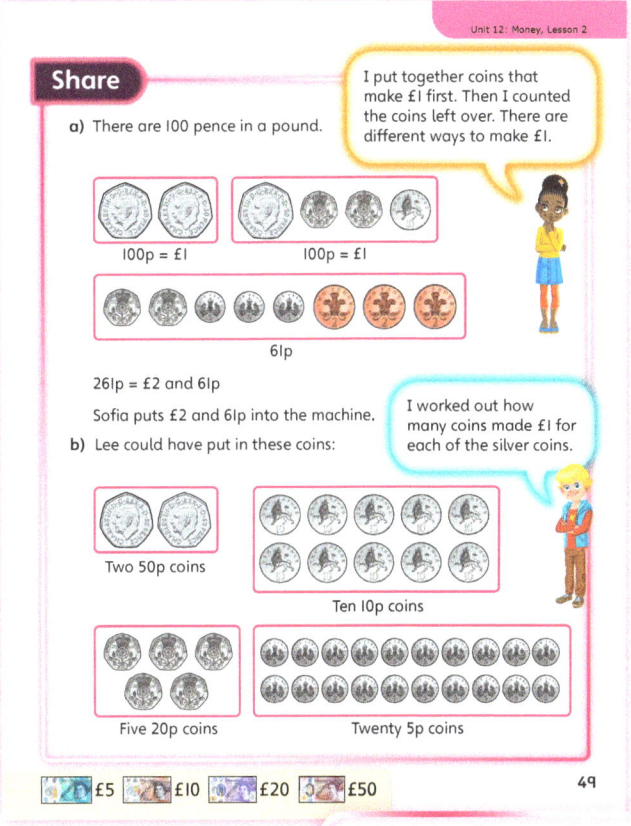

PUPIL TEXTBOOK 3C PAGE 48

PUPIL TEXTBOOK 3C PAGE 49

87

Unit 12: Money, Lesson 2

Think together

WAYS OF WORKING Whole class teacher led (I do, We do, You do)

ASK
- Question ①: *What coins are shown here? Can you count which coins make 100 pence? How can you count the coins?*
- Question ②: *How have the part-whole models partitioned the total amount? How many pence are there in £1? How many pence in £2?*
- Question ③: *How many pence are in £1? How many pence do you have if you add 100p to 52p? How many pounds are in 500p?*

IN FOCUS Question ① looks at which coins can be grouped together to make £1. Encourage children to make their own representations for each set of coins. Ensure children are aware they need to make 100 pence from each set of coins. Question ② introduces the part-whole model representation for money. Explain how these can be used to partition an amount into the pounds and pence. Discuss with children multiples of 100 pence and how many pounds these are worth. You might want children to first split the amounts into 200p and 86p and then £2 and 86p.

In question ③ a) and c), children convert pounds and pence to pence. In question ③ b) and d), children convert pence to pounds and pence. Encourage children to describe to a partner what they need to do in each part before they start, so they are clear what the question is asking them. There are also many ways to show each amount using coins, so encourage children to check one another's work.

STRENGTHEN Encourage children to use plastic or paper coins to model the problems.

DEEPEN For question ③, ask children to write some pounds and pence or just pence amounts of their own and then convert them. Ask: *What happens when you reach £10 and above? How many pence do you have then?*

ASSESSMENT CHECKPOINT Can children explain that 100 pence is equivalent to £1 and recognise groups of coins that make £1? Use question ③ to assess whether children can convert between pounds and pence and pence, or vice versa.

ANSWERS

Question ① a): The first two sets make £1. The two 50p coins make £1. The five 20p coins also make £1. The last set is £1 and 10p.

Question ① b): Holly has £5 and 32p.

Question ② a): £2 and 86p

Question ② b): 809p

Question ③ a): 152p

Question ③ b): £5 and 68p

Question ③ c): 475p

Question ③ d): £3 and 7p

Question ③ a–d): Children may draw different combinations of coins to make up each total. For example, for a), they might draw a £1 coin then a 50p and a 2p coin.

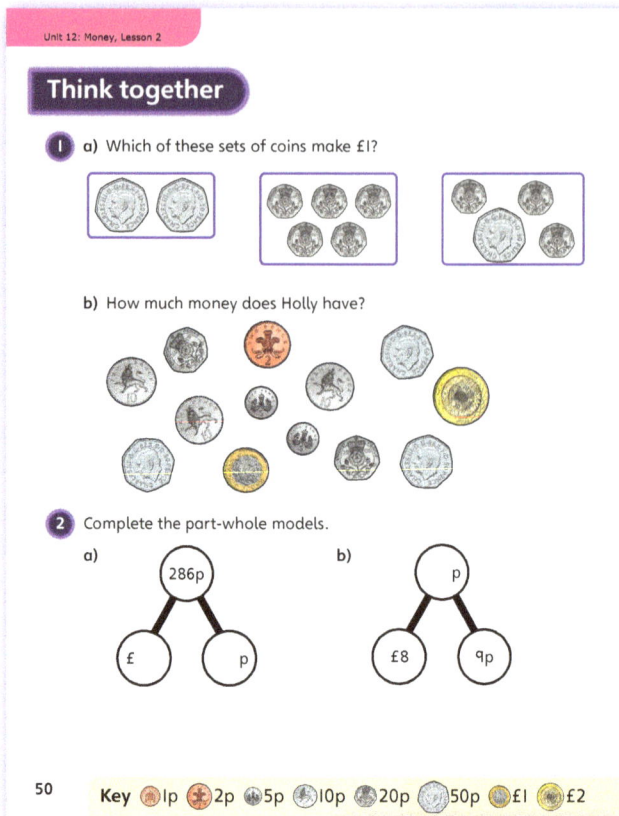

PUPIL TEXTBOOK 3C PAGE 50

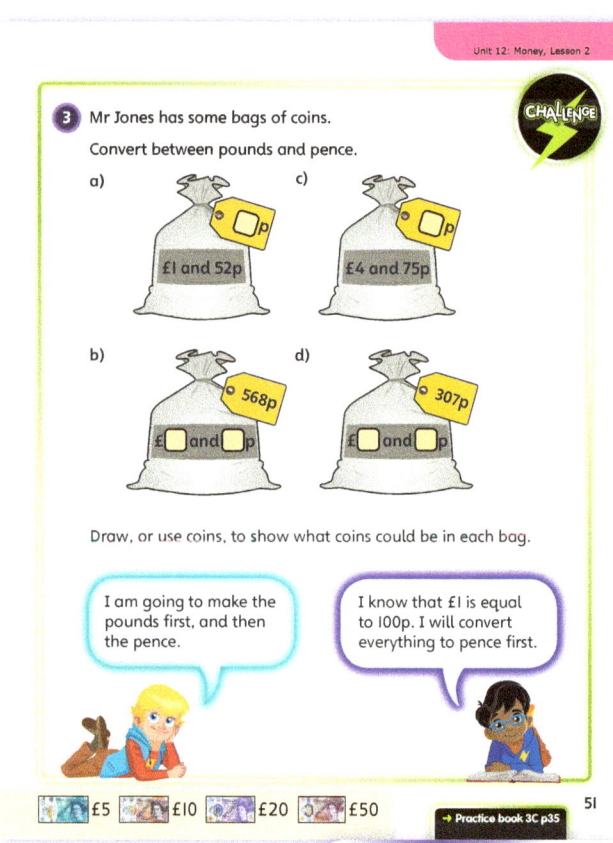

PUPIL TEXTBOOK 3C PAGE 51

Unit 12: Money, Lesson 2

Practice

WAYS OF WORKING Independent thinking

IN FOCUS Question ❷ encourages children to identify the total money from coins that are all given in pence. Ask children to think about how they could make £1 from the coins and to write an answer in just pence and then in pounds and pence. Questions ❺ and ❻ are more abstract with no pictorial representations given, so ensure plastic or paper coins are available to children as they may need to make their own representations of the questions.

STRENGTHEN Encourage children to use play money to support their understanding and to make their own representations of the problems.

DEEPEN Questions ❹ b) and c) can be explored further by asking children if there are any other ways they could complete the part-whole models.

THINK DIFFERENTLY In question ❹, part-whole models are used in order to emphasise the equivalence between pounds and pence.

ASSESSMENT CHECKPOINT Children should be confident in converting pence into pounds given a pictorial or abstract representation.

ANSWERS Answers for the **Practice** part of the lesson can be found in the *Power Maths* online subscription.

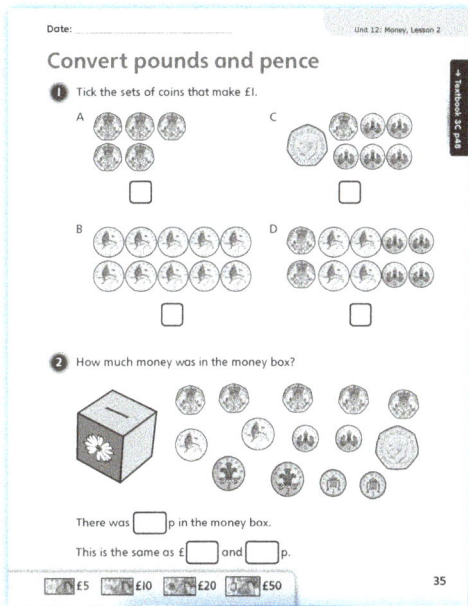

PUPIL PRACTICE BOOK 3C PAGE 35

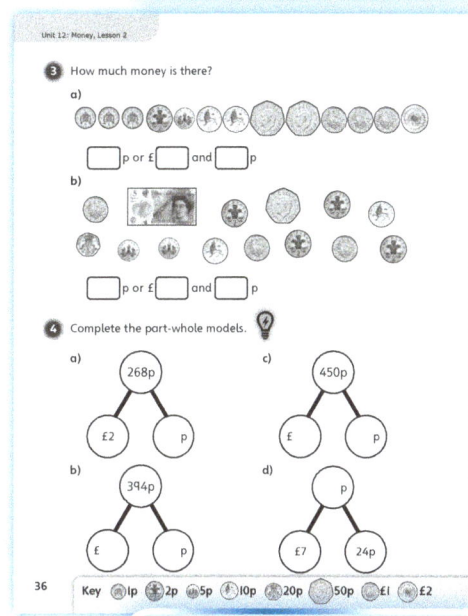

PUPIL PRACTICE BOOK 3C PAGE 36

Reflect

WAYS OF WORKING Independent thinking

IN FOCUS This activity checks children's understanding that £1 and 100 pence are equivalent and that they can reason that £2 and 200 pence are equivalent.

ASSESSMENT CHECKPOINT Ensure children can explain why £2 and 72 pence and 272p are equivalent.

ANSWERS Answers for the **Reflect** part of the lesson can be found in the *Power Maths* online subscription.

After the lesson ⏸

- Can children convert pence to pounds and represent amounts on a part-whole model?
- Can children write the amount shown from an image of pounds and pence?

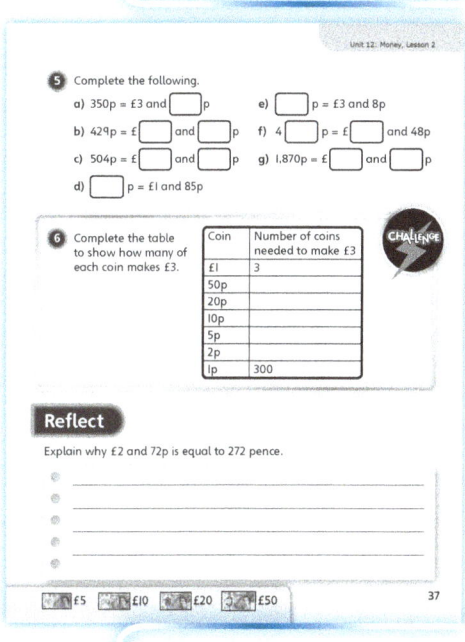

PUPIL PRACTICE BOOK 3C PAGE 37

Unit 12: Money, Lesson 3

Add money

Learning focus
In this lesson, children will add amounts of money that are given in pounds and pence.

Before you teach
- Can children convert pence to pounds?

NATIONAL CURRICULUM LINKS

Year 3 Measurement – money

Add and subtract amounts of money to give change, using both £ and p in practical contexts.

ASSESSING MASTERY

Children can add amounts of money given in pounds and pence that also involve converting pence to pounds.

COMMON MISCONCEPTIONS

Children may not add the correct numbers. For example, they may add £1 and 88 pence together and get 89 as they have added 1 and 88. Children need to understand the importance of the units. Ask:
- *Which are the pounds and which are the pence?*

Children may add the pounds and pence separately but leave the pence answer greater than £1. For example, when finding the total of £1 and 89 pence and £2 and 62 pence, children may answer £3 and 151 pence. Ask:
- *How many pence are in one pound? Can you write this another way?*

STRENGTHENING UNDERSTANDING

Encourage children to add the pounds together and then the pence together separately, then combine their answer.

GOING DEEPER

Ask children to add more than two amounts of money. Give them a total amount of money and ask how many different ways they can make the amount. For example, £2 and 57 pence + ☐ + ☐ = £5. Encourage them to represent their thinking on a part-whole model. This will also help children to link adding money to subtraction.

KEY LANGUAGE

In lesson: add, cost, total, pounds, pence

Other language to be used by the teacher: convert

STRUCTURES AND REPRESENTATIONS

Column addition

RESOURCES

Mandatory: plastic or paper coins and notes

 In the eTextbook of this lesson, you will find interactive links to a selection of teaching tools.

Quick recap

It is important that children can add two 2-digit numbers and also two 3-digit numbers. Give children questions that involve adding numbers and check they are confident. Include questions where they need to cross the hundreds and/or tens.

Discover

WAYS OF WORKING Pair work

ASK

- Question 1 a): *How much does a cup of tea cost? How much does a slice of cake cost? How could you add these amounts together? How many pounds; how many pence? What is the total?*
- Question 1 b): *How much does the juice cost? How much does a toastie cost?*

IN FOCUS Question 1 a) is used to get children to add amounts of money that do not require them to convert pence to pounds. Encourage children to add the pounds and pence separately. Question 1 b) asks children to add amounts of money that also involve converting pence to pounds. Children need to know 100 pence makes one pound.

PRACTICAL TIPS Ask children to price classroom items and open a 'shop'. Provide plastic money or diagrams of coins so children can go shopping.

ANSWERS

Question 1 a): The tea and cake cost Sofia £3 and 52p in total.

Question 1 b): A juice is 145p, a toastie is 280p.
145 + 280 = 425p
The juice and toastie cost Lee £4 and 25p in total.

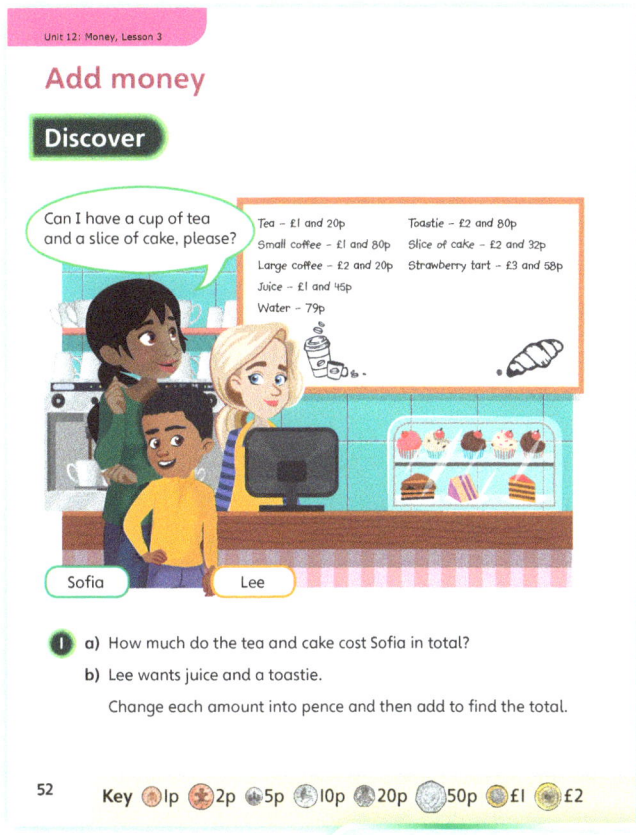

PUPIL TEXTBOOK 3C PAGE 52

Share

WAYS OF WORKING Whole class teacher led

ASK

- Question 1 a): *Can you add the pounds together? What is the total? Can you add the pence together? What is the total? What is the overall total? How much does it cost Sofia?*
- Question 1 b): *What is the total of the pounds? Can you add the pence together? What is the total? How many pence are in £1? Can you convert the pence to pounds? What is the overall total? How much does it cost Lee?*

IN FOCUS Question 1 a) prompts children to look at the diagram of the coins and encourages them to add the pounds and pence separately. Ask children to explain why the total cost is £3 and 52 pence.

Question 1 b) encourages children to notice what is the same and what is different when the pounds and pence are added separately. Ask children to explain how to convert the pence amount into pounds and pence. Emphasise the equivalence of 100 pence and £1. Use a part-whole model to show how the pence total can be separated into pounds and pence. Ask children to explain why the total is £4 and 25 pence.

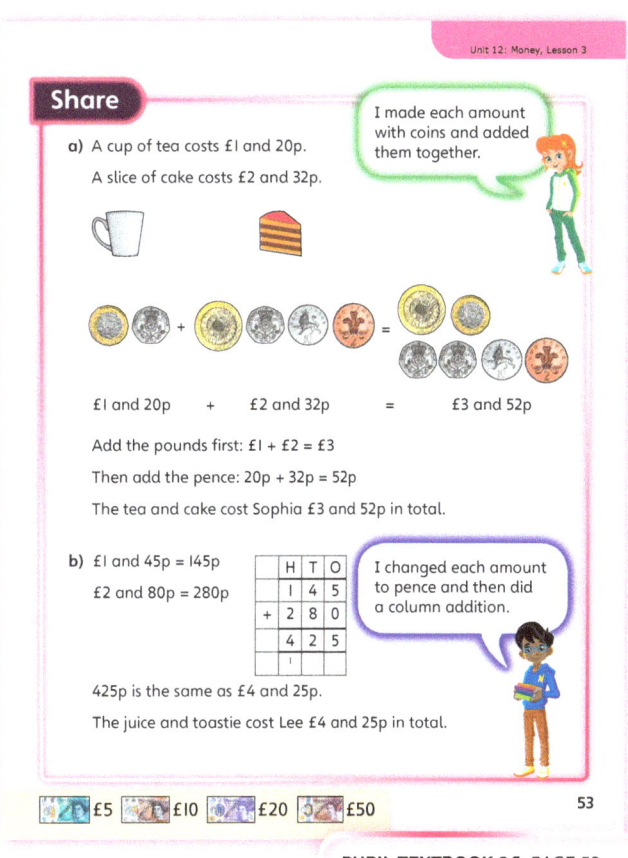

PUPIL TEXTBOOK 3C PAGE 53

Think together

WAYS OF WORKING Whole class teacher led (I do, We do, You do)

ASK
- Question ❶: *Add the pounds together, what is the total? Add the pence together, what is the total?*
- Question ❷: *What is the total cost in pence? What is the total cost in pounds and pence?*
- Question ❸: *Which items could you add together to get these totals? Could you use more than two items?*

IN FOCUS Question ❸ asks children to explore different amounts of money that add to the totals shown. Encourage children to look at the last number in each price and use this to help them without having to carry out the full calculation. For example, for a total of £5 and 12 pence, the end numbers need to add to make 2 or 12. So the items bought cannot be a Large coffee and a Toastie (£2 and 20 pence and £2 and 80 pence) because the end number would be 0 + 0 = 0.

STRENGTHEN Provide children with plastic or paper coins to support understanding. For questions ❶ and ❷, encourage children to add the pounds and pence separately then put their answer together to get the total. Ensure children are secure in the knowledge that 100 pence is equivalent to £1.

DEEPEN In question ❸, ask children to work out the totals for other combinations of items that involve adding more than two items.

ASSESSMENT CHECKPOINT Can children represent their answers correctly in pounds and pence?

ANSWERS

Question ❶: £2 + £3 = £5
20p + 58p = 78p
The total cost is £5 and 78p.

Question ❷: £1 and 80p = 180p
£1 and 45p = 145p
The total cost is 325p or £3 and 25p.

Question ❸: A tea is £1 and 20p and a water is 79p.
£1 and 20p + 79p = £1 and 99p.
The first customer bought tea and water.

A toastie is £2 and 80p and a slice of cake is £2 and 32p.
£2 and 80p + £2 and 32p = £5 and 12p.
The second customer bought a toastie and a slice of cake.

A slice of cake is £2 and 32p, a strawberry tart is £3 and 58p and a large coffee is £2 and 20p.
£2 and 32p + £3 and 58p + £2 and 20p = £8 and 10p.
The third customer bought a slice of cake, a strawberry tart and a large coffee.

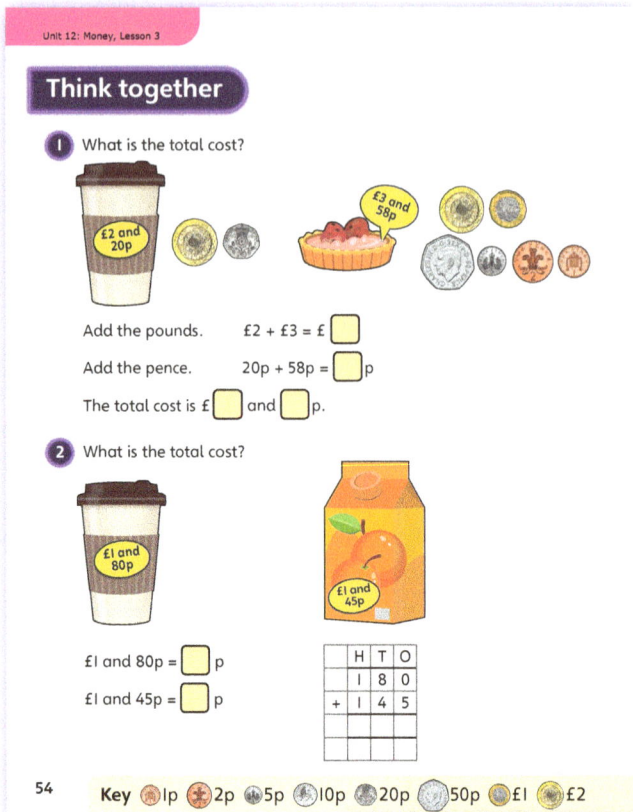

PUPIL TEXTBOOK 3C PAGE 54

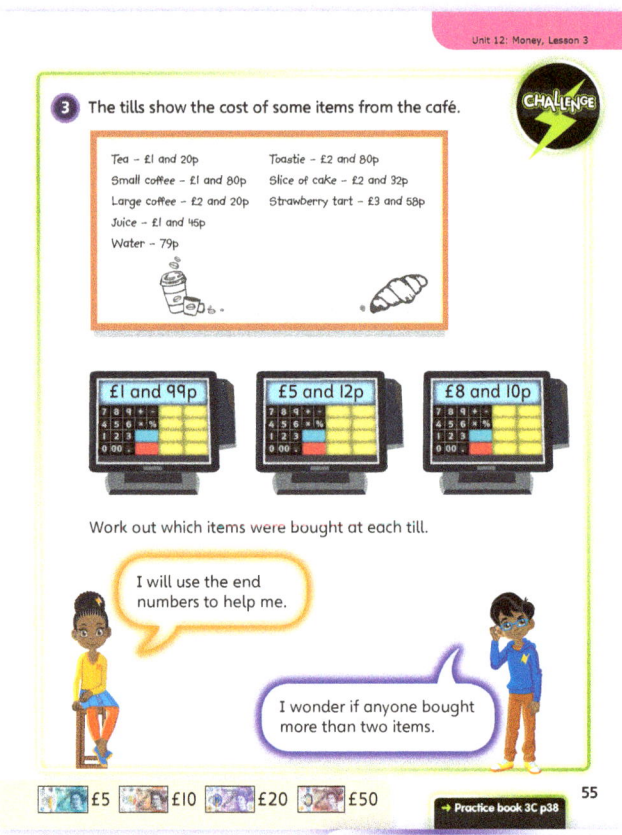

PUPIL TEXTBOOK 3C PAGE 55

Unit 12: Money, Lesson 3

Practice

WAYS OF WORKING Independent thinking

IN FOCUS Questions 1, 2 and 4 consolidate children's understanding of adding two amounts of money where the information is represented through diagrams of coins, price tags and lists of prices. Encourage children to add the pounds and pence separately and to convert the pence into pounds if necessary. Question 6 requires children to explore ways in which amounts of money can be added together to make the given totals.

STRENGTHEN Encourage children to use plastic or paper coins to support their understanding and, when diagrams are not given, to make their own representations.

DEEPEN Question 6 can be explored further by asking children what other totals they can make. Encourage children to look at adding more than two items together.

ASSESSMENT CHECKPOINT By the end of this part of the lesson, children should be confident in converting pence into pounds given a pictorial or abstract representation.

ANSWERS Answers for the **Practice** part of the lesson can be found in the *Power Maths* online subscription.

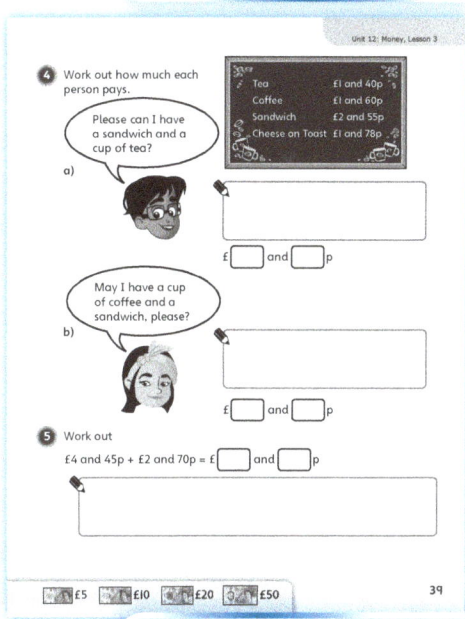

PUPIL PRACTICE BOOK 3C PAGE 38

PUPIL PRACTICE BOOK 3C PAGE 39

Reflect

WAYS OF WORKING Independent thinking

IN FOCUS This activity checks children's understanding of adding two amounts of money together where they also have to convert pence to pounds. Reinforcing the equivalence of 100 pence and £1 will help children tackle this problem.

ASSESSMENT CHECKPOINT Can children explain how they have found the total of the two amounts? Do they give their answer in the correct amount of pounds and pence?

ANSWERS Answers for the **Reflect** part of the lesson can be found in the *Power Maths* online subscription.

PUPIL PRACTICE BOOK 3C PAGE 40

After the lesson

- Can children add two amounts of money together?
- Are children confident converting pence to pounds and then writing the total in pounds and pence?

93

Unit 12: Money, Lesson 4

Subtract money

Learning focus
In this lesson, children will subtract amounts of money that are given in pounds and pence. Children will also find the difference between two amounts of money.

Before you teach
- Can children convert pence to pounds?
- Can children subtract simple whole numbers using column subtraction?

NATIONAL CURRICULUM LINKS

Year 3 Measurement – money

Add and subtract amounts of money to give change, using both £ and p in practical contexts.

ASSESSING MASTERY

Children can subtract amounts of money given in pounds and pence.

COMMON MISCONCEPTIONS

Children may not be confident with their number bonds to 100. For example, they may understand that £1 is equivalent to 100 pence but think that 67 pence subtracted from £1 is 43 pence rather than 33 pence. Ask:
- How many pence are there in total? Can you add the 10s and then the 1s? Can you use number bonds to 10 to help you?

STRENGTHENING UNDERSTANDING

Strengthen understanding of finding the difference between two amounts of money by encouraging children to use a number line and count on.

GOING DEEPER

Ask children to add and subtract in one question. For example, give them an amount of money and ask how much money would be left if they bought two or three different items rather than just one item. This could be represented on a part-whole model.

KEY LANGUAGE

In lesson: subtract, difference, total, less, more, pounds, pence

Other language to be used by the teacher: convert, change

STRUCTURES AND REPRESENTATIONS

Number lines

RESOURCES

Mandatory: plastic or paper coins and notes

 In the eTextbook of this lesson, you will find interactive links to a selection of teaching tools.

Quick recap

On the board write the question: 438 – 195. Ask children to attempt this question. Discuss the methods that they could use including a column method, subtraction using a number line and counting on. Children need to be confident subtracting 3-digit numbers from 3-digit numbers.. Practise until children remember the method.

94

Unit 12: Money, Lesson 4

Discover

WAYS OF WORKING Pair work

ASK
- Question 1 a): *How much money does Sofia have in her hand? How much does a cupcake cost?*
- Question 1 a): *How could you find out how much money she has left?*
- Question 1 b): *How much does the loaf of bread cost? How much does the pack of bread rolls cost?*
- Question 1 b): *How could you find the difference between these amounts?*

IN FOCUS Question 1 a) requires children to subtract one amount from another and to work systematically, first by subtracting the pounds and then the pence. Question 1 b) asks children to find the difference between two amounts of money and encourages them to use a number line to count on to find the difference.

PRACTICAL TIPS Encourage children to role-play the transaction with the correct coins and change. Children could devise their own menu for a class bakery.

ANSWERS

Question 1 a): £2 and 50p – £1 and 10p = £1 and 40p
Sofia has £1 and 40p left.

Question 1 b): £2 and 50p – £1 and 89p = 61p
A loaf of bread is 61p cheaper than a pack of bread rolls.

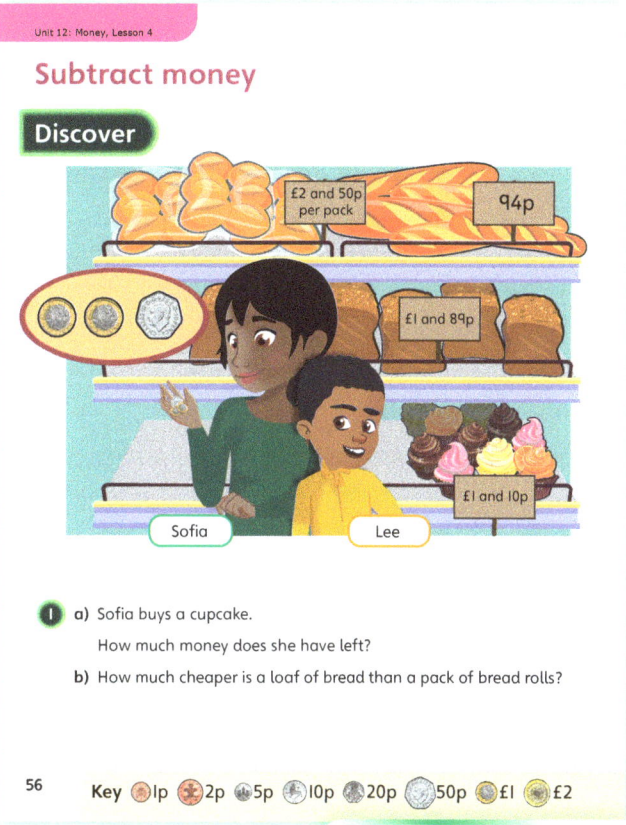

PUPIL TEXTBOOK 3C PAGE 56

Share

WAYS OF WORKING Whole class teacher led

ASK
- Question 1 a): *Look at the coins in Sofia's hand: how much money does she have in total? How much does the cupcake cost? Subtract the £1 first; how much is left? How much more do you need to subtract? How much does Sofia have left?*
- Question 1 b): *How much does each item cost? How many pence from £1 and 89 pence up to £2? How do you know? How many pence from £2 up to £2 and 50 pence? What could you do with your two answers to find the difference?*

IN FOCUS Question 1 a) prompts children to use the diagrams of coins to work out how much money Sofia has. Can children explain why Sofia has £2 and 50 pence in total? Ask children how much is left if £1 is subtracted and identify this in the diagram where the pound is crossed out. Show the diagrams of 50 pence being exchanged for two 20p coins and one 10p coin so the remaining 10 pence can be easily subtracted. Show children the diagram of the coins that are left. Can children explain why Sofia has £1 and 40 pence left?

STRENGTHEN In question 1 b), ensure children can use the number line and count aloud together from £1 and 89 pence up to £2. Can children explain why this is 11 pence? Show the difference from £2 to £2 and 50 pence on the number line. Can children explain why the difference is 50 pence? Explain to children that adding these amounts together will find the total difference.

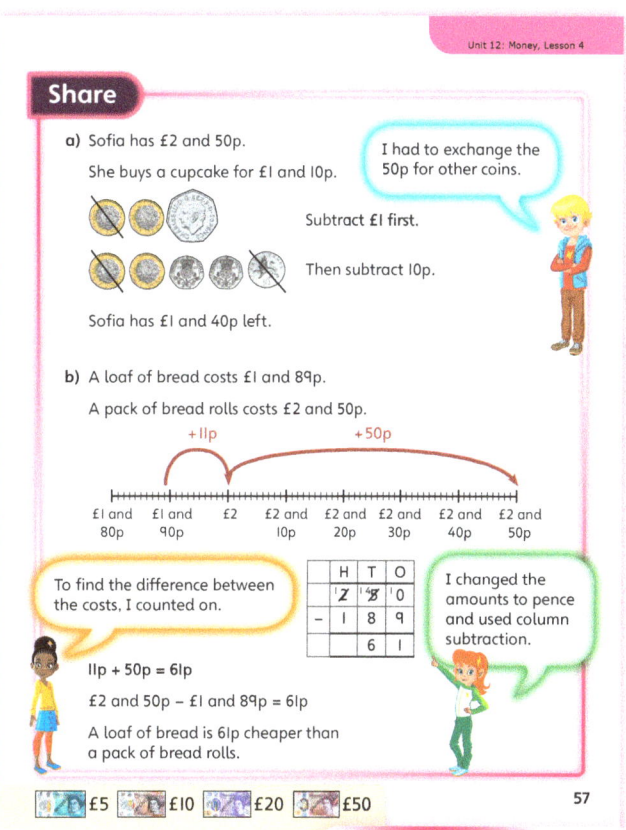

PUPIL TEXTBOOK 3C PAGE 57

95

Unit 12: Money, Lesson 4

Think together

WAYS OF WORKING Whole class teacher led (I do, We do, You do)

ASK
- Question ❶: *What is the difference between 94 pence and £1? What is the difference between £1 and £1 and 89 pence? What is the difference altogether?*
- Question ❷: *How much money does Lee have? How much does a custard tart cost? How can you work out how much Lee has left?*

IN FOCUS Question ❸ encourages children to make predictions first before working out the actual answers. Encourage children to mentally subtract the pounds first, then to think about whether subtracting the pence part would give an answer more or less than £1. Children can use a number line to work out the actual answers.

STRENGTHEN Children can use plastic or paper coins to support their understanding. Encourage children to draw number lines or provide them with pre-made number lines to support their calculations.

DEEPEN Question ❷ can be explored further by asking children to work out the maximum number of custard tarts that Lee could buy with the money he has.

ASSESSMENT CHECKPOINT Can children subtract and find the difference and represent their answers correctly in pounds and pence?

ANSWERS

Question ❶: 6p + 89p = 95p
A loaf of bread costs 95p more than a breadstick.

Question ❷: Lee has £4 and 86p left.

Question ❸ a): Check children's predictions. They should recognise that the answers £1 and 95p – £1 and 42p as well as £4 and 45p – £3 and 88p will have an answer of less than £1.

Question ❸ b): Using Ash's suggestion of changing the amounts into pence:
195p – 142p = 53p
530p – 150p = 380p or £3 and 80p
218p – 64p = 154p or £1 and 54p
445p – 388p = 57p

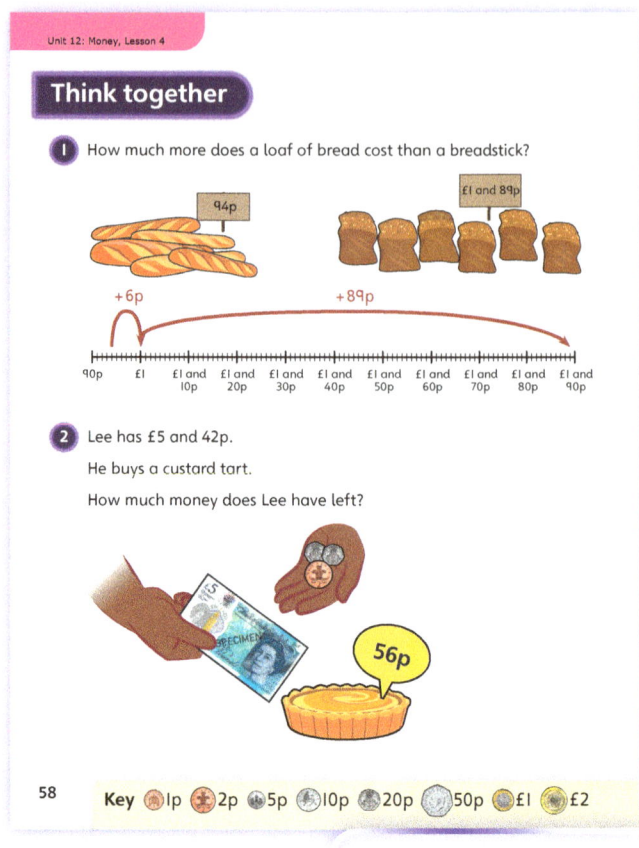

PUPIL TEXTBOOK 3C PAGE 58

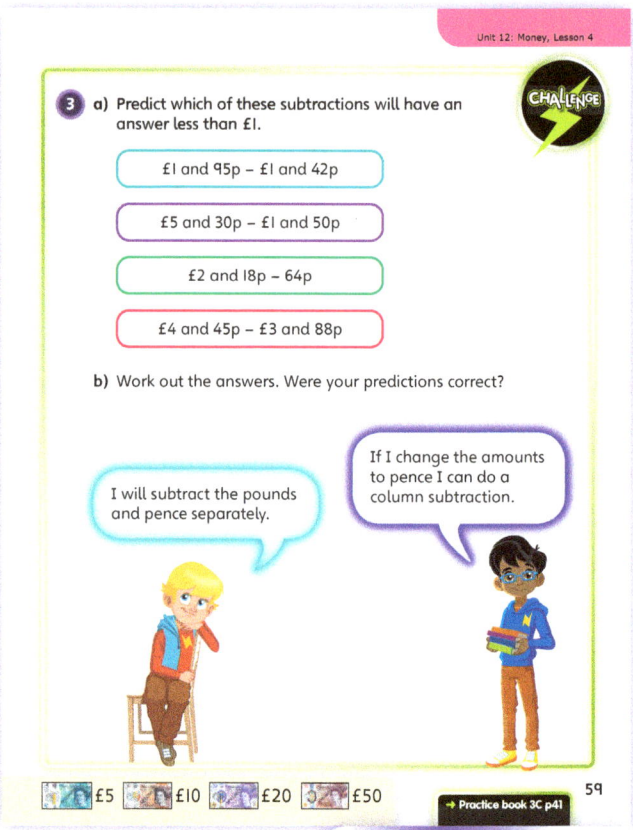

PUPIL TEXTBOOK 3C PAGE 59

Unit 12: Money, Lesson 4

Practice

WAYS OF WORKING Independent thinking

IN FOCUS Questions ❶ to ❹ consolidate children's understanding of subtracting amounts of money or finding the difference where the information is represented through diagrams of coins, price tags or just numbers. Encourage children to look at the diagrams and cross out the coins or use a number line to help with their calculations. Question ❺ encourages children to explore the best way to find the difference between two amounts of money. Encourage children to discuss their method and why they chose it.

STRENGTHEN Encourage children to use plastic or paper coins to support their understanding and draw their own number lines where necessary (they may need guidance on where to start and end their number lines).

DEEPEN Question ❺ can be explored further by asking children what has been subtracted if the answer is a certain amount. For example, ask: *What has been subtracted from £2 and 34p if the answer is 78p?*

ASSESSMENT CHECKPOINT Children should be confident at subtracting one amount of money from another and finding the difference between two amounts.

ANSWERS Answers for the **Practice** part of the lesson can be found in the *Power Maths* online subscription.

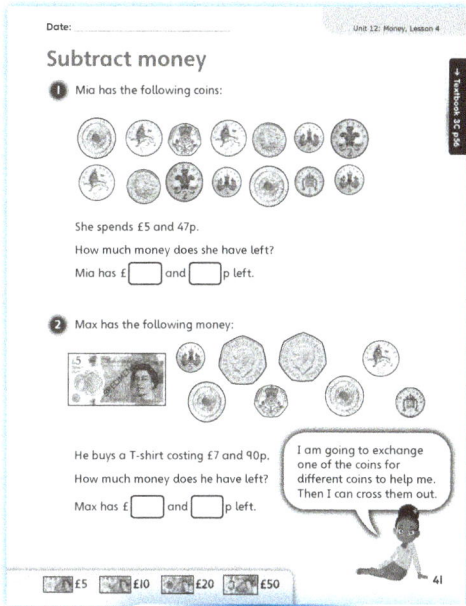

PUPIL PRACTICE BOOK 3C PAGE 41

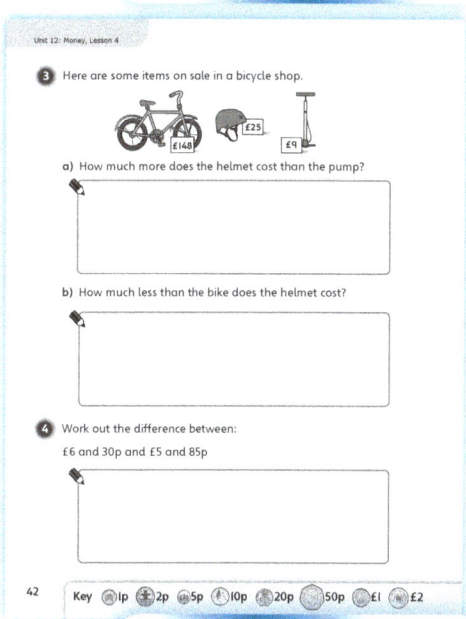

PUPIL PRACTICE BOOK 3C PAGE 42

Reflect

WAYS OF WORKING Independent thinking

IN FOCUS This checks children's understanding of subtracting one amount of money from another amount. Reinforce the usefulness of using a number line and counting on to show their method.

ASSESSMENT CHECKPOINT Can children explain how they have subtracted the amount?

ANSWERS Answers for the **Reflect** part of the lesson can be found in the *Power Maths* online subscription.

After the lesson ⏸

- Can children subtract and find the difference between two amounts of money and then write the total in pounds and pence?

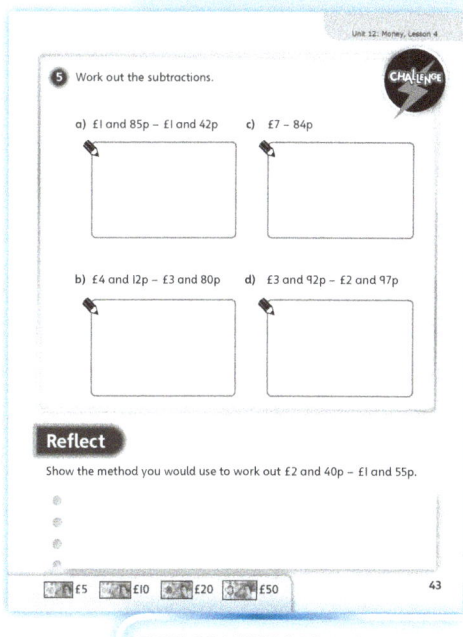

PUPIL PRACTICE BOOK 3C PAGE 43

Unit 12: Money, Lesson 5

Find change

Learning focus
In this lesson, children will find the change from a given coin or note.

Before you teach
- Are children confident using counting on methods to find a difference?
- Check that children can work out a 3-digit by 3-digit subtraction.
- Can children count a mixed set of coins?

NATIONAL CURRICULUM LINKS

Year 3 Measurement – money
Add and subtract amounts of money to give change, using both £ and p in practical contexts.

ASSESSING MASTERY

Children can find the change from a given note used to pay for items. They understand the concept of change. Children start to realise that the cost and the change together make up the note or coin with which they paid.

COMMON MISCONCEPTIONS

Children may forget to make the next pound first when asked to find the change from a given note. For example, for an item that costs £7 and 10p, children may think that the change from a £10 note is £3 and 90p. In actual fact, they first need to make the next pound to £8 and then they need £2 more. Use a number line to support their understanding. Ask:
- *How much do you need to make the next £? How many more pounds do you need to make £10?*

STRENGTHENING UNDERSTANDING

Children may find it much easier to use coins alongside a number line. Using actual coins will help children make the next pound and then work out how many more pounds are needed to make the note with which they paid. Children may need to count on in smaller jumps to get the answer. This is acceptable, but prompt children to consider using bigger jumps if they can. The number line and coins will help them understand the concept, rather than just going straight to column methods.

GOING DEEPER

Ask children to solve problems that require a mixture of calculations, not just subtracting. For example, they find the total of two or more items and then find the change. Alternatively, ask: *How many pencils costing 30p can you buy for £2? How much change would you get?*

KEY LANGUAGE

In lesson: pounds, pence, change, subtract
Other language to be used by the teacher: cost, total

STRUCTURES AND REPRESENTATIONS

Number line, column, methods, place value counters if necessary

RESOURCES

Mandatory: plastic or paper coins and notes
Optional: place value grids

 In the eTextbook of this lesson, you will find interactive links to a selection of teaching tools.

Quick recap
On the board, draw a blank number line. Mark 0 and 100 at either end. Make a jump of + 20 from 0. Ask children how much more they need to jump to make 100. Repeat for different starting jumps: + 35, + 49 and + 72. Discuss with children whether they can do it in one jump or whether they need to make more jumps.

Unit 12: Money, Lesson 5

Discover

WAYS OF WORKING Pair work

ASK

- Question 1 a): *How much does the shopping cost? What note is Sofia paying with? How can you work out the change by counting on? How much do you need to make the next pound?*
- Question 1 b): *How many pence are there in £5? How many pence is £2 and 35 pence worth? How can you find the change using column methods?*

IN FOCUS In this **Discover** activity, children use two different methods to calculate the change. In question 1 a) they use the counting on method to work out how much they need to add on to make the next pound and then how much more to get to £5. Encourage use of a number line. In question 1 b), children have to first convert each amount to pence and then use column subtraction methods. Encourage children to double check their answers to make sure they are accurate. For both questions have coins and place value grids available for children.

PRACTICAL TIPS Set the classroom up as a shop to help children engage with the activity. Children can sell items to each other, each with different prices. When a child pays for an item, the shopkeeper has to work out the change. Encourage use of both the count on method and the column subtraction method. Discuss with children which method they prefer and why.

ANSWERS

Question 1 a): Sofia received £2 and 65p change.

Question 1 b): 500p – 235p = 265p, which is £2 and 65p.

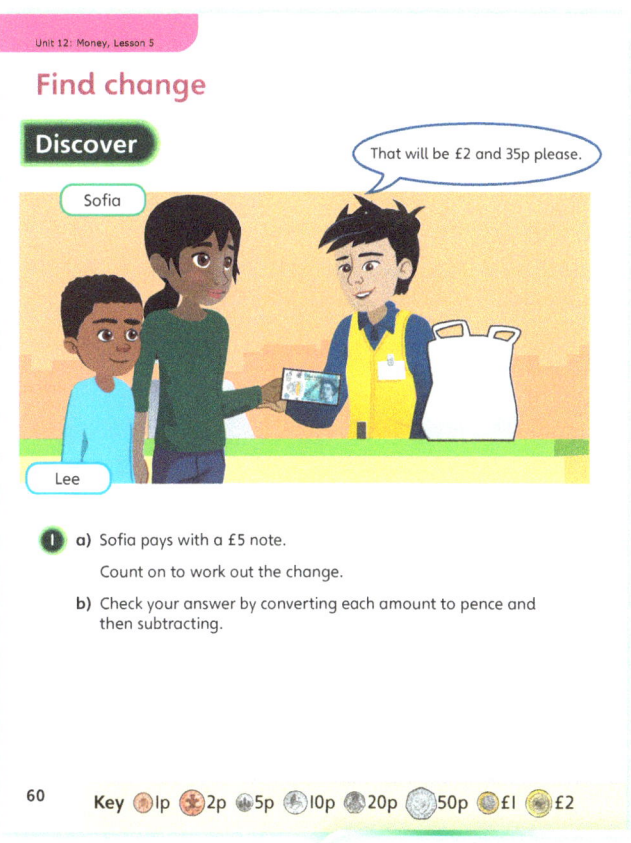

PUPIL TEXTBOOK 3C PAGE 60

Share

WAYS OF WORKING Whole class teacher led

ASK

- Question 1 a): *What intervals does the number line go up in? How much do you need to add on to make £3? How much more do you need to add on to make £5? What is the total change?*
- Question 1 b): *How many pence are there in £5? How many pence is £2 and 35p worth? What steps do you need to go through to work out the answer?*

IN FOCUS In question 1 a), ask children to draw an open number line with you. If they cannot work out straight away that they need to add on 65p to make £3, then break it down into a jump from 35 to 40 and then from 40 to the next pound. Explain there are different jumps they can make to the next pound, but encourage a single jump from £3 to £5. In question 1 b), discuss the column subtraction method and go through it step-by-step. You may want to use a place value grid and coins to support children's understanding. Discuss how 500 – 235 is the same as 499 – 234 and how this is an easier calculation.

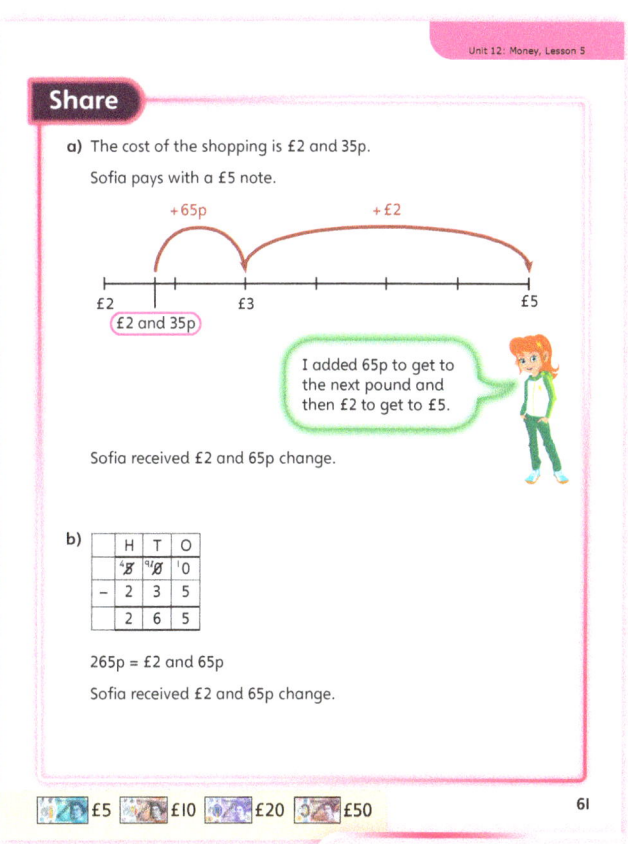

PUPIL TEXTBOOK 3C PAGE 61

99

Unit 12: Money, Lesson 5

Think together

WAYS OF WORKING Whole class teacher led (I do, We do, You do)

ASK
- Question ①: *How much does the till show? How much is the customer paying with? What methods to you know to find the change?*
- Question ②: *What mistake has been made? Why do you think Sofia has made this mistake?*
- Question ③: *How is this question similar to finding the change? How can you work out the cost of the chocolate?*

IN FOCUS In question ①, children are given amounts and a note and they have to work out the change. Discuss the different methods that they can use to work out the answer (as used in **Share**). Discuss which method they prefer and check that all methods give the same answer.

Question ② demonstrates a common misconception where children see the £6 and 80p and think they have to make £6 up to £10 and so get £4 change. Use a number line to remind children that they first need to get to £7, meaning there will only be £3 and 20p change.

Question ③ provides an example where children are given the change but need to work out the cost. They should realise that this is a similar question to finding the change. Can children explain why a subtraction is still needed?

STRENGTHEN Throughout these questions, children may find it much easier to use coins alongside a number line. Using actual coins will help children make the next pound and then work out how many more pounds are needed to reach the value of the note used for payment. Children may need to count on in multiple smaller jumps to get the answer. This is acceptable, but prompt children to consider using bigger jumps if they can. The number line and coins will help them understand the concept of counting on.

DEEPEN Work through examples where children need to first add the costs of two or more items and then find the change. To make it harder, the items might be priced with a mix of pence, and pounds and pence. For example, ask: *A bat cost £6 and a pack of balls £3 and 50p. How much change will you get from £20?*

ASSESSMENT CHECKPOINT Can children solve problems involving money that include adding, subtracting, multiplying and dividing?

ANSWERS

Question ① a): 500p – 382p = 118p, which is £1 and 18p change.

Question ① b): 1,000p – 618p = 382p, which is £3 and 82p change.

Question ②: Sofia is incorrect.
£10 – £4 and 80p = £5 and 20p

Question ③: £2 – £1 and 37p = 63p
Alternatively, Ambika could add on from £1 and 37p, to reach £2.
The bar of chocolate cost 63p.

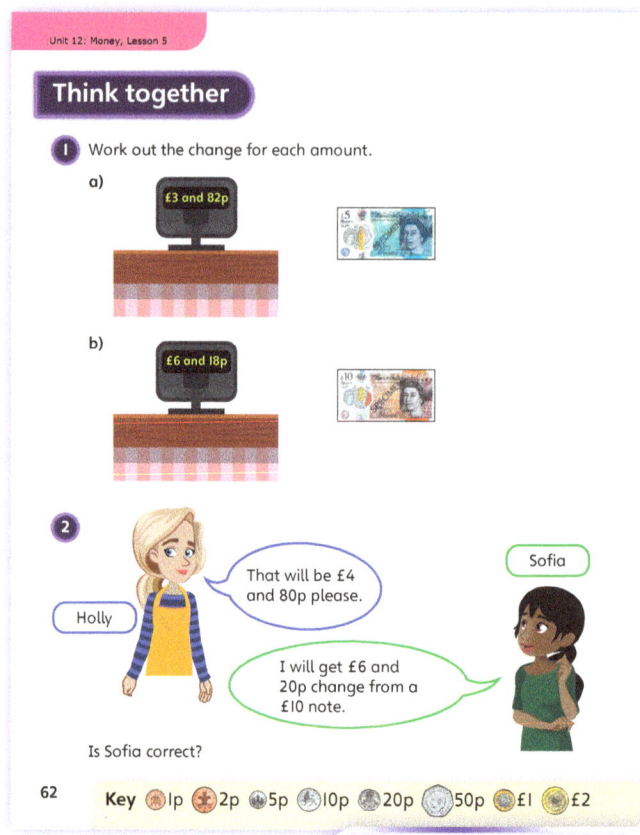

PUPIL TEXTBOOK 3C PAGE 62

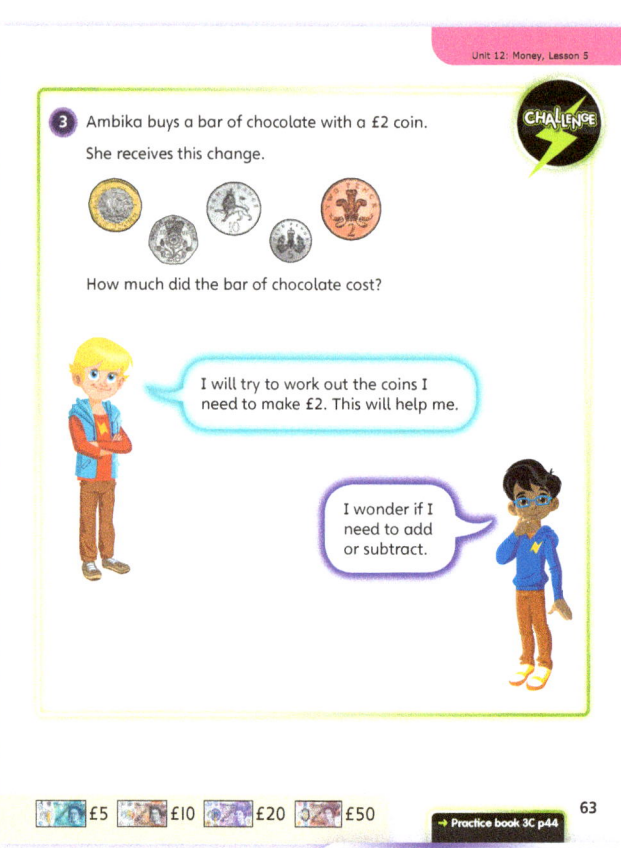

PUPIL TEXTBOOK 3C PAGE 63

Unit 12: Money, Lesson 5

Practice

WAYS OF WORKING Independent thinking

IN FOCUS In questions ❶ and ❷, children either use counting on or the column method to find the change from the cost of some items. Check that they use an accurate method. Question ❸ can be approached in different ways. Children could work out the change expected and then work out if the coins add up to this amount. Alternatively, children could count on from the cost using the coins and check that it makes £10.

STRENGTHEN Throughout these questions, children may find it much easier to use coins alongside a number line. Using actual coins will help children make the next pound and then work out how many more pounds are needed to reach the value of the note used for payment. Children may need to count on in multiple smaller jumps to get the answer. This is okay, but prompt children to consider using bigger jumps if they can. The number line and coins will help them understand the concept of counting on to make a total.

DEEPEN Ask children to work through examples where they need to first add the cost of two or more items and then find the change. Question ❺ provides an example, and similar questions could be asked.

THINK DIFFERENTLY In question ❹, children are given the amount of change and have to use subtraction to find the cost of the item. They should notice that it feels very similar to finding change.

ASSESSMENT CHECKPOINT By the end of the **Practice** part of the lesson children should be confident in problem solving with money.

ANSWERS Answers for the **Practice** part of the lesson can be found in the *Power Maths* online subscription.

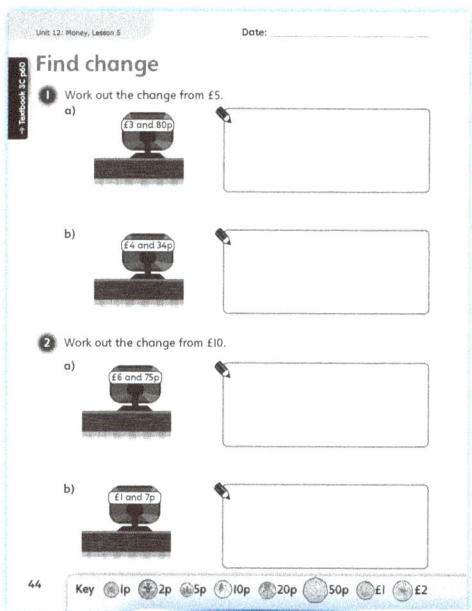

PUPIL PRACTICE BOOK 3C PAGE 44

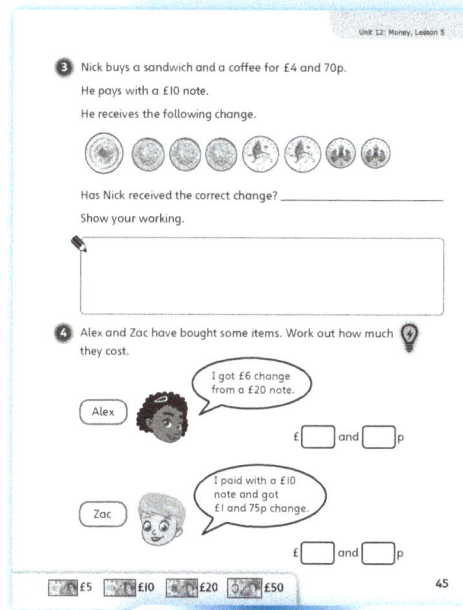

PUPIL PRACTICE BOOK 3C PAGE 45

Reflect

WAYS OF WORKING Pair work

IN FOCUS This checks children's understanding of a method they can use to find the change from a note, when given an amount. Listen for children explaining the method and check that they use the correct language. Ask: *How could you check your answers?*

ASSESSMENT CHECKPOINT Children explain a method to each other that allows them to find the change. This method may be using a number line or column method.

ANSWERS Answers for the **Reflect** part of the lesson can be found in the *Power Maths* online subscription.

After the lesson ⏸

- Can children use a number line to count on to find the change?
- Can children use column subtraction methods to find the change?

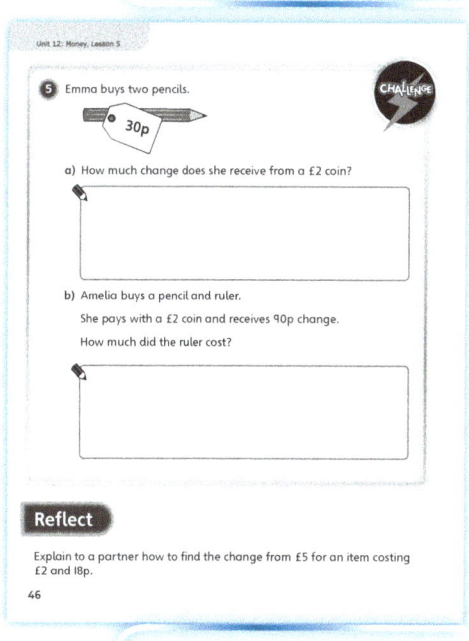

PUPIL PRACTICE BOOK 3C PAGE 46

Unit 12: Money

End of unit check

Don't forget the unit assessment grid in your *Power Maths* online subscription.

WAYS OF WORKING Group work adult led

IN FOCUS
- These questions cover the full range of topics within the unit.
- The diagnostic questions in this unit cover the key misconceptions that children may come across. Question ❶ assesses whether children can distinguish between pounds and pence. Question ❷ assesses whether children can convert properly from pence to pounds. Question ❺ assesses whether children can find change.
- Children need to be confident converting between pounds and pence as this is the foundation for the work across other year groups. This knowledge of converting should help children cross the pounds when adding and subtracting.
- Question ❻ is a SATs-style question that is solvable and can be reasoned in different ways. Some children may, for example, add the amounts and compare. Others may subtract what Olivia has, and what her mum gives her, from the price of the cap and show that there is some money left so Olivia does not have enough. Some children may round up and show that these whole amounts added together would not make £12.

ANSWERS AND COMMENTARY

By the end of the unit, children will be able to convert amounts from pounds and pence to just pence and vice versa. They should also be able to solve addition and subtraction problems (including finding change).

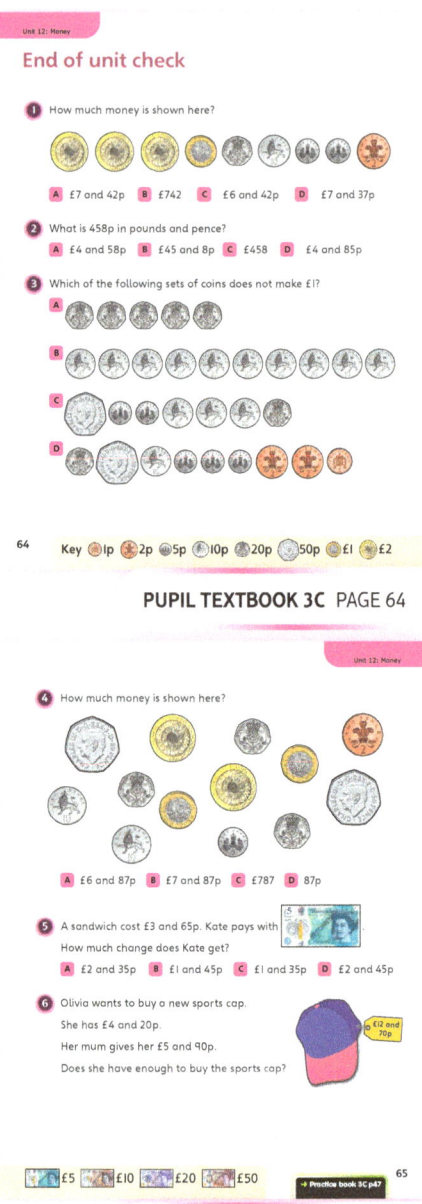

PUPIL TEXTBOOK 3C PAGE 64

PUPIL TEXTBOOK 3C PAGE 65

Q	A	WRONG ANSWERS AND MISCONCEPTIONS	STRENGTHENING UNDERSTANDING
1	A	B indicates children have just used the £1 symbol and not thought about the pence. C or D shows that children have miscounted the coins. This often happens when there is more than one of a particular coin.	A common mistake is that children miscount, count too many or do not count enough coins. Provide plastic or paper coins for them to use throughout to strengthen understanding. Also encourage them to use a number line and/or cross coins out or cover them up once they have counted them. Ask: • How many pounds are there? • How many pence are there? • How many pence are in one pound? • How much in total? How can you check? Children may find it useful to represent problems with a bar model to help them see the structure of the problem and work out which amounts they need to find.
2	A	B indicates an incorrect conversion, perhaps thinking that there are 10p in £1. C suggests children have just changed the pence to pounds. D indicates a reading error as children have transposed the last two digits.	
3	C	A or B suggest that children do not recognise common amounts that make £1. D indicates inaccuracy of counting.	
4	B	D suggests that the children may have counted correctly but forgotten to write down the pounds alongside the pence.	
5	C	A shows that children have made £1 too many because they forgot that they made £1 when they added 65p and 35p together.	
6	No	Children may misread the question and simply subtract the amounts saved from the price of the hat and so think Olivia has enough.	

Unit 12: Money

My journal

WAYS OF WORKING Independent thinking

ANSWERS AND COMMENTARY A large part of what children have to do in real life, and also in maths assessments, is to solve number problems that involve money. Typically these problems involving finding change or finding totals by adding where children have to cross £1. This activity asks children to make up their own stories involving money to fit the pictures.

The story could be: Sofia goes into a shop with a £5 note. She wants to buy an ice cream for £2 and 45p and a bottle of water for £2 and 72p. Does she have enough money?

£2 and 45p + £2 and 72p = £5 and 17p

No, Sofia needs another 17p to be able to afford both an ice cream and a water.

Power check

WAYS OF WORKING Independent thinking

ASK

- Can you convert an amount in pence to pounds and pence?
- How confident are you in working out change when shopping?
- Do you feel unsure about the value of any of the coins or notes we have looked at?

Power puzzle

WAYS OF WORKING Independent thinking

IN FOCUS This puzzle provides a real-life example for children to add amounts of money together and then find change. Aki needs to buy the correct amount of the ingredients for his recipe. Some children may just add the amounts shown, however some will realise that they need to work out the cost of each ingredient first. Once children have done this, they can explore methods for adding up several amounts. Do they add two at a time. Or use coins and then add the pounds and pence? Allow children to explore different approaches.

ANSWERS AND COMMENTARY
- 400 g of butter = £1 and 75p
- 2 eggs = 40p
- 400 g sugar = £1 and 40p: Children may double and double again to get this amount. Allow them to select coins as necessary.
- 400 g of flour = £1 and 4p: Children could double and double again.
- 50 g of cocoa = 90p: Children need to find half of £1 and 80p. Some may simply halve £1 to get 50p and halve 80p to get 40p; others may need coins to help them.
- Sprinkles = 87p

Total cost is £6 and 36 pence, so change from £10 is £3 and 64 pence.

PUPIL PRACTICE BOOK 3C PAGE 47

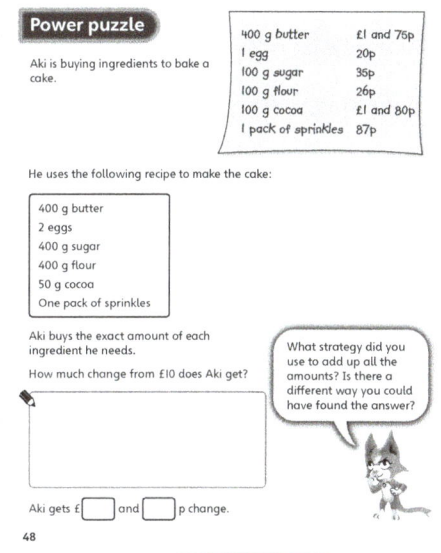

PUPIL PRACTICE BOOK 3C PAGE 48

After the unit

- Can children find the total amount of money, including crossing £1?
- Can children solve simple addition and subtraction problems involving money?
- Can children convert between pence and pounds and pence?

Strengthen and **Deepen** activities for this unit can be found in the *Power Maths* online subscription.

Unit 13
Time

Mastery Expert tip! 'Throughout children's work on time, look for real-life opportunities to reinforce the learning. Focus on telling the time at different moments in the day and regularly discussing where children may see times shown on analogue and digital clocks in their school or home environments.'

Don't forget to watch the Unit 13 video!

WHY THIS UNIT IS IMPORTANT

This unit is important because it will help to develop children's understanding of the length of a day, and their awareness of times of day, of which they may have little real-life experience. These concepts will be used for reading, estimating and measuring time, and in problem-solving contexts. Children will build on their knowledge of reading clocks, including those that feature Roman numerals. They will tell the time to the nearest minute, using 'past' and 'to' language. They will read times from digital clocks and understand when a time is am and when a time is pm. Finally children will find durations of events, including those that cross the hour.

WHERE THIS UNIT FITS

→ Unit 12: Money
→ **Unit 13: Time**
→ Unit 14: Angles and properties of shapes

In this unit, children will begin with an introduction to Roman numerals, and afterwards will recap their understanding of time from Year 2. They will develop a deeper understanding of the length of a year, a month, a day, an hour, a minute and a second, and will use this to solve problems involving reading and measuring time.

Before they start this unit, it is expected that children:
- know the number of minutes in an hour, and can read and write times on a clock to five minutes
- know the months of the year and key dates (including everyday usage)
- have some prior knowledge of everyday usage of time and the o'clock times that occur throughout the day
- are familiar with moving from a start time through a duration to an end time.

ASSESSING MASTERY

Children who have mastered this unit know the number of days and months in a year, and the number of days in each month. They can explain what a leap year is and apply this in different contexts. They understand that there are 24 hours in a day, 60 minutes in an hour and 60 seconds in a minute, and can use this to estimate times. Children can tell times on analogue clocks to 5 minutes and to the nearest minute. They use am and pm appropriately and can read time from analogue and digital clocks, including those that feature Roman numerals.

COMMON MISCONCEPTIONS	STRENGTHENING UNDERSTANDING	GOING DEEPER
Children think that one day begins when they wake up and ends when they go to bed.	Discuss how sunrise and sunset differ between countries. Ask: *When is it light or dark in the evenings? Does it change in the summer holidays?*	Ask children to discuss if they notice the sunset and sunrise times changing during the different seasons.
Children may think that the hour hand always points directly to a number.	Show four school activities that do not all happen at o'clock times. Children draw minute and hour hands on clocks to match and put the activities in order.	Give children four clocks, each with just the hour hand showing a different time within an hour. Children should put the clocks in order. Ask: *Where could the minute hand be?*
If a start time is given as 12 minutes to 4 and the duration of the activity is 10 minutes, children may try to find the end time by adding 10 minutes to 12 minutes and say that the end time is 22 minutes to 4.	Provide real examples of durations (such as cooking instructions). Ask children what the end time might be, given a specific start time involving minutes to the hour.	Look at a local bus timetable. Ask: *Where can you travel to if you leave now and need to arrive at another stop before ☐ am?*

Unit 13: Time

UNIT STARTER PAGES

Go through the unit starter pages of the Textbook and introduce the unit focus. Use the characters to explore different ways of working.

STRUCTURES AND REPRESENTATIONS

Analogue and digital clocks: These models are used regularly to represent 12-hour times; some analogue clocks use Roman numerals, and digital clocks can also show 24-hour times. Children will complete analogue clock faces with no hands, to demonstrate their understanding.

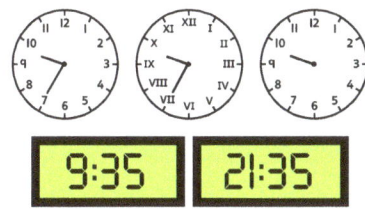

Number line: This model helps children to visualise the order of numbers. It can help them to count on and back in minutes from a given start time, and to identify patterns within the count. In this unit, a number line will be used to represent minutes within an hour, so will go from 0 to 60.

Bar model: This model helps children to find the time left in problem-solving questions.

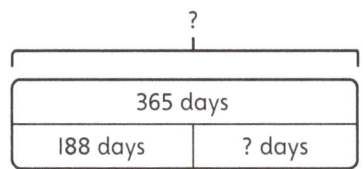

KEY LANGUAGE

There is some key language that children will need to know as part of the learning in this unit:

→ month, year, leap year
→ January, February, March, April, May, June, July, August, September, October, November, December
→ day, hour, minute, second
→ midnight, midday/noon
→ hour hand, minute hand, past, to, half past, o'clock, quarter past, quarter to, numerals, Roman numerals
→ longer, shorter, the same, units, last, convert, passed, fastest, slowest
→ digital clock
→ start time, end time, start, to, duration, how long?, how long left?, time taken, finish, forwards, backwards, twice
→ daytime, night time, around the clock, am, pm
→ morning, afternoon, evening, night

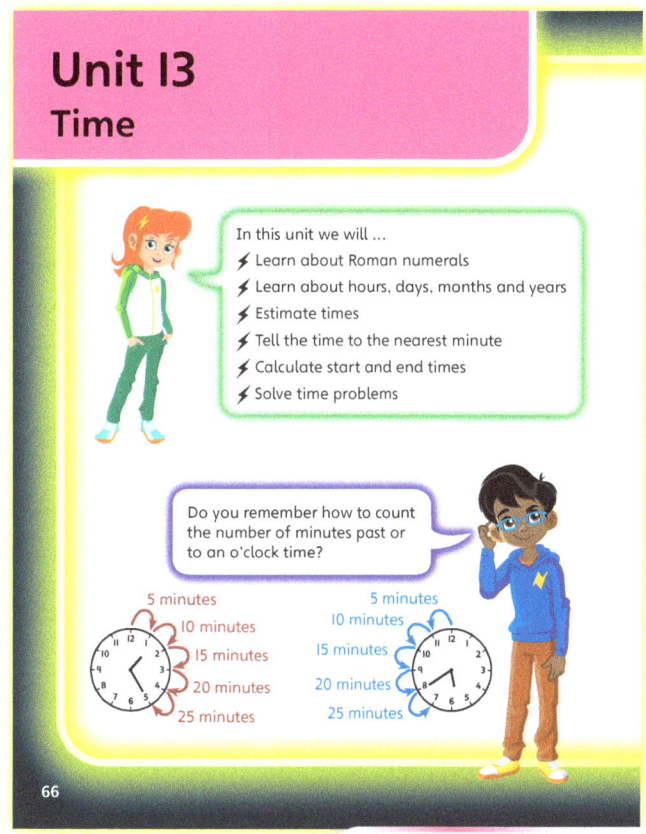

PUPIL TEXTBOOK 3C PAGE 66

PUPIL TEXTBOOK 3C PAGE 67

105

Unit 13: Time, Lesson 1

Roman numerals to 12

Learning focus
In this lesson, children will learn the Roman numerals from 1 to 12 and use this knowledge to read clock faces that have Roman numerals.

Before you teach
- Can children read o'clock and half past times?
- Can children draw o'clock and half past times?
- Can children read quarter past and quarter to times?

NATIONAL CURRICULUM LINKS

Year 3 Measurement – time

Tell and write the time from an analogue clock, including using Roman numerals from I to XII, and 12-hour and 24-hour clocks.

ASSESSING MASTERY

Children know the number of days and months in a year and can say the number of days in each month. Children can explain what a leap year is and apply the concept in different contexts.

COMMON MISCONCEPTIONS

One of the biggest challenges for children reading simple Roman numerals are numbers such as 4 and 9. They may think that IIII represents 4. Acknowledge that this is a possible way to form the number, but explain to children that once beyond half-way between 1 and 5, the Romans decided, for ease, to write 4 as IV as they thought of it as '1 less than 5'. The same applies to IX representing 9. Discuss with children the difference between IV and VI and the placement of the I, encouraging them to think about it as 1 less and 1 more.

STRENGTHENING UNDERSTANDING

Reading the time from a clock can be a difficult skill for young children, even without Roman numerals. If children are struggling to read the clock times in **Think together**, take a step back to focus their learning on reading o'clock and half past times using standard clock faces. Then move back to clocks with Roman numerals. Children may want to make the times on clocks themselves.

GOING DEEPER

Ask children to continue reading Roman numerals to 20, explaining any patterns that they see. Then ask them if they can then go beyond 20 themselves, using their knowledge of the numbers they already know. Ask children to talk about generalisations they can see and how numbers are formed.

KEY LANGUAGE

In lesson: I, V, X, o'clock, half past, quarter past, quarter to, Roman numerals

Other language to be used by the teacher: tell the time

RESOURCES

Mandatory: clock face with Roman numerals

Optional: printed copies of clocks

 In the eTextbook of this lesson, you will find interactive links to a selection of teaching tools.

Quick recap
Before embarking on this lesson, it is important that children can confidently read simple times from clocks. Display some clocks on the board and ask children to write down the times shown. Feature times that involve o'clock, half past, quarter past and quarter to. You could ask children to make or draw some other times on their own clocks.

Unit 13: Time, Lesson 1

Discover

WAYS OF WORKING Pair work

ASK

- Question 1 a): *Does anyone know who the Romans were? What do you know about them? What do you notice about the Roman numerals for 1 to 7? Can you spot any patterns?*
- Question 1 b): *Have you seen these numbers anywhere before?*

IN FOCUS Set the scene for **Discover** by asking children about the Romans and if any of them know anything about their history. Explain that in Roman times they had a different way of writing numbers to the one that we use now. They used letters. Explain that the table (when complete) shows the first ten numbers. Give them the opportunity to look at these first ten numbers and see if they notice any patterns. For question 1 b), discuss as a class where they might have seen Roman numerals before. This might include on clock faces, on statues and memorials, or in the name of a king or queen (King Charles III).

PRACTICAL TIPS This is a great opportunity to bring in links with History and to really set the scene of what it was like in Ancient Rome. You might want children to research aspects of the Roman era.

ANSWERS

Question 1 a): VIII represents 8.
IX represents 9.

Question 1 b): Clock faces, film or book titles and coins.

PUPIL TEXTBOOK 3C PAGE 68

Share

WAYS OF WORKING Whole class teacher led

ASK

- Question 1 a): *What patterns did you notice? Why do you think 2 and 3 are written that way? What happens when you get to 4?*
- Question 1 b): *Have you seen any Roman numerals in real life? Has anyone seen them on a clock before? Where else have you seen them?*

IN FOCUS It is important to draw out in your class discussion the patterns children may have noticed with Roman numerals. Talk through each number and the letters that represent them. Focus on I, V and X first as the key letters that make up the Roman numerals. Children will understand why II and III represent 2 and 3 but they may struggle more when they get to IV for 4. Compare IV and IX as '1 less than 5' and '1 less than 10'. Look especially at the difference between IV (4) and VI (6). Help children understand that the position of the I is very important. They can think of IV as '1 *less* than 5, because the I comes *before* the V' and of VI as '1 *more* than 5, because the I comes *after* the V'. In question 1 b), discuss where children may have seen Roman numerals before. Explain that throughout this unit they are going to look at them on clocks.

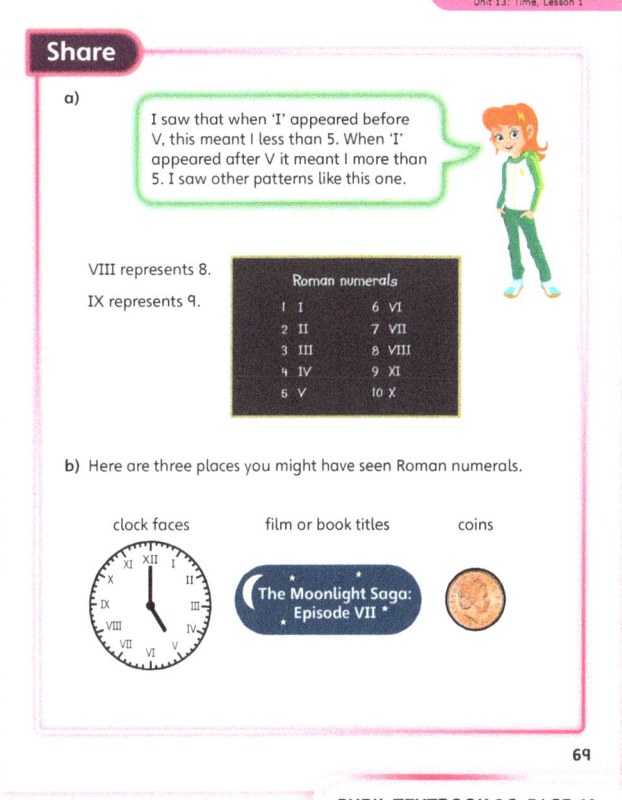

PUPIL TEXTBOOK 3C PAGE 69

107

Think together

WAYS OF WORKING Whole class teacher led (I do, We do, You do)

ASK

- Question ①: *What do you think these are the titles of? What Roman numerals can you see? What does each one mean? What happens when you put them together?*
- Question ②: *What is different about these clocks to other analogue clocks you have worked with before? What times can you see? How do you know?*
- Question ③: *What times can you see on these clocks? How do you know if they are half past, quarter past or quarter to times?*

IN FOCUS Question ① focuses on children learning the Roman numerals for 11 and 12. Discuss with them the letters they can see, what numbers they represent, and how they make the numbers 11 and 12 when they are put together. Question ② provides the Roman numerals on a clock with some o'clock times to read, whereas in question ③, children have a mix of other times to read. They would have met these in Year 2 when they used standard clock faces.

STRENGTHEN Reading the time from a clock can be a difficult skill for young children, even without Roman numerals. If children are struggling to read the clock times in questions ② and ③, take a step back to focus learning on o'clock and half past times using standard clock faces. Then move on to learning quarter past and quarter to times on standard clock faces, before revisiting clocks with Roman numerals. Children may want to make the times on clocks themselves.

DEEPEN Children could explore Roman numerals in more depth. Ask them what patterns they think continue to 20. Ask children if they can then go beyond 20. Children should see how the same rules and principles apply.

ASSESSMENT CHECKPOINT Questions ② and ③ check children's understanding of reading times from clocks that feature Roman numerals.

ANSWERS

Question ① a): 11

Question ① b): 12

Question ② a): 5 o'clock

Question ② b): 9 o'clock

Question ③ a): Half past 1

Question ③ b): Quarter past 4

Question ③ c): Half past 8

Question ③ d): Quarter to 11

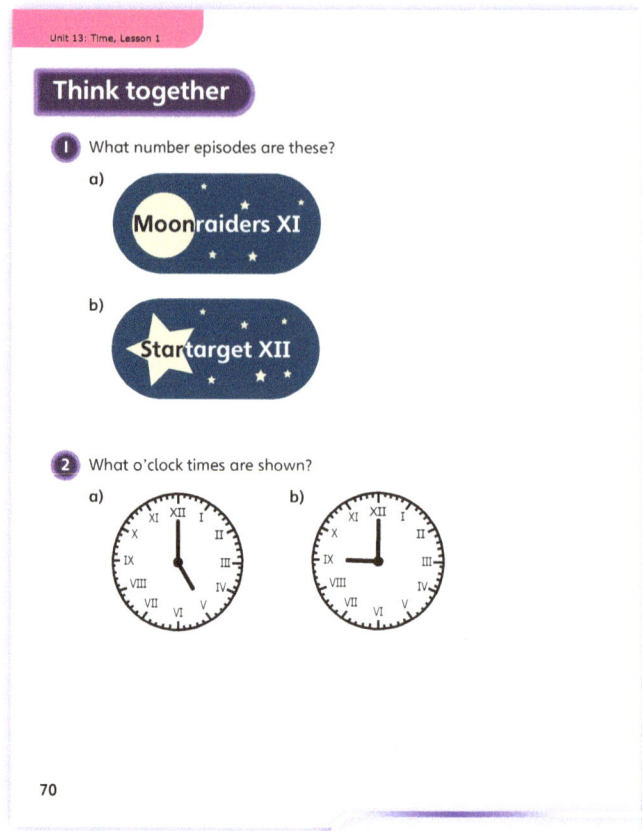

PUPIL TEXTBOOK 3C PAGE 70

PUPIL TEXTBOOK 3C PAGE 71

Unit 13: Time, Lesson 1

Practice

WAYS OF WORKING Independent thinking

IN FOCUS In question ①, children write down the Roman numerals that they see. In question ②, children write the numbers that are represented by Roman letters. Question ③ asks children to match clocks with the correct o'clock times. The clocks in this lesson all feature Roman numerals. In questions ④ and ⑤, children demonstrate that they can tell the time to a mixture of half past, quarter past and quarter to.

STRENGTHEN For questions ① and ②, you may want to display the numbers I, V and X on the board and show that they represent 1, 5 and 10. These are the key letters that are used in smaller Roman numerals. This would give children a starting point to work out some of the numbers on the doors. For question ③ onwards, children may find it easier to write the numbers 1 to 12 next to the Roman numerals to help them tell the time.

DEEPEN Ask children to challenge their partners to tell the time from clocks with Roman numerals.

ASSESSMENT CHECKPOINT Use questions ① and ② to check that children know their Roman numerals to 12. Use questions ③ to ⑤ to check that children can accurately read o'clock, half past, quarter past and quarter to times.

ANSWERS Answers for the **Practice** part of the lesson can be found in the *Power Maths* online subscription.

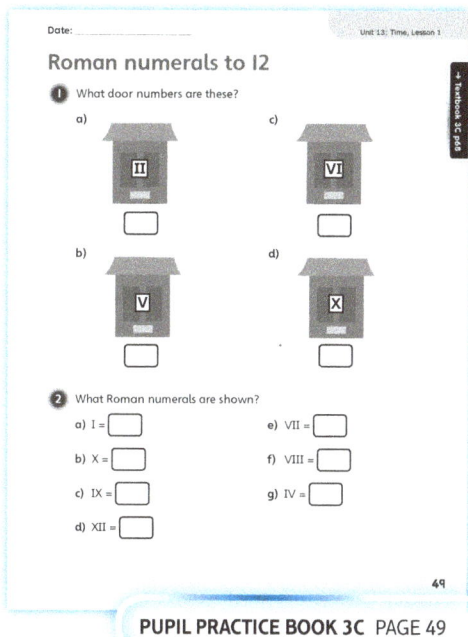

PUPIL PRACTICE BOOK 3C PAGE 49

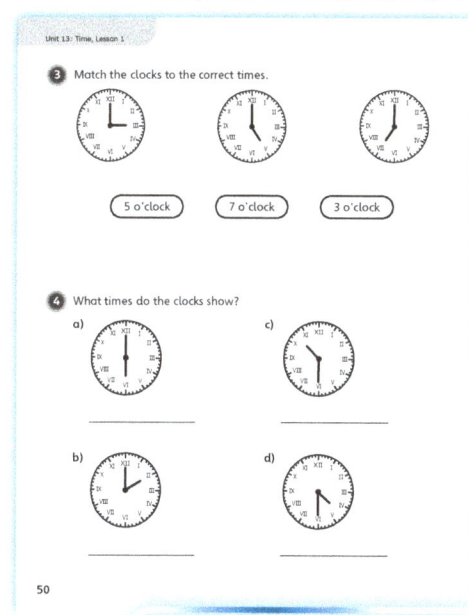

PUPIL PRACTICE BOOK 3C PAGE 50

Reflect

WAYS OF WORKING Pair work

IN FOCUS Children show off what they know about telling time with o'clock and half past times. Encourage children to not just focus on o'clock times. Discuss with children what they notice about all the o'clock times and the half past times. Children could also challenge each other to show different times.

ASSESSMENT CHECKPOINT Check that children can accurately label a clock with the o'clock and half past times and also read clocks with Roman numerals.

ANSWERS Answers for the **Reflect** part of the lesson can be found in the *Power Maths* online subscription.

After the lesson

- Do children know the Roman numerals from 1 to 12?
- Can children read o'clock, half past, quarter past and quarter to times from a clock that has Roman numerals?

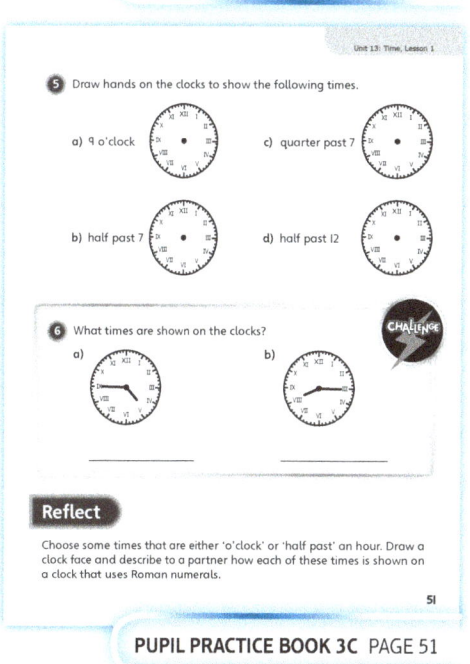

PUPIL PRACTICE BOOK 3C PAGE 51

Unit 13: Time, Lesson 2

Tell the time to 5 minutes

Learning focus
Children will continue to develop their ability to tell the time to 5 minutes and link this to prior knowledge of reading analogue clocks by reading the 5-minute intervals.

Before you teach
- Are children confident in counting in 5s?
- What other practical resources will you provide?

NATIONAL CURRICULUM LINKS

Year 3 Measurement – time

Tell and write the time from an analogue clock, including using Roman numerals from I to XII, and 12-hour and 24-hour clocks.

ASSESSING MASTERY

Children can tell times displayed on analogue clocks to 5 minutes (including all 5 minutes past and to the hour). Children will be able to transfer this knowledge when using clocks with Roman numerals.

COMMON MISCONCEPTIONS

Some children will mix up the roles of the hour and minute hands. They think '10 to' means the hour hand points to the number 10. Ask:
- It is 10 o'clock. Where is the hour hand? Where is the minute hand? What do they show us?

STRENGTHENING UNDERSTANDING

Show children an analogue clock and ask them to read around the clock in steps of minutes. For example, *5 past, 10 past*, etc. Change the time on the clock. Repeat with a mixture of 'past' and 'to' times.

GOING DEEPER

Ask: *What different activities can you do within an hour?* For example, wake up, brush teeth, get ready for school, have breakfast. Show these as multiples of 5-minute times on the analogue clock. Begin with times past the hour, then times to the hour. Ask: *Can you make a timetable of the activities you chose?*

KEY LANGUAGE

In lesson: hour hand, minute hand, minutes to, o'clock, Roman numerals

Other language to be used by the teacher: minutes past, quarter past, quarter to, half past

RESOURCES

Mandatory: analogue clock tool, key language flashcards ('o'clock', 'half past', 'quarter past', 'quarter to'), flashcards showing analogue clocks and written times

 In the eTextbook of this lesson, you will find interactive links to a selection of teaching tools.

Quick recap

Ask children to draw or make an o'clock time. Ask five children to show their answers. Ask: *What is the same and what is different about the clocks?* Do the same with half past and quarter past times. It is important children are confident reading these times.

Discover

WAYS OF WORKING Pair work

ASK

- Question 1 a): *How many minutes are in an hour? How many minutes are between each number on the clock? How many minutes past 11 o'clock is it? How did you work it out?*
- Question 1 b): *What is the next hour? How many minutes are there to the next hour? How did you work it out?*

IN FOCUS It is important to remind children that there are 60 minutes in an hour. This is knowledge from Year 2. In question 1 a), ask children to work out how many minutes there are between each number on a clock. Discuss how they can work this out using the knowledge of 12 × 5 = 60. They could then count in 5s to 55 minutes past 11 o'clock.

In question 1 b), some children may say things like '11 minutes to 12' as the minute hand is pointing to 11. Correct children by reminding them that the interval between each number represents 5 minutes, so there are only 5 minutes to go until 12 o'clock. Children might also use their answers from question 1 a) to subtract 55 minutes from 60 minutes, leaving 5 minutes until 12 o'clock.

PRACTICAL TIPS Show examples of simple bus or train timetables. Ask: *Why is it important for trains and buses to leave at the same time each weekday? Why is it useful to know that they leave at the same time?*

ANSWERS

Question 1 a): It is 55 minutes past 11 o'clock.

Question 1 b): The time is 5 minutes to 12.

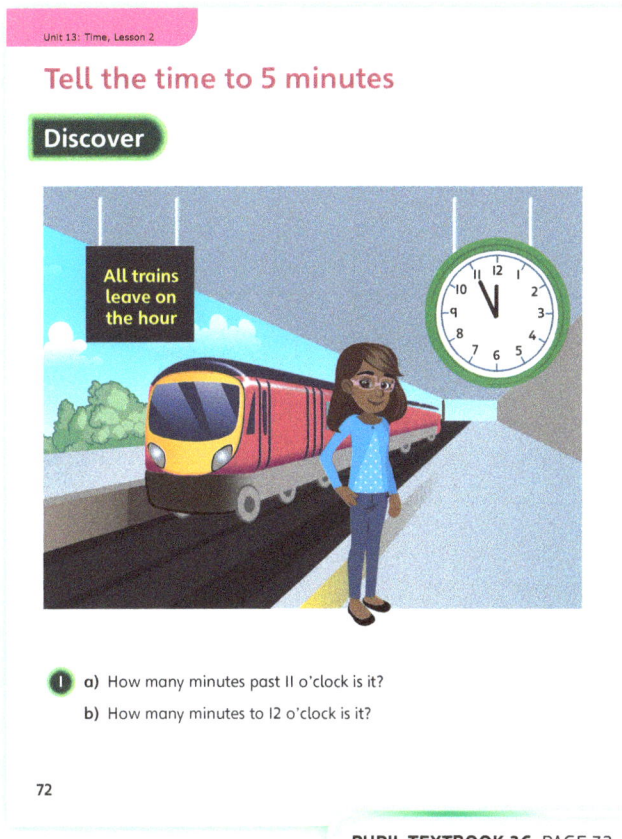

PUPIL TEXTBOOK 3C PAGE 72

Share

WAYS OF WORKING Whole class teacher led

ASK

- Question 1 a): *How did Ash work out how many minutes there are between each number on a clock? How do we show the minutes? How can you use your knowledge of times-tables to help?*
- Question 1 b): *Can you count back in 5 minute intervals from 12 o'clock?*

IN FOCUS In question 1 a), children understand that there are 5 minutes between each of the numbers on a clock. They can see this on the clocks with the little 1 minute marks that were not there on the clock in **Discover**. Use their knowledge of the 5 times-table to calculate 11 × 5 = 55 minutes.

In question 1 b), practise counting back in 5 minute intervals from 12 o'clock so children are reminded of all the 'to' times. Finally, emphasise that when the minute hand is on the 11, it is 5 minutes to the next hour.

STRENGTHEN Point at each number around a clock face, starting at 1. Ask children to say each number as you point to it. Then repeat, this time asking children to count up in 5s as you go around the clock face, from 5 to 60. Repeat again, but now ask children to say *five past, ten past, quarter past, …, half past, twenty-five to, …, quarter to, ten to, …, o'clock.* Continue until children can get all the way round the clock with no mistakes.

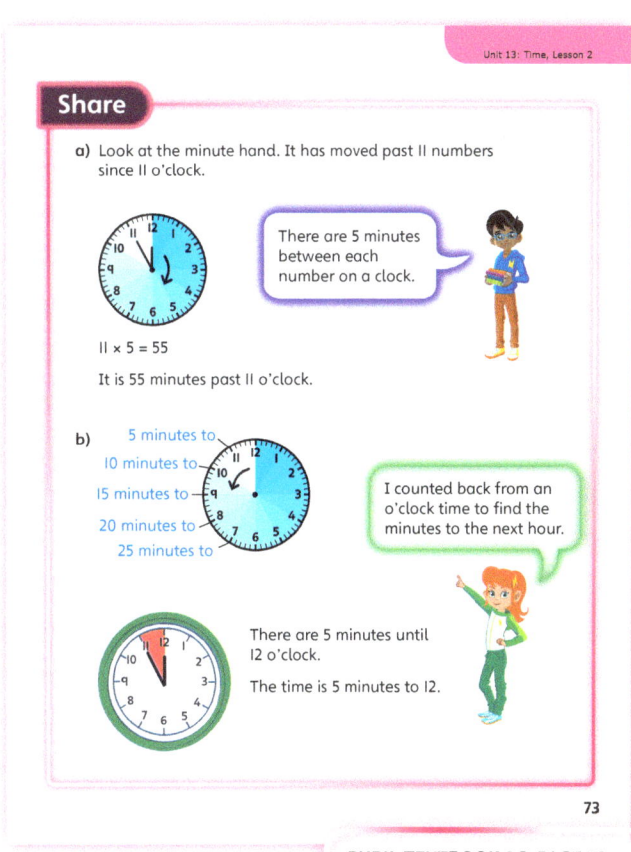

PUPIL TEXTBOOK 3C PAGE 73

111

Think together

WAYS OF WORKING Whole class teacher led (I do, We do, You do)

ASK
- Question ❶: *What time does each clock show? How do you work it out? Is it a 'past' or 'to' time?*
- Question ❷: *What do you notice about the numbers on this clock? What numbers do they represent?*

IN FOCUS In question ❶, children read some 'past' and 'to' times to the nearest 5 minutes. Work through question ❶ a) with children counting around the clock. In question ❶ c), children may say this as a 'past' time, but discuss how they may say it as a 'to' time. Discuss when it might be better to say 'past' and 'to' times (after half-way around the clock, we say 'to' as the minute hand is then closer to the next hour). In question ❷, apply children's understanding of times to 5 minutes on a clock that shows Roman numerals. They met these types of clocks in the previous lesson.

STRENGTHEN You could draw out a large clock in the playground and ask children to walk around it, counting the 5-minute intervals. Make sure to reinforce the 'past' and 'to' elements of the clock face. In question ❸, strengthen children's understanding by asking them to draw all the times they can think of. Ask: *Where would the minute hand be? Where would the hour hand be? How do you know that you are correct?*

DEEPEN Challenge children in question ❸ by asking them to find as many answers as they can within 5 minutes. This will help them practise reading the time. It will also give them an opportunity to estimate and get a sense of what 5 minutes could be. To deepen their understanding even further, ask them to draw the times on a clock that uses Roman numerals.

ASSESSMENT CHECKPOINT At this point in the unit, children should be more confident in telling the time to 5 minutes, by counting forwards and backwards in 5-minute intervals. They should be increasingly fluent in using 'minutes past' and 'minutes to' and in reading the time using analogue clocks (including those with Roman numerals).

ANSWERS

Question ❶ a): 10 minutes past 3

Question ❶ b): 2 minutes past 7

Question ❶ c): 10 minutes to 3 or 50 minutes past 2

Question ❷: 40 minutes past 3 or 20 minutes to 4

Question ❸ a): Children's answers will vary, but they should refer to the minute hand being in the first half of the clock, and the hour hand pointing between the 6 and the 7.

Question ❸ b): Children's answers will vary, but they should refer to the minute hand being in the second half of the clock, and the hour hand pointing between the 9 and the 10.

Question ❸ c): Children's answers will vary, but they should refer to the minute hand pointing to the 8 on the clock, and the hour hand pointing more than half-way between any two numbers.

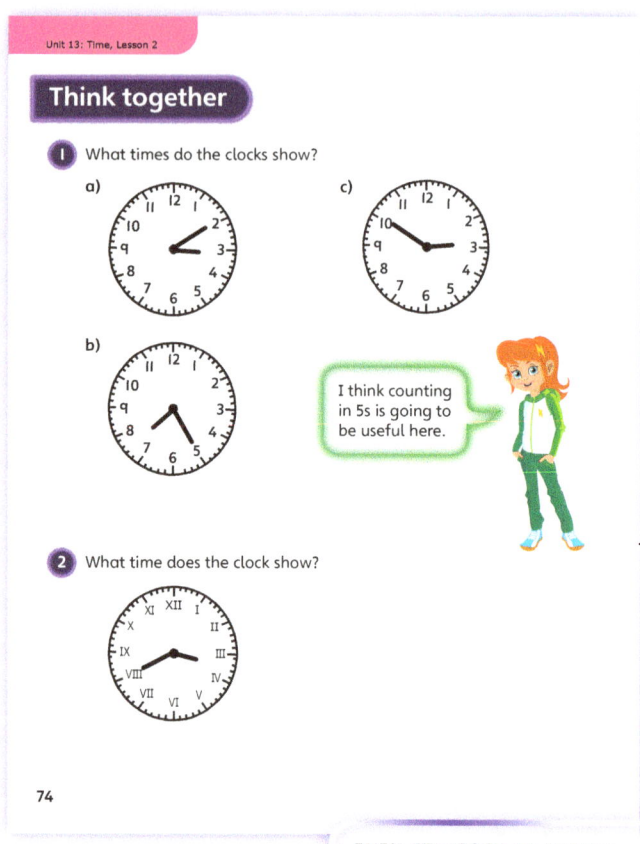

PUPIL TEXTBOOK 3C PAGE 74

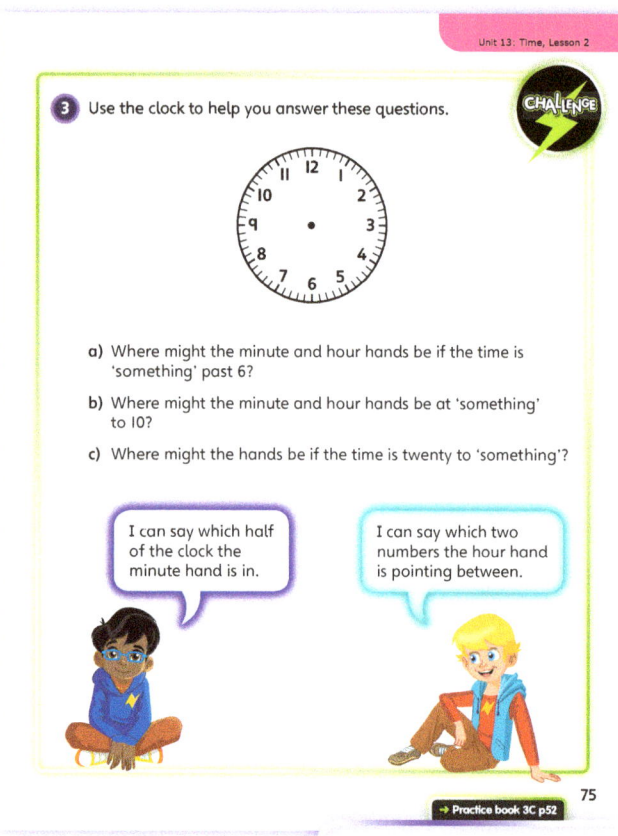

PUPIL TEXTBOOK 3C PAGE 75

Unit 13: Time, Lesson 2

Practice

WAYS OF WORKING Independent thinking

IN FOCUS Questions 1 and 2 scaffold children's understanding of reading an analogue clock in 5-minute intervals. It will help children to develop their fluency if you also use shaded representations of time and clocks with Roman numerals and standard numbers.

STRENGTHEN For children who find question 5 challenging, it may help to use a clock with moving hands. Children could make 6 o'clock, then move the minute hand a full turn. Ask: *What happens to the hour hand? Does it move? Is the time more than 6 o'clock? How do you know?*

DEEPEN Use question 4 to deepen children's explanation and reasoning skills. This question addresses the misconception of mixing up the minute hand with the hour hand. Ask: *What part of the lesson did Lexi not understand? What would you say to explain her mistake? What should the answer have been? Can you shade the clock to show the journey of the minute hand from 1 o'clock until 5 minutes to 2?*

THINK DIFFERENTLY In question 5, children are given a description of a clock and are asked to work out the time. To support children, provide them with a clock so that they can make the time. Some children may be able to work out the time by creating a visual image of the clock in their head. You could extend this activity by asking children to think of another time and describe to a partner the different positions of the hands on the clock face. Their partner has to try to draw the time being described.

ASSESSMENT CHECKPOINT Children should be able to tell the time confidently. They should understand that the minute hand travels around the clock once every hour, and that it tells you how many minutes are past an hour or to the next hour. They should know that each number on the clock face represents an interval of 5 minutes. They should be able to recognise and record different times on an analogue clock, including clocks with Roman numerals.

ANSWERS Answers for the **Practice** part of the lesson can be found in the *Power Maths* online subscription.

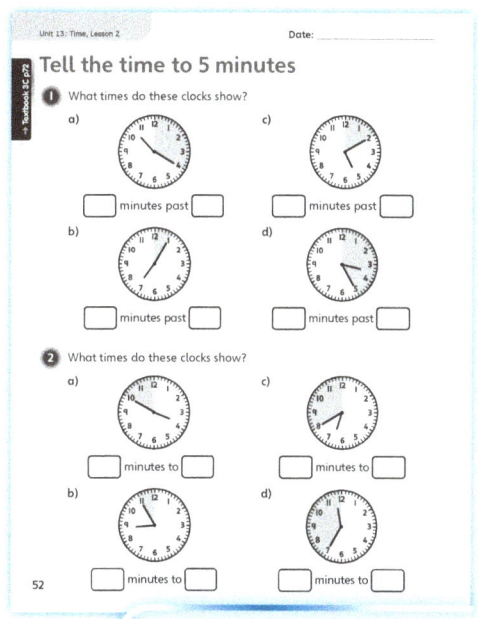

PUPIL PRACTICE BOOK 3C PAGE 52

PUPIL PRACTICE BOOK 3C PAGE 53

Reflect

WAYS OF WORKING Pair work

IN FOCUS Pay particular attention to children's reasoning when it comes to 'twenty-five past' and 'twenty-five to'. Ask: *Does a clock showing twenty-five minutes to 4 also show twenty-five minutes past 4? Can you draw both times? Can you explain how these times differ?*

ASSESSMENT CHECKPOINT Children should recognise that at 'twenty-five to', the minute hand points to the 7. Children should be able to explain their reasoning clearly. They may use a concrete representation or picture to justify their answer.

ANSWERS Answers for the **Reflect** part of the lesson can be found in the *Power Maths* online subscription.

After the lesson

- Have children recognised that, for the hour hand, the numbers on the clock face each represent 1 hour, whilst, for the minute hand, each number on the clock face represents 1 minute.
- Do they understand how these two hands work together to tell us the time in hours and minutes?
- How will you reinforce this link by bringing telling the time into other subjects?

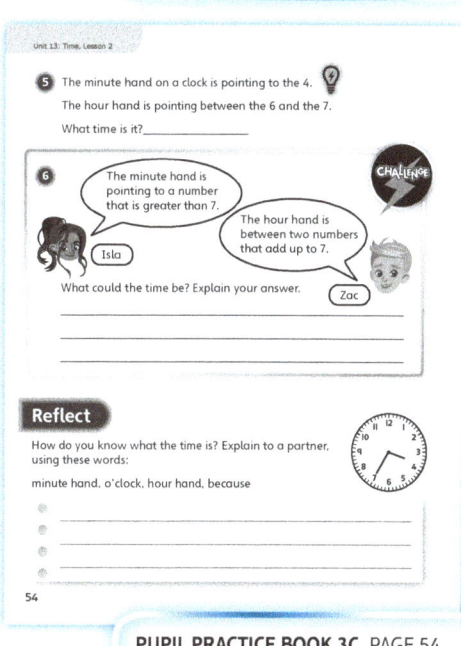

PUPIL PRACTICE BOOK 3C PAGE 54

Unit 13: Time, Lesson 3

Tell the time to the minute

Learning focus
In this lesson, children will tell the time using 'minutes past' and 'minutes to' and using the 12-hour analogue clock. They will read and describe times to the nearest minute.

Before you teach
- Can children see the time easily in the classroom?
- Can you provide a variety of analogue clocks?

NATIONAL CURRICULUM LINKS

Year 3 Measurement – time

Tell and write the time from an analogue clock, including using Roman numerals from I to XII, and 12-hour and 24-hour clocks.

Estimate and read time with increasing accuracy to the nearest minute; record and compare time in terms of seconds, minutes and hours; use vocabulary such as o'clock, am/pm, morning, afternoon, noon and midnight.

ASSESSING MASTERY

Children can tell times displayed on analogue clocks to the minute. Children recognise that there are 60 minutes in an hour, and are able to use this information when giving times past and to the hour.

COMMON MISCONCEPTIONS

Children may think that the minute hand moves straight from number to number. Ask:
- *How many minutes in an hour? How does the minute hand show this?*

Children may read 'five to 3' as' five to 2' because the hour hand is between the 2 and the 3. Ask:
- *What o'clock time has just happened? What o'clock time will come next?*

STRENGTHENING UNDERSTANDING

Show '50 minutes past 3' on an analogue clock. Ask: *What time is it? Show me the time 10 minutes later. Has the hour hand moved?*

GOING DEEPER

Compare two clocks showing 'five to' times, with one hour difference. Ask: *What times do they show? Which is later? How do they differ?* Repeat for 6 o'clock, half past 12, quarter past 9 and quarter to 3. Encourage children to use 'minutes past' and 'minutes to', and to describe the position of the hour hand and minute hand.

KEY LANGUAGE

In lesson: time, past, to, o'clock

Other language to be used by the teacher: hour hand, minute hand, analogue

STRUCTURES AND REPRESENTATIONS

Number line

RESOURCES

Mandatory: analogue clock tool

Optional: flashcards showing analogue clock times, number cards from 1 to 12, large pieces of paper

 In the eTextbook of this lesson, you will find interactive links to a selection of teaching tools.

Quick recap

On the board write the times '10 minutes past 5', '10 minutes past 7' and '10 minutes past 9'. In groups of three, ask children to draw these times. Encourage them to work together and to discuss what is the same and what is different. Repeat for three similar 'to' times.

Unit 13: Time, Lesson 3

Discover

WAYS OF WORKING Pair work

ASK

- Question 1 a): *What does the picture show? What do the hands mean on the clock? How many minutes are there in an hour?*
- Question 1 b): *If the minute hand was at number 6, what would the time have been? How many minutes past number 6 is the minute hand showing?*

IN FOCUS Use the picture to recap and briefly assess children's understanding. Begin by reinforcing the idea that the minute hand does not jump from one number to the next, but travels gradually around the clock.

PRACTICAL TIPS Place a large hoop on the floor on a large piece of paper. Ask children to place number cards from 1 to 12 inside the hoop to make a clock. Using a pen, divide the gap between each pair of numbers into five, so that there are 60 divisions altogether on the clock. Children can make their own clocks. Encourage them to put the numbers in the correct places. They can set their own clock-based problems, telling the time to the nearest minute.

ANSWERS

Question 1 a): The photo was taken at 33 minutes past 10.

Question 1 b): Another way to say this time is 27 minutes to 11.

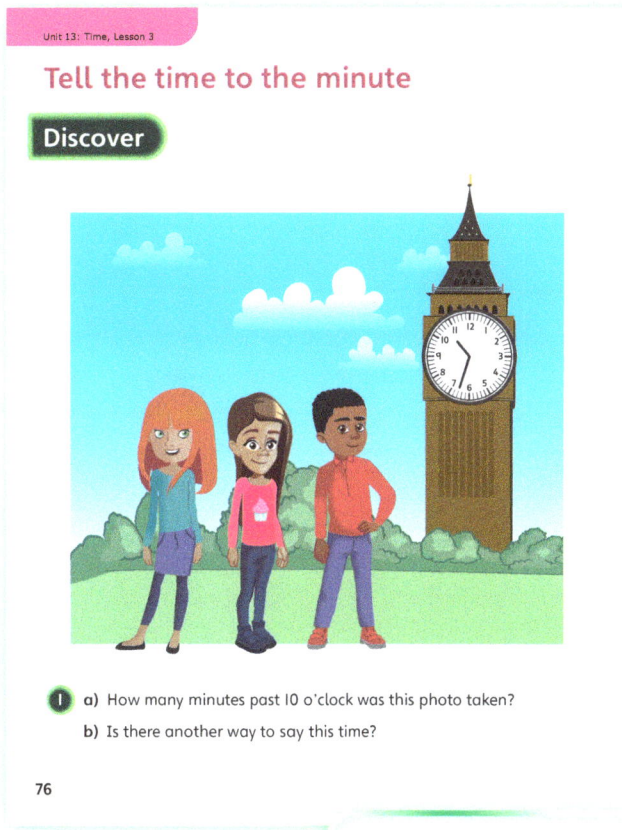

PUPIL TEXTBOOK 3C PAGE 76

Share

WAYS OF WORKING Whole class teacher led

ASK

- Question 1 a): *How can you use a clock face to prove that there are 60 minutes in an hour?*
- Question 1 a): *If 35 minutes have gone by since the last o'clock time, how many minutes remain until the next hour? How many minutes have gone by if it is 10 minutes to the hour?*
- Question 1 b): *How can we say the time in two different ways? Which is the more common way?*

IN FOCUS Use this opportunity to address the misconception that the hands jump from one number to the next. Allow children to show their understanding of times – for example, by using clock manipulatives, drawing times on blank clock faces or calling out times when shown different clocks.

STRENGTHEN Provide two large classroom clocks that children can see clearly during the day; one clock should show the individual minutes, while the other does not. Ask children to read the time from each clock. Ask one group to read the time approximately (*It's about …*). Ask another group to read it exactly to the nearest minute (*It's exactly …*). Ask: *Do all clocks show each individual minute on their faces? What do you notice? Is it easier to read the time to the minute with or without the minute marks?*

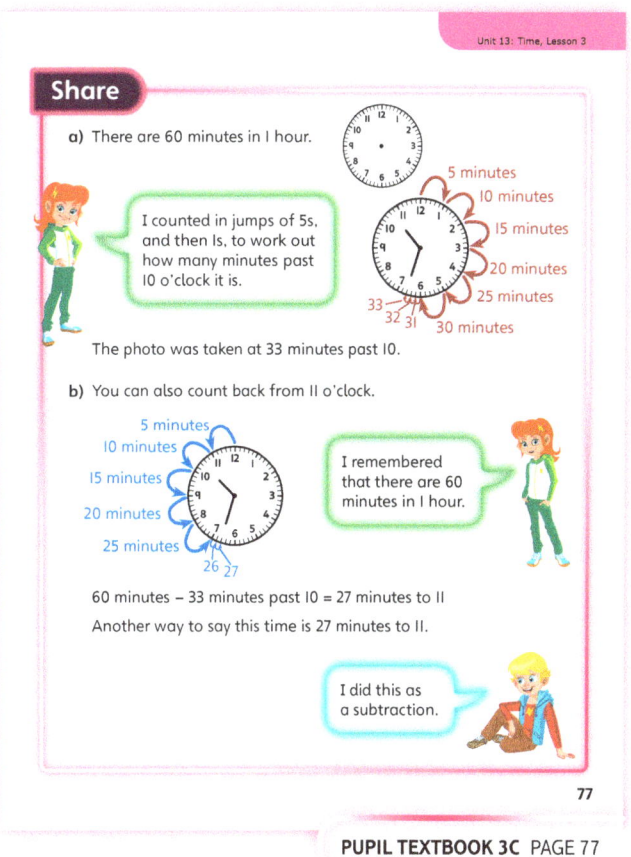

PUPIL TEXTBOOK 3C PAGE 77

115

Unit 13: Time, Lesson 3

Think together

WAYS OF WORKING Whole class teacher led (I do, We do, You do)

ASK
- Question ❶: *Why are we counting in 5s and 1s? How can you tell what hour it is? Which hand should you look at?*
- Question ❷: *What is different about the clocks in a) and b)? What would you be doing at these times in the afternoon?*

IN FOCUS Questions ❶ and ❷ give children an opportunity to practise their recognition of 'minutes past' and 'minutes to'. Discuss with children the clues they use to identify whether the clock is showing 'minutes past' or 'minutes to'. Ask: *Has the minute hand gone past the number 6 (half-way through an hour)?* In question ❷ a), ask: *Has the hour hand gone past the number 4 or is it travelling towards the number 4?* Encourage discussion and share ideas.

STRENGTHEN For these questions, encourage children to make and to write in words all the times shown. Ask: *Can you think of two ways to say this time?* Questions ❶ to ❸ provide opportunities for children to estimate times. Ask and discuss: *Why is estimating time important in everyday life? Do you think you are more likely to hear 'about twenty to 3' or '21 minutes to 3' in real life? Does that mean reading the time exactly is not important?*

DEEPEN Ask children to organise the times they have seen in this lesson into two groups: 'minutes to' and 'minutes past'. Ask: *In which group will this time go? How do you know you have organised the times correctly? Can you draw a clock showing each time?*

ASSESSMENT CHECKPOINT Children should be able to identify clocks that show 'minutes to' and clocks that show 'minutes past'. Assess if children can recognise and explain how the position of the hour hand changes as the time moves from one hour to the next.

ANSWERS

Question ❶ a): 18 minutes past 5

Question ❶ b): 21 minutes to 12

Question ❷ a): 9 minutes past 4

Question ❷ b): 12 minutes to 5

Question ❷ c): 4 minutes to 9

Question ❷ d): 28 minutes past 1

Question ❸: 41 minutes past 9 or 19 minutes to 10

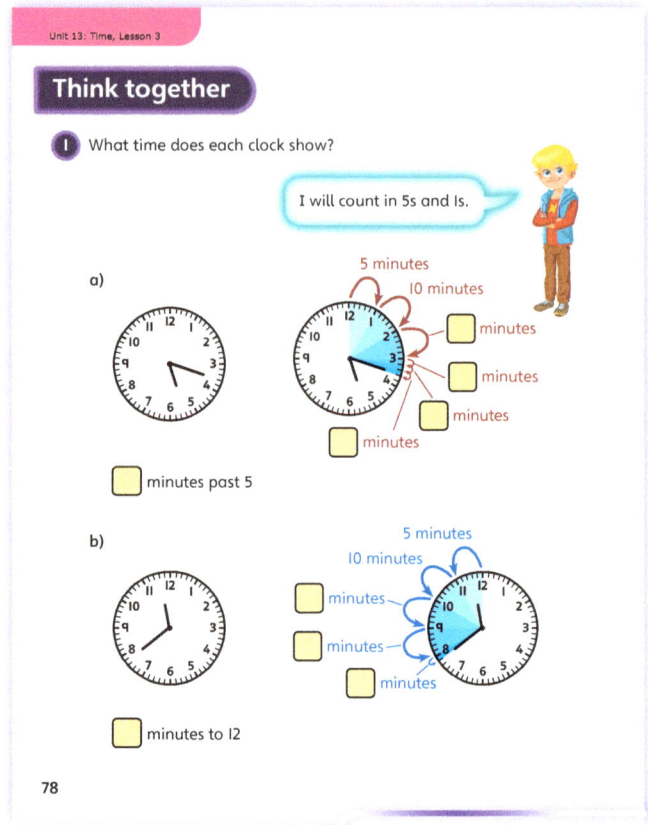

PUPIL TEXTBOOK 3C PAGE 78

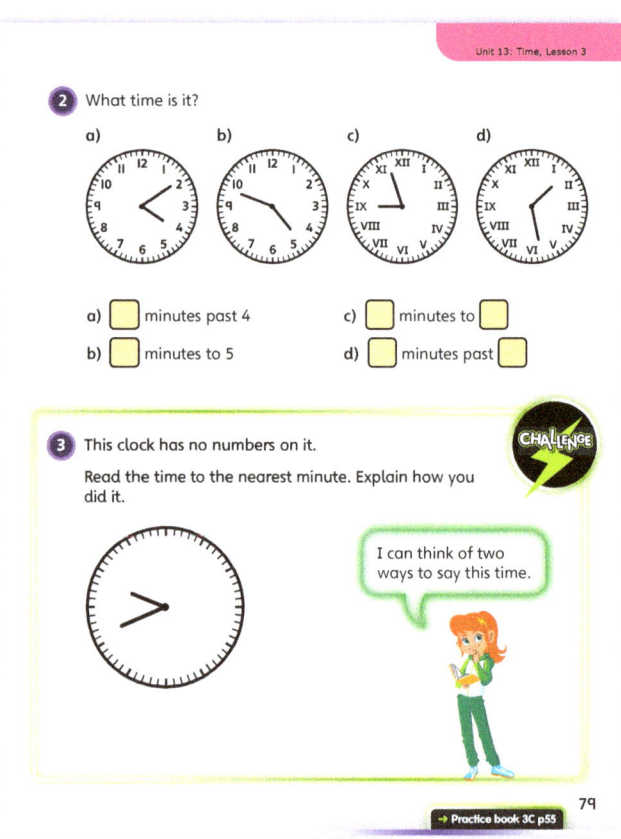

PUPIL TEXTBOOK 3C PAGE 79

Unit 13: Time, Lesson 3

Practice

WAYS OF WORKING Independent thinking

IN FOCUS Question ① links the concepts from this lesson with children's real-life experiences. It may be beneficial to provide blank clock faces, and the statements written on slips of paper, so children can draw the right time on each clock and match it with the corresponding activity. Question ② gives children the opportunity to develop fluency in recognising and recording different times.

STRENGTHEN For children struggling to read the 'minutes past' and the 'minutes to' times in question ③, it may be beneficial to use a number line. Ask: *How many minutes have gone by? How many minutes are there until the next hour?*

DEEPEN Use question ④ to deepen children's reasoning and explanation skills. Ask children what part of the lesson they think Kate did not understand. Ask: *What would you say to help her understand? Can you explain where she went wrong? Can you use a resource or picture to help explain the mistake to her?*

ASSESSMENT CHECKPOINT At this point in the unit, children should be able to confidently recognise and record times to the nearest minute. They should be able to use 'minutes past' and 'minutes to' and fluently describe the times they read.

ANSWERS Answers for the **Practice** part of the lesson can be found in the *Power Maths* online subscription.

Reflect

WAYS OF WORKING Independent thinking

IN FOCUS Give children an opportunity to develop their own line of thinking. They should be able to link the vocabulary they have learnt in the past with the vocabulary in this lesson to justify their reasoning.

ASSESSMENT CHECKPOINT Children should show a confident understanding of estimating and reading time to the nearest minute. They should be able to explain confidently and accurately the importance of being able to estimate times and of being able to read the time to the nearest minute.

ANSWERS Answers for the **Reflect** part of the lesson can be found in the *Power Maths* online subscription.

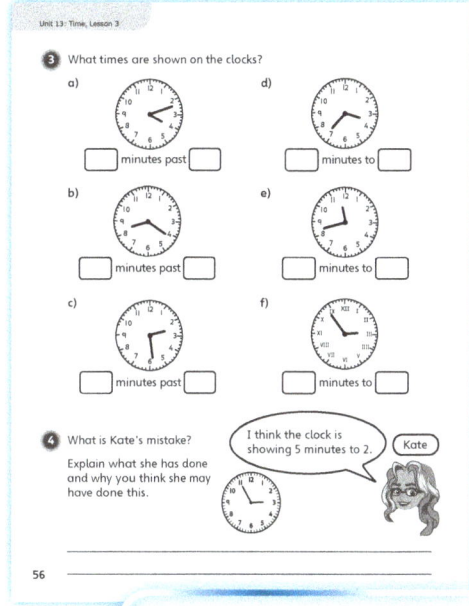

PUPIL PRACTICE BOOK 3C PAGE 55

PUPIL PRACTICE BOOK 3C PAGE 56

PUPIL PRACTICE BOOK 3C PAGE 57

After the lesson

- Are children able to confidently estimate times?
- Are children able to read the time accurately?
- How will you build in more opportunities to practise these skills throughout the school day?

Unit 13: Time, Lesson 4

Read time on a digital clock

Learning focus
In this lesson, children will read times from digital clocks. They will be able to read the time as 'minutes past' and as 'minutes to'.

Before you teach
- Are children able to read 'past' times from an analogue clock?
- Are children able to read 'to' times from an analogue clock?
- Can children convert 'past' times to 'to' times for simple examples?

NATIONAL CURRICULUM LINKS

Year 3 Measurement – time

Estimate and read time with increasing accuracy to the nearest minute; record and compare time in terms of seconds, minutes and hours; use vocabulary such as o'clock, am/pm, morning, afternoon, noon and midnight.

Tell and write the time from an analogue clock, including using Roman numerals from I to XII, and 12-hour and 24-hour clocks.

ASSESSING MASTERY

Children can read times from a digital clock, and they should be able to convert digital 'past the hour' times to 'to the hour times' using their understanding that there are 60 minutes in an hour.

COMMON MISCONCEPTIONS

When converting times children may find the complement to 60 incorrectly. For example children may think that 45 minutes past 4 is the same as 25 minutes to 5, as they think that 25 + 45 makes 60. Ask children to use a number line to support their understanding of number bonds to 60. Ask:
- *From 45, how many minutes are there to the closest 10 minutes? How many more 10s do you need to add to get to 60?*

STRENGTHENING UNDERSTANDING

Children should use a number line from 0 to 60 to help them convert between 'past' and 'to' times. They should be encouraged to show two jumps on the number line. The first jump is from the minutes to the next 10; the second is a jump to 60. Each time, ask them to record the size of the jump above the number, and this will tell them the complement to 60.

GOING DEEPER

Children could be encouraged to discuss different parts of the day and what a digital clock might show. Children could also start to convert analogue clock times into digital clock times.

KEY LANGUAGE

In lesson: time, past, to, hour, digital, analogue,

Other language to be used by the teacher: clock, complement

STRUCTURES AND REPRESENTATIONS

Number lines

RESOURCES

Mandatory: analogue and digital clocks, dice

 In the eTextbook of this lesson, you will find interactive links to a selection of teaching tools.

Quick recap
Roll some dice and use the digits to make a two-digit number less than 60. (Roll again if you roll two 6s). Ask children to work out the number that they need to add on to make 60, working as fast as they can. Discuss the strategies that they used. Make sure that number lines are available for the children to use.

Unit 13: Time, Lesson 4

Discover

WAYS OF WORKING Pair work

ASK
- Question 1 a): *Has anyone got numbers like this on their watch or clock? What do you think the first number tells you? What about the second number?*
- Question 1 b): *How can you make or draw this time on an analogue clock? How many minutes past the hour is it? How many minutes to the next hour is it?*

IN FOCUS This may be the first time some children have seen times shown on a digital watch or clock. It is important to discuss where children might come across them – for example, on an adult's phone, an alarm clock or the microwave at home. Ask children to think about what the first number might tell them and what the second number might tell them. Provide clocks for children to make the time on an analogue clock.

PRACTICAL TIPS Provide images showing examples of where children might have seen digital clock times.

ANSWERS

Question 1 a): The time is 35 minutes past 11.

Question 1 b):

It is 25 minutes to 12.

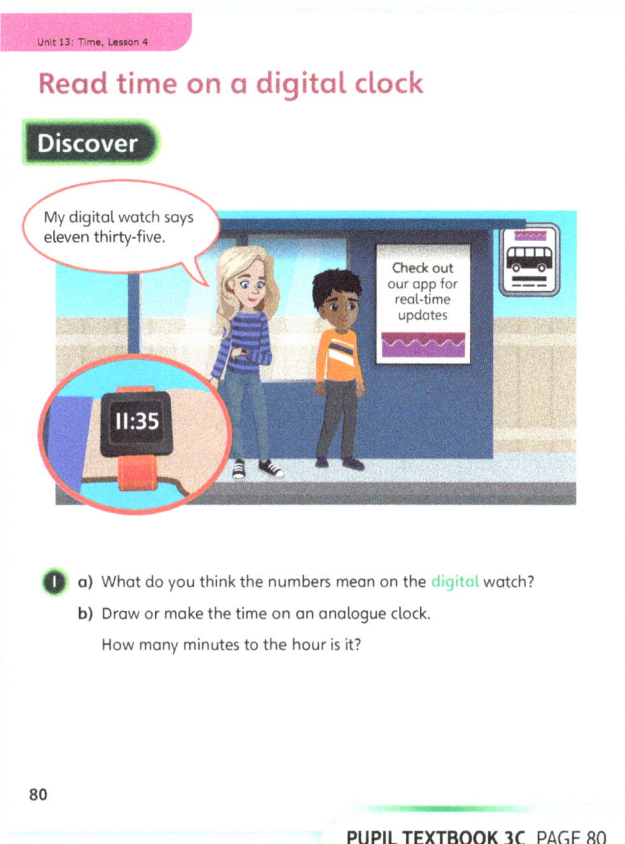

PUPIL TEXTBOOK 3C PAGE 80

Share

WAYS OF WORKING Whole class teacher led

ASK
- Question 1 a): *What do you think the first number is? What about the second number? What do you think is the maximum the numbers could be? How do you think you read the time?*
- Question 1 b): *What number does the minute hand point to? How many minutes past 11 o'clock does this represent? How many minutes to 12 o'clock does this represent? What do the number of minutes past 11 and the number of minutes to 12 add up to?*

IN FOCUS In question 1 a), discuss children's thinking about the two numbers on the clock. Explain to children how you read the times, for example, '35 minutes past 11'. Explain that we don't usually change this to minutes to the hour. Sometimes we say the time is 'eleven thirty-five'. Discuss why the second number can never go above 59 and that, when it needs to, it restarts at 00. For question 1 b), ask children to show this time on an analogue clock and then to use this to work out how many minutes to the hour it is. Children may use a number line to support them. Explain that the numbers past the hour and the numbers to the next hour should add to make 60, because there are 60 minutes in an hour. In this case, 35 minutes past + 25 minutes to = 60 minutes.

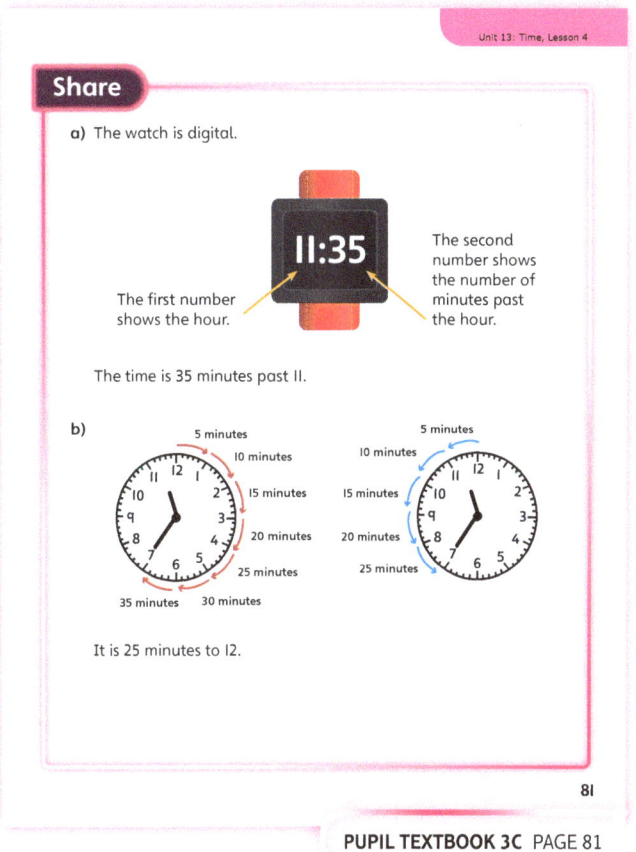

PUPIL TEXTBOOK 3C PAGE 81

119

Think together

WAYS OF WORKING Whole class teacher led (I do, We do, You do)

ASK

- Question ❶: *How can you read these times? What is the difference between the two times shown in each part of this question?*
- Question ❷: *What times are shown? How can you say these as 'to the hour' times? What did you do to convert?*
- Question ❸: *What time is shown? How many minutes to the hour is this?*

IN FOCUS Question ❶ focuses children on reading times. They may first read them as 'two twenty', but encourage children to say 'twenty past two'. The clocks below have am and pm on. Although this is more of a focus for the next lesson, take the time now to highlight that some clocks will show this feature and that it tells you about what part of the day it is. In question ❷, ask children to say the times 'past' the hour and ask them to convert these to 'to' the hour times by working out how many minutes to make 60. They may use a number line to help them.

STRENGTHEN Children should use a number line from 0 to 60 to help them to convert between 'past' and 'to' times. They should be encouraged to show two jumps on the number line. The first jump is from the minutes to the next 10, followed by a jump to 60. Each time they should record the size of the jump above the number, and this will tell them the complement to 60.

DEEPEN Ask children to say some 'to' times and then ask them to convert to digital time (for example, '32 minutes to 5'). What time will this show on a digital clock?

ASSESSMENT CHECKPOINT Use question ❶ to check that children can accurately read a time on a digital clock and use question ❷ to check that children can convert a 'past the hour' time to a 'to the hour' time.

ANSWERS

Question ❶ a): 20 minutes past 2

Question ❶ b): 29 minutes past 10

Question ❶ c): 43 minutes past 7

Question ❷ a): 10 minutes to 3

Question ❷ b): 24 minutes to 6

Question ❸ a): Ebo is correct.

Question ❸ b): 8:26 am

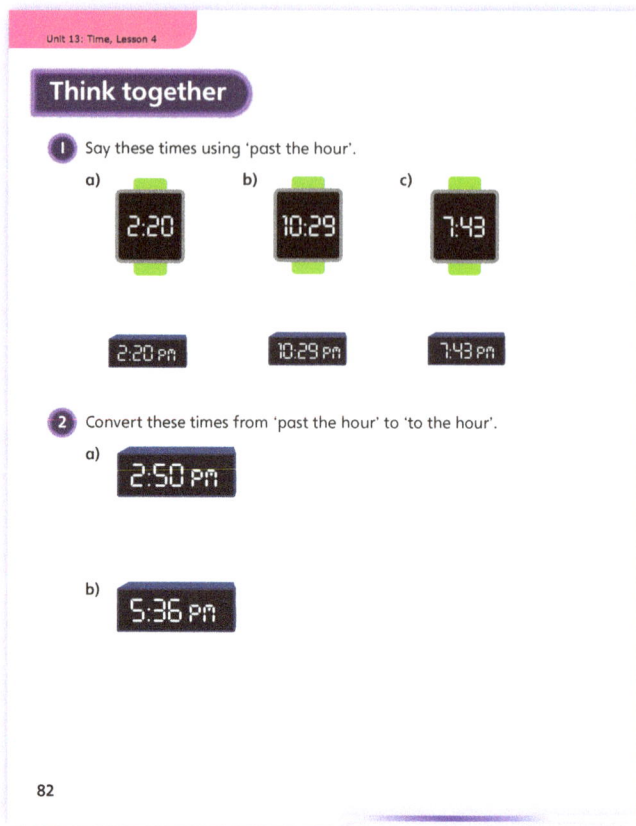

PUPIL TEXTBOOK 3C PAGE 82

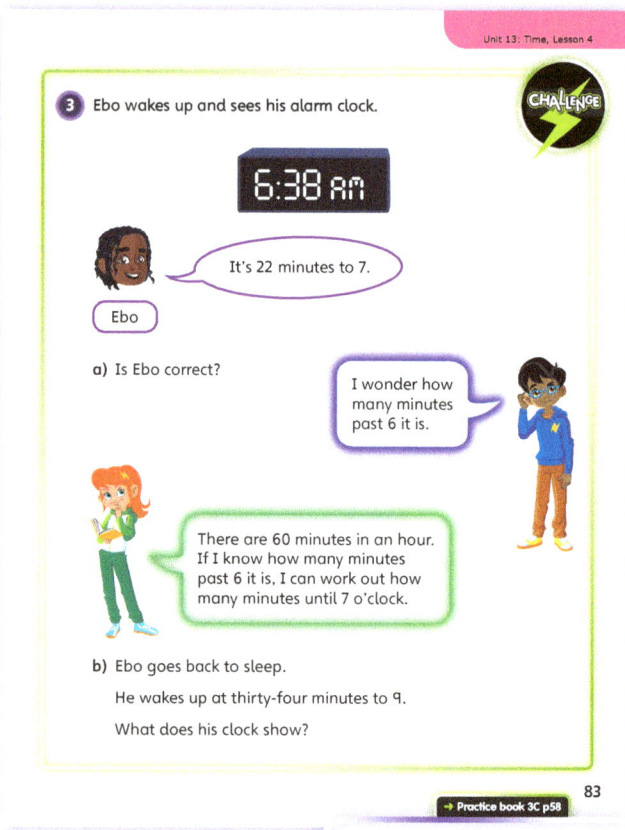

PUPIL TEXTBOOK 3C PAGE 83

Unit 13: Time, Lesson 4

Practice

WAYS OF WORKING Independent thinking

IN FOCUS In question ❶, children match the digital clock times to the times that children may say out loud. In question ❷, they convert simple digital 'past' times to 'to' times by thinking about how many minutes to the hour there are left. In question ❹, children must convert the times that are said to a digital clock time. In question ❺, encourage children to say the times out loud in different ways. Can they say each time in three different ways?

STRENGTHEN Children should use a number line from 0 to 60 to help them convert between 'past' and 'to' times. They should be encouraged to show two jumps on the number line. The first jump is from the minutes to the next 10, followed by a jump to 60. Each time they should record the size of the jump above the number, and this will tell them the complement to 60.

DEEPEN Ask children to convert between analogue and digital clock times. Also ask children to say some 'to' times and to convert these to digital times. (For example, '32 minutes to 5'. Ask: *What time will this show on a digital clock?*)

ASSESSMENT CHECKPOINT Use the questions to check that children can accurately read a time in different ways from a digital clock.

ANSWERS Answers for the **Practice** part of the lesson can be found in the *Power Maths* online subscription.

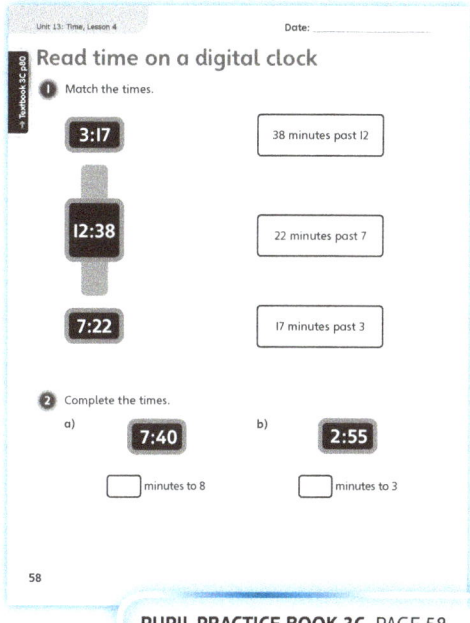

PUPIL PRACTICE BOOK 3C PAGE 58

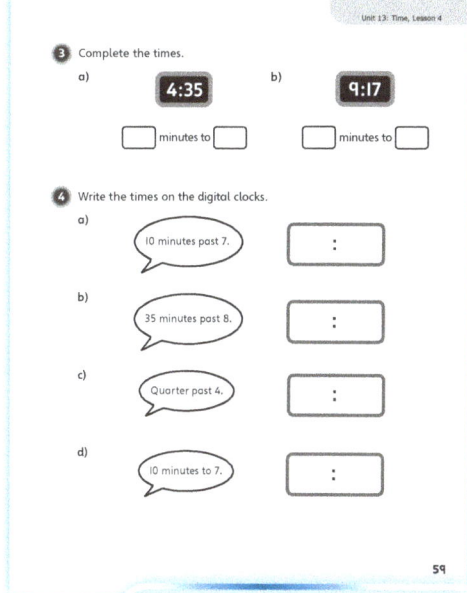

PUPIL PRACTICE BOOK 3C PAGE 59

Reflect

WAYS OF WORKING Pair work

IN FOCUS Ask children, in pairs, to discuss what time it is now. Where can they see the time? How can they read and represent this time in different ways?

ASSESSMENT CHECKPOINT Check the different ways children say, write or draw the current time. Discuss the different ways they might do this including digital and analogue, 'to' and 'past' the hour times, and so on.

ANSWERS Answers for the **Reflect** part of the lesson can be found in the *Power Maths* online subscription.

PUPIL PRACTICE BOOK 3C PAGE 60

After the lesson ⏸

- Can children read a time from a digital clock?
- Are children able to convert a 'past the hour' time to a 'to the hour' time?

Unit 13: Time, Lesson 5

Use am and pm

Learning focus
Children will read times using analogue and digital clocks. They will recap their learning about measuring time and describe time using am and pm, or morning and afternoon or evening.

Before you teach
- Are children confident using am and pm?
- Can children easily see the time in the classroom?

NATIONAL CURRICULUM LINKS

Year 3 Measurement – time

Estimate and read time with increasing accuracy to the nearest minute; record and compare time in terms of seconds, minutes, hours; use vocabulary such as o'clock, am/pm, morning, afternoon, noon and midnight.

Tell and write the time from an analogue clock, including using Roman numerals from I to XII, and 12-hour and 24-hour clocks.

ASSESSING MASTERY

Children can read times displayed on analogue clocks to the minute and use the terms am and pm appropriately. Children can read 12-hour digital times and make links between digital and analogue times.

COMMON MISCONCEPTIONS

Children may think that digital clock numbers are the same as analogue clock numbers – for example, that 3:10 means the hour hand points to the 3, and the minute hand points to the 10. Ask:
- *What does the analogue clock show? What does this mean? How many hours? How many minutes?*

Children may think that am means the hours of daylight and pm means the hours of darkness. Ask:
- *Is it sometimes dark at 7 am? Is it ever dark at 1 pm?*

STRENGTHENING UNDERSTANDING

Offer children opportunities to look at pictures of the same clock at different times in one hour (for example, 9 o'clock, 9:15, 9:20, 9:30, 9:40, 9:45, 9:55 and 10 o'clock). Ask them to describe the journey of the minute hand within this one hour, and then to describe the journey of the hour hand within this one hour.

GOING DEEPER

Challenge children to look at one channel in an online TV guide. Ask them to record each programme in a table with two columns: am and pm. Give children a set amount of time to do this – for example, 15 minutes. This will develop their ability to recognise the times. It will also help to develop their awareness of how long a certain number of minutes is.

KEY LANGUAGE

In lesson: past, to, digital, digit, **ante meridiem (am)**, **post meridiem (pm)**, **midday**, **midnight**, morning, evening, time, minute, hour

Other language to be used by the teacher: hour hand, minute hand, o'clock, consecutive, analogue

RESOURCES

Mandatory: analogue clock tool, digital clock showing am and pm

Optional: flashcards showing written times

 In the eTextbook of this lesson, you will find interactive links to a selection of teaching tools.

Quick recap
Check that children can read a digital time. Display a time for 3 seconds on a digital clock and ask children to tell you the time. Discuss whether it could be in the morning or evening.

122

Unit 13: Time, Lesson 5

Discover

WAYS OF WORKING Pair work

ASK

- Question 1 a): *What clocks can you see in this picture? How many minutes are there in an hour? Is the time shown closer to 8 o'clock or 9 o'clock?*
- Question 1 b): *What information can we use to decide whether it is morning, afternoon or evening?*

IN FOCUS This picture provides children with experience of using different representations of time. Use this picture as an opportunity to discuss similarities and differences between digital and analogue time. Address the potential misconception that the hour and minute hands point to the numbers shown on a digital clock, to ensure children do not think this later.

PRACTICAL TIPS If possible, provide a digital and an analogue clock on each table, so children can observe how the time changes throughout the day. At regular intervals, ask children to tell the time on both clocks. Ask: *How do the clocks differ?*

ANSWERS

Question 1 a): The clock on the wall should show the time like this:

Question 1 b): It is morning because the digital clock says 'am'.

Share

WAYS OF WORKING Whole class teacher led

ASK

- Question 1 a): *The whole hour is 60 minutes. Can you use this to find out how many minutes until the hour is complete?*
- Question 1 a): *Can you count around the clock face and show where the minute hand should be? How would this time look on a digital clock?*
- Question 1 b): *How do you decide whether it is going to be am or pm?*

IN FOCUS Give children different times in the day, and ask them to think of activities they might be doing at these times. Reinforce what am and pm mean. Ask children to think of mistakes that they think might happen when using am and pm, such as thinking that am means the hours of daylight and pm means the hours of darkness. Clarify any misconceptions, reinforcing that am means before midday and pm means after midday.

STRENGTHEN Have a class 'clock day' where children can bring a clock from home. Also have different types of watches and clocks in the classroom. Ask: *Which one do you prefer? Why? How do they differ?* Ask children throughout the day to find out what the time is.

PUPIL TEXTBOOK 3C PAGE 84

PUPIL TEXTBOOK 3C PAGE 85

123

Think together

WAYS OF WORKING Whole class teacher led (I do, We do, You do)

ASK
- Question ❶: *How will you read the analogue clocks? How will you read the digital clocks? What clues can you look for when reading the time on each clock?*
- Question ❷: *Do you use am or pm for a time in the evening? Where is the hour hand pointing on the analogue clock?*

IN FOCUS Questions ❶ and ❷ make the link between analogue and digital times. Make sure children are confident when reading a digital time and are able to decide whether the time shown is morning or afternoon/evening.

STRENGTHEN In questions ❶ and ❷, strengthen the link between the times shown in the question and the correct use of the mathematical vocabulary 'am', 'pm', 'minutes past' and 'minutes to'. In question ❸, ask children to explain how they will use the fact that there are 60 minutes in one hour to help solve the problem. Discuss what Flo says: what ways could they work out how many minutes it is to the hour? Children may suggest either counting on from 49 to 60, or instead subtracting 49 from 60.

DEEPEN Encourage children to use previous learning and the number line to help solve question ❸. Ask: *Can you think of a way to prove that your answer is right?*

ASSESSMENT CHECKPOINT At this point in the unit, children should be more confident in reading analogue and digital times and explaining when to use 'minutes past' and 'minutes to'. Look for confident and fluent understanding and use of 'am' and 'pm'.

ANSWERS

Question ❶ a):

Analogue	Digital
clock 1	6:20 pm
clock 2	10:46 am
clock 3	4:30 am
clock 4	4:06 pm

Question ❶ b): 10:46 am and 4:30 am show times in the morning.

Question ❷: Digital clock D

Question ❸: 11 minutes to 1

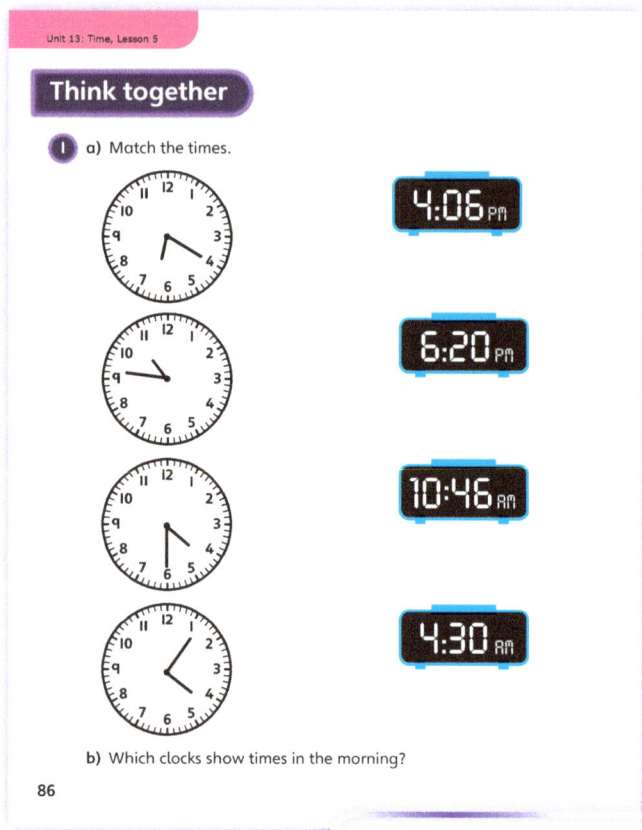

PUPIL TEXTBOOK 3C PAGE 86

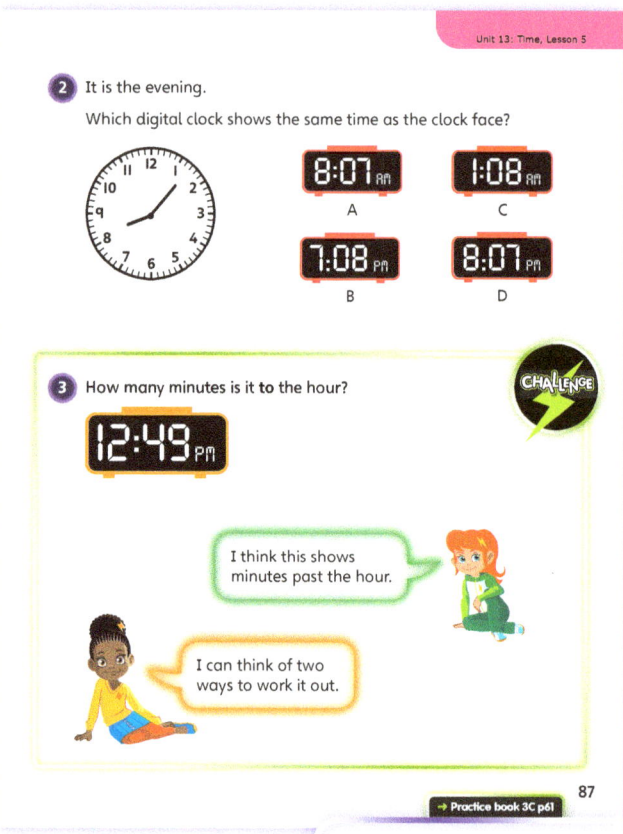

PUPIL TEXTBOOK 3C PAGE 87

Unit 13: Time, Lesson 5

Practice

WAYS OF WORKING Independent thinking

IN FOCUS Question ❶ checks that children know the difference between am and pm. In questions ❷ and ❸, children need to draw digital times on analogue clock faces and vice versa. Check that children understand what each number in a digital clock time represents and how this will be shown on an analogue clock. Ask: *Does this time show 'minutes past' or 'minutes to'? How will you write or draw that? Which half of the clock will the minute hand be in?* Explain that analogue clocks do not show 'am' or 'pm'.

STRENGTHEN Question ❹ requires children to practise reading the vocabulary of time. Remind them to draw on what they have learnt in previous lessons.

DEEPEN In question ❼, challenge children to find all the possible times that could fit the criteria. Ask children to prove their ideas and explain how they ensured that they found all the possible times. What systematic approach did they take?

ASSESSMENT CHECKPOINT Children should be showing a confident understanding of what 'am' and 'pm' mean. They should be able to read the time clearly using a digital or analogue clock. They should be able to explain clearly how to read the time using 'minutes past', and how to calculate 'minutes to' if the time shown is more than half past the hour.

ANSWERS Answers for the **Practice** part of the lesson can be found in the *Power Maths* online subscription.

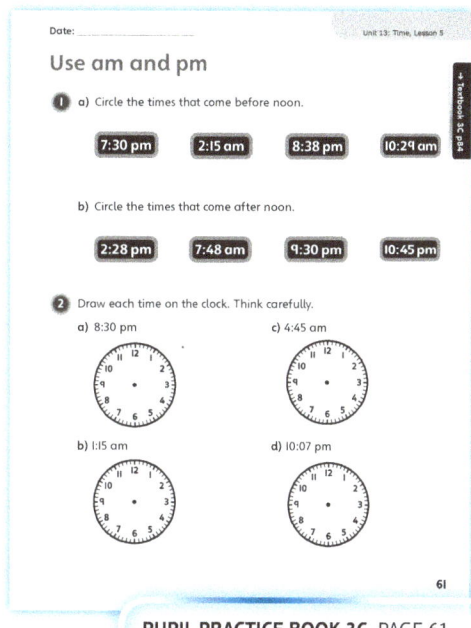

PUPIL PRACTICE BOOK 3C PAGE 61

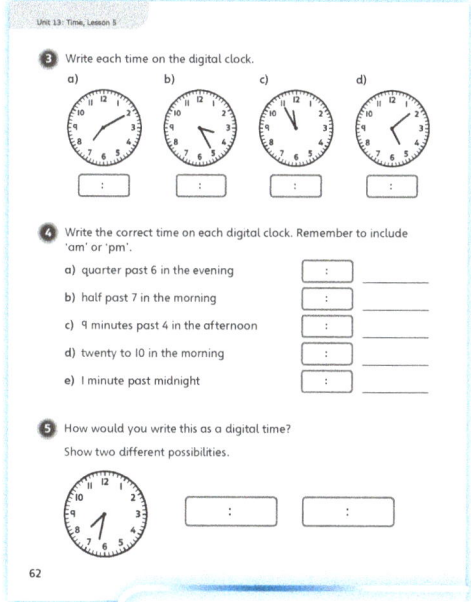

PUPIL PRACTICE BOOK 3C PAGE 62

Reflect

WAYS OF WORKING Pair work

IN FOCUS Children can think of their own ideas about when to use 'am' and 'pm', and share their ideas with their partners. Ask: *If you have different ideas, can you explain each other's ideas? Why did your partner think that?*

ASSESSMENT CHECKPOINT Children should clearly refer to the fact that 'am' and 'pm' are not related to whether it is dark or light outside. They should draw on previous knowledge that the day starts at 12 o'clock midnight and is 24 hours long.

ANSWERS Answers for the **Reflect** part of the lesson can be found in the *Power Maths* online subscription.

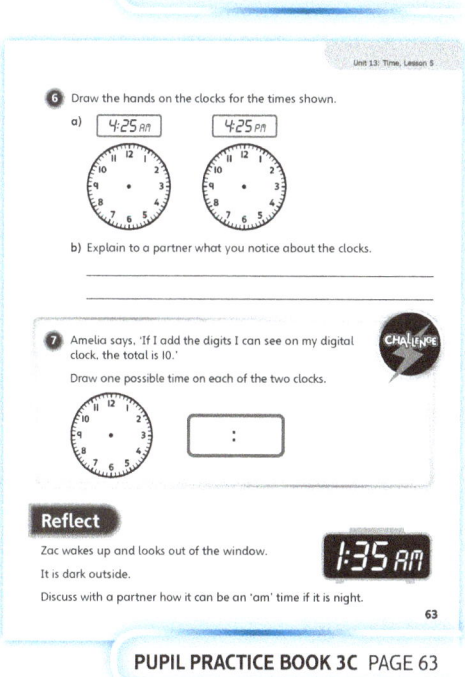

PUPIL PRACTICE BOOK 3C PAGE 63

After the lesson
- Are children able to confidently explain the differences between 'am' and 'pm'?
- Are children able to confidently explain how analogue and digital clocks differ?
- How will you build in more opportunities to practise reading the time throughout the school day?

125

Unit 13: Time, Lesson 6

Years, months and days

Learning focus
In this lesson, children will learn what a year is and be able to explain why there is a leap year every four years. They will learn the number of days in each month.

Before you teach
- Can children name the months of the year?
- Can children name the days of the week?

NATIONAL CURRICULUM LINKS

Year 3 Measurement – time

Know the number of seconds in a minute and the number of days in each month, year and leap year.

ASSESSING MASTERY

Children know the number of days and months in a year and can say the number of days in each month. Children can explain what a leap year is and apply the concept in different contexts.

COMMON MISCONCEPTIONS

Children may not see the difference between calendar facts that are fixed each year (months in the year, days in each month – apart from February) and those that change (the days of the week that individual dates fall on). Ask:
- *True or false? 1. All years have 365 days. 2. All years have 12 months.*

STRENGTHENING UNDERSTANDING

If children cannot remember how many days are in each month, ask them to make fists of both hands. From the left, point to knuckles and dips between knuckles as we say the months (ignore the thumbs). Any month we say when we point to a knuckle has 31 days; other months have 30 days, apart from February (28, or 29 if it is a leap year).

GOING DEEPER

Encourage children to think about what would happen if there wasn't a leap year every 4 years. Since the earth takes $365\frac{1}{4}$ days to travel around the sun, months of the year would, very slowly, start to shift 'backwards' so that, for example, December would become summer time in the northern hemisphere. The summer and winter months would not be fixed.

KEY LANGUAGE

In lesson: month, year, day, week, calendar, January, February, March, April, May, June, July, August, September, October, November, December, leap year, adjust, **end**

Other language to be used by the teacher: common year (a year with 365 days)

STRUCTURES AND REPRESENTATIONS

Bar model

RESOURCES

Mandatory: calendars

 In the eTextbook of this lesson, you will find interactive links to a selection of teaching tools.

Quick recap
Before the lesson begins, ask children what they already know about months, days and years. Determine any misconceptions children may have so that you can address them during the lesson.

Discover

WAYS OF WORKING Pair work

ASK

- Question 1 a): *What does the picture show? How many months are shown in the picture?*

IN FOCUS Question 1 a) brings attention to the number of days in each month. The picture should encourage children to look for the date of the last day in each month, as this will show how many days are in that month. For example, January's last day is Wednesday the 31st. January has 31 days.

PRACTICAL TIPS Make sure each pair or group has a copy of the same calendar, so they can discuss the number of days in each month.

ANSWERS

Question 1 a): There are 12 months in a year.
January, March, May, July, August, October and December have 31 days.
April, June, September and November have 30 days. February has 28 days.

Question 1 b): There are 365 days in a year.

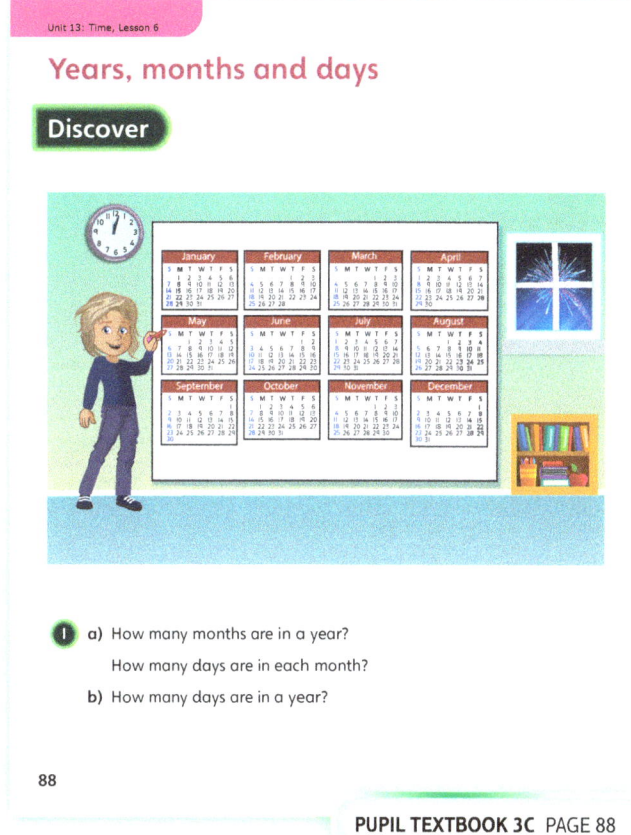

PUPIL TEXTBOOK 3C PAGE 88

Share

WAYS OF WORKING Whole class teacher led

ASK

- Question 1 a): *Do you know any of the months of the year? What order do they come in? Try counting the months along your knuckles. Do you notice a pattern? How can you check how many days there are in one year?*

IN FOCUS This may be a good opportunity to clarify any misconceptions that children have.

Ensure children are confident that every year has 12 months, and every week has 7 days. Some children may think they need to count the number of days in each of the 12 months individually. Remind them that the calendar makes it clear how many days are in each month. Ensure children understand that leap years have 366 days, as February has an extra day (29 days) in these years.

STRENGTHEN Discuss special days. Ask: *What dates of the year are special to you?* Talk to children about dates they look forward to every year. Ask: *Can you find these dates in the calendar?* Provide a printout of a calendar for children to explore. Ask them to circle the last day of each month in the calendar. Then ask them to write down how many days each month has, next to the name of the month. Ask: *What do you notice?*

Provide children with calendars for different years. Include a calendar for a leap year – for example, 2016 or 2020. Ask: *If the years are different, what do you notice? Is this year a leap year?*

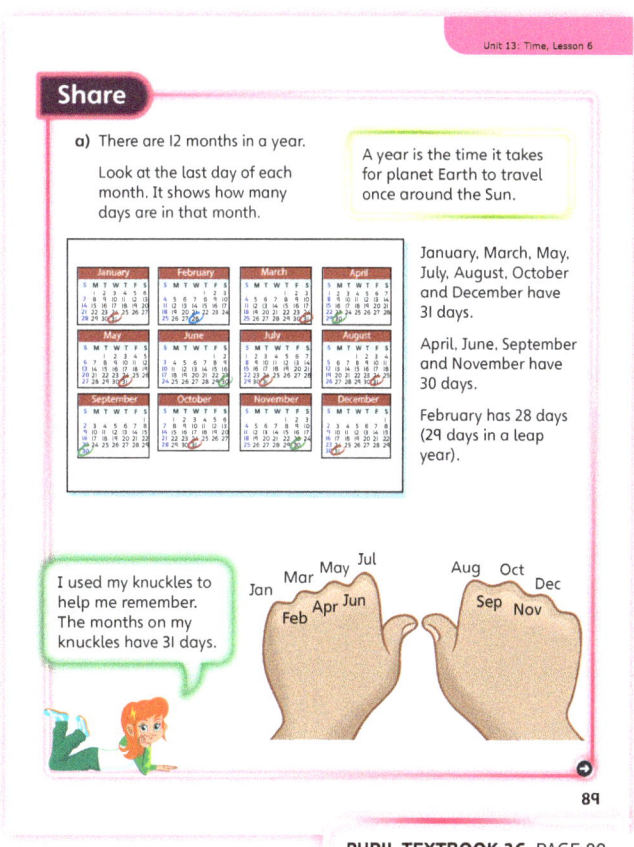

PUPIL TEXTBOOK 3C PAGE 89

127

Unit 13: Time, Lesson 6

Think together

WAYS OF WORKING Whole class teacher led (I do, We do, You do)

ASK
- Question ❶: *How many days are there in a week? If today is Monday 1st, what date will the next Monday be? If today is the 20th, what date is it tomorrow? What date was it yesterday?*
- Question ❷ a): *How many days are there in a week? How can you work out how many days there are in 5 weeks?*
- Question ❷ b): *How many days are there in a common year? How can you use the bar model to work out how many days are left?*

IN FOCUS In question ❷, highlight the link between the number of days in the year and the bar model.

Question ❸ asks children to use their knowledge of fractions to explain why leap years usually occur every 4 years. The bar model provides a visual connection between four $\frac{1}{4}$s and one whole.

STRENGTHEN For questions ❷ and ❸, strengthen children's understanding of what a year is by showing a picture of the Earth travelling around the Sun, so that children can visualise the journey. In question ❸, encourage children to think about what would happen if there wasn't a leap year every 4 years. Since the earth takes $365\frac{1}{4}$ days to travel around the sun, very slowly months of the year would start to shift 'backwards' so that, every 4 years, 1st January would happen one day earlier than it did the previous year. After hundreds of years, this would mean that it would be summer in November or December! So leap years are used in order to keep each month in the correct season.

DEEPEN When children have solved question ❸, deepen their understanding by explaining that $365\frac{1}{4}$ days is not exact, so some adjustments are made in deciding when there is a leap year and when there is not. If a year does not end in '00' and is a multiple of 4, then it is a leap year (for example, 2212). If a year ends in '00', to be a leap year the first 2 digits must be a multiple of 4. For example, 2400 is a leap year because 24 is a multiple of 4. However, 2100 is not a leap year because 21 is not a multiple of 4. Ask children to predict years that are leap years, and years that are not. Encourage them to explain their reasoning clearly.

ASSESSMENT CHECKPOINT Children know the number of days in a year. They can identify which months have 30 days and which have 31 days. They can differentiate between a leap year and a common year.

ANSWERS

Question ❶ a): Children should count on: 15 April, 16 April, 17 April, 18 April, 19 April, 20 April, 21 April.

Question ❶ b): Children should count back: 23 July, 16 July, 9 July, 2 July.

Question ❷ a): There are 35 days in 5 weeks.

Question ❷ b): There are 177 days left in the year.

Question ❸: Four $\frac{1}{4}$ days make 1 whole day. So one extra day is added to the year in every leap year (29 February).

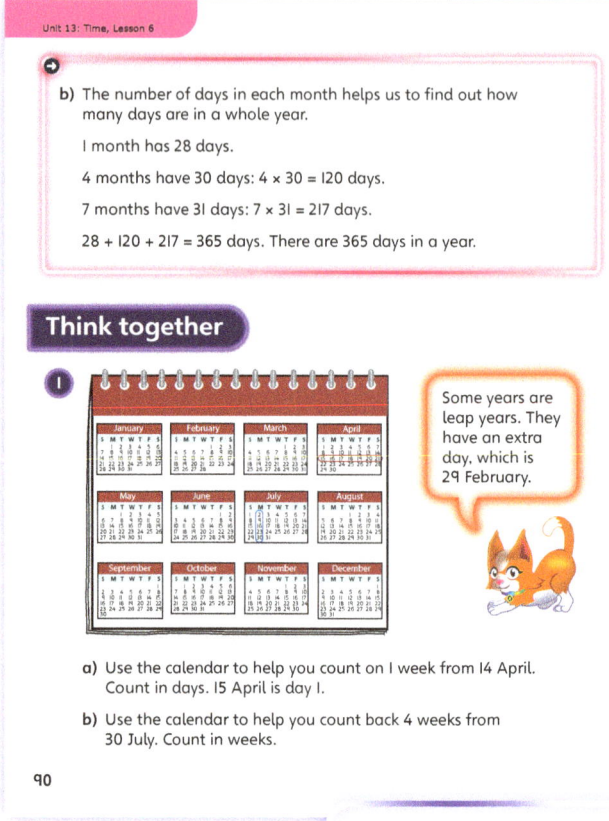

PUPIL TEXTBOOK 3C PAGE 90

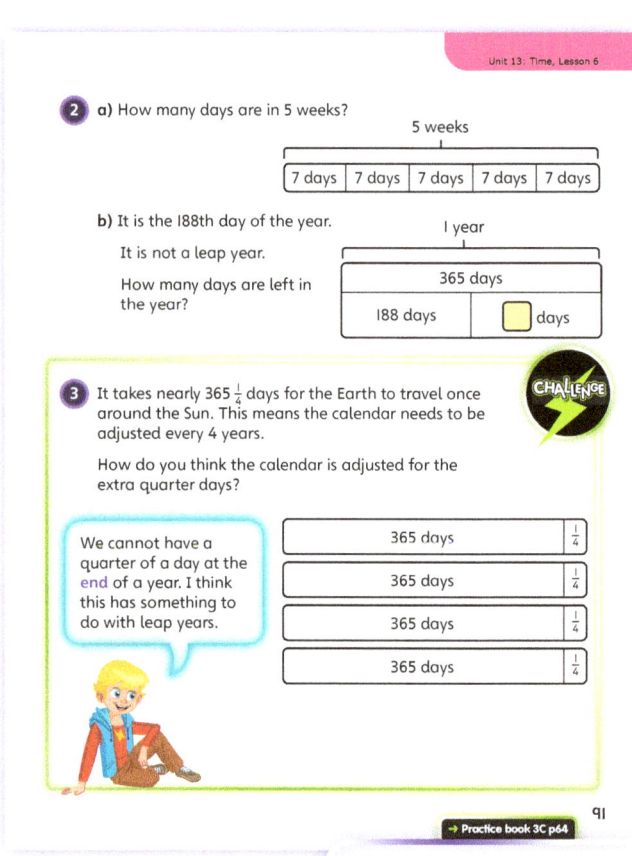

PUPIL TEXTBOOK 3C PAGE 91

Unit 13: Time, Lesson 6

Practice

WAYS OF WORKING Independent thinking

IN FOCUS When working independently on the questions in this section, children should continue to secure their understanding of how many days there are in a week, month and year. In question ❶, encourage children to circle the starting date where relevant.

STRENGTHEN If children are finding it difficult to calculate how many days are left in question ❹, ask them to use a bar model similar to the one used in the lesson. Ask: *How many days are there in a leap year? Where in the bar model will you write that number?*

DEEPEN Ask children to solve problems such as: *What is the date 2 weeks after 7th March? What is the date 10 days after 29th June?* Provide a variety of problems, with answers that span months and possibly a year.

THINK DIFFERENTLY In question ❹, children need to work out how many days there are left in the year from a particular date. They need to know how many days there are in a leap year – they can use a bar model to help them.

ASSESSMENT CHECKPOINT Children's knowledge of the number of days in each month should be secure. Children can relate the number of days in a year to the journey of the Earth orbiting the Sun, and understand why leap years exist.

ANSWERS Answers for the **Practice** part of the lesson can be found in the *Power Maths* online subscription.

Reflect

WAYS OF WORKING Pair work

IN FOCUS Give children time to discuss with a partner whether the statement is true or not. Ask: *How can you be certain that the statement is true or false? Is there a way to check?* Once children have discussed their methods, give them time to write their explanations.

ASSESSMENT CHECKPOINT Look for clarity in children's explanations. They should make reference to the fact that leap years have an extra day, as February has 29 days in a leap year. Additionally, they may give examples of years that are leap years, and years that are not leap years.

ANSWERS Answers for the **Reflect** part of the lesson can be found in the *Power Maths* online subscription.

After the lesson

- Do children know how many days there are in each month?
- Do children know how many days there are in a leap year?
- Can children explain why there are leap years and common years?

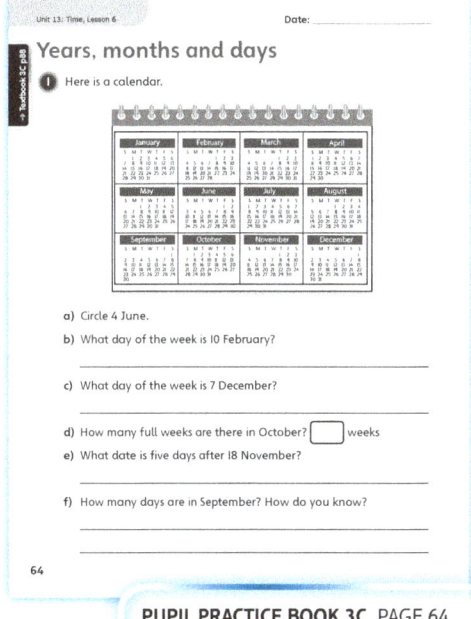

Unit 13: Time, Lesson 7

Days and hours

Learning focus
Children will be introduced to the 24 hours in a day including noon and midnight. They will think about the activities they spend time doing during a typical day, and the length of time they spend doing them.

Before you teach
- Do children use o'clock times to describe events?
- How can you give real-life experiences of durations of time?

NATIONAL CURRICULUM LINKS

Year 3 Measurement – time

Estimate and read time with increasing accuracy to the nearest minute; record and compare time in terms of seconds, minutes and hours; use vocabulary such as o'clock, am/pm, morning, afternoon, noon and midnight.

Tell and write the time from an analogue clock, including using Roman numerals from I to XII, and 12-hour and 24-hour clocks.

ASSESSING MASTERY

Children understand that there are 24 hours in a day and that a day runs from midnight until midnight. Children are able to apply this in different contexts.

COMMON MISCONCEPTIONS

Some children think that one day only includes the daytime, or that it begins when they wake up and ends when they go to bed. Ask:
- *How many hours in a day? How many hours have you counted? Do these two amounts match up?*

Some children think that the hour hand goes around the clock once a day. Ask:
- *What number does the hour hand on the clock point to when you wake up in the morning? What number does it point to when you go to bed? How many times has it travelled around the clock in this time?*

STRENGTHENING UNDERSTANDING

Give opportunities to discuss how the times of sunrise and sunset differ from winter to summer, and between different countries. Ask: *When is it light/dark in the evenings? Is it lighter at the beginning of your school day in summer or winter? Are the days long or short in the summer holidays?*

Encourage discussion about how the duration of 'one day' is not linked with the hours of daylight or the hours of darkness.

GOING DEEPER

Ask children to investigate how many hours are between sunrise and sunset today. Draw a timeline from midnight to midnight labelling each hour. Emphasise that there are 24 hours in a day including noon and midnight. Ask children to think of 24 activities, one for each hour in a one-day period, and to place them on their timeline.

KEY LANGUAGE

In lesson: day, hour, midnight, midday/noon, morning, night, hour hand, minute hand

Other language to be used by the teacher: sunset, sunrise, afternoon, evening, twice, night time, duration

STRUCTURES AND REPRESENTATIONS

Bar model, number line

RESOURCES

Optional: analogue clock manipulatives, pictures of sunset and sunrise, laminated pictures of clock faces, pictures of activities for each hour in the day, soft toys for role play

 In the eTextbook of this lesson, you will find interactive links to a selection of teaching tools.

Quick recap
Ask children to describe what they do in a day and ask them to estimate how long each activity takes. You might want to ask children at what times they go to bed and get up, and see whether they can work out how many hours it takes to do certain things.

Discover

WAYS OF WORKING Pair work

ASK

- Question 1 a): *What does the picture show? Which of the children is correct?*

IN FOCUS The answers the children give offer an opportunity to discuss when the day starts and ends, and how many hours there are in a day. Emphasise that one of the children says the day ends at midnight the next day, to encourage children to realise that 24 hours have passed.

PRACTICAL TIPS Try to give each pair or group access to an analogue clock. Provide pictures of motorway workers, street cleaners, nurses and doctors. Ask: *What time are their working hours?* Include examples of working hours throughout the 24 hours of a day.

ANSWERS

Question 1 a): The start of the day is 12 o'clock at night. This is called midnight. The end of the day is the next midnight. This is when a new day begins.

Question 1 b): There are 24 hours in one day.

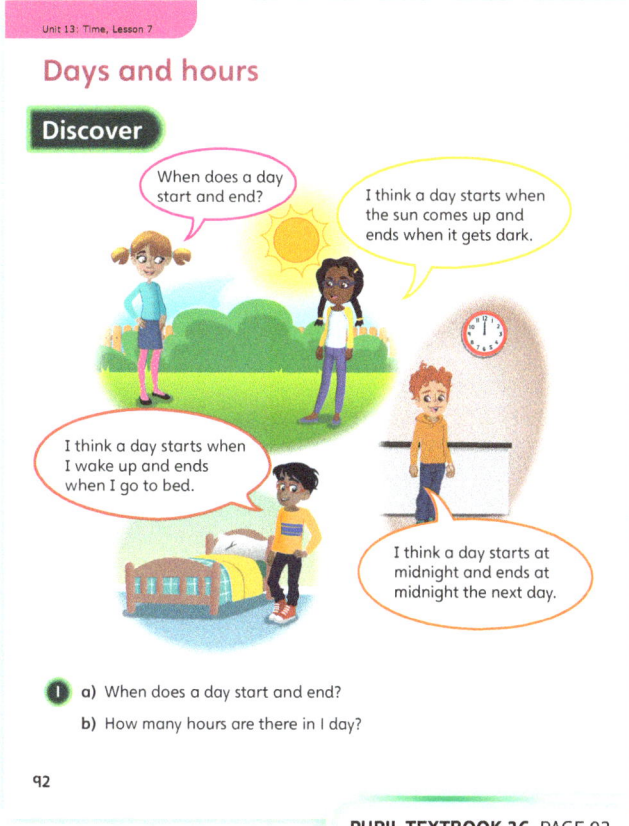

PUPIL TEXTBOOK 3C PAGE 92

Share

WAYS OF WORKING Whole class teacher led

ASK

- Question 1: *Where have you come across the idea of 24 hours before?*
- Question 1: *Can you think of an example from real-life concerned with a new day beginning at midnight?* [For example, New Year's Eve or the fairy tale *Cinderella*.]

IN FOCUS Use this opportunity to correct potential misconceptions and to give children the opportunity to show their understanding of the hours in a day. Ask them to explain where each child has gone wrong with their reasoning. What real-life examples can children give to justify their answers?

STRENGTHEN Challenge children to explore hints from everyday life that show there are 24 hours in a day. For example, a 24-hour petrol station and a 24-hour emergency vet.

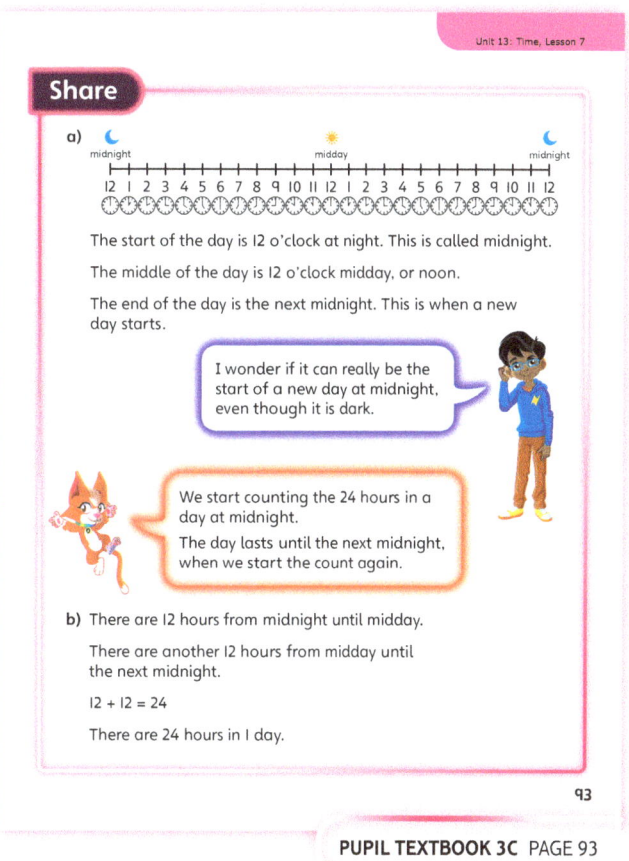

PUPIL TEXTBOOK 3C PAGE 93

Think together

WAYS OF WORKING Whole class teacher led (I do, We do, You do)

ASK
- Question ❶: *How can you work out the answers to some of Tim's questions? How many hours are in a day? How many times does the hour hand go around the clock in a day? How many times in the day is it 8 o'clock?*
- Question ❷ a): *What time do you think is the end of the day? How many hours are there from 7 pm to midnight?*
- Question ❷ b): *Is there a way you can work out the answer using your answer to part a)?*
- Question ❸: *How many hours in total are shown doing the activities? How can you add them up quickly? What do you need to do to work out how much time is left?*

IN FOCUS Questions ❶ and ❷ offer a good opportunity to reinforce the understanding that a day is made up of 24 hours, which is twice round the clock. Use a model to show how the hands move around a clock face. Encourage children to use analogue clocks to move through 24 hours, counting the hours as they do so. Ask children to think about the number of minutes in an hour. Ask: *How many times does the minute hand go around the clock in one day? What about the hour hand?*

STRENGTHEN Use a clock to support children who are struggling with some of the questions. In question ❷, for example, count the hours as the clock goes around from 7 pm to 12 pm.

DEEPEN Ask children to think about the number of days in one week and the number of hours in one day. They could use a bar model to show how to work out the number of hours in one week. Ask: *How many hours is it from 1 o'clock in the morning on Tuesday, to 3 o'clock in the morning next week on Wednesday? How did you come to that solution?*

ASSESSMENT CHECKPOINT Children should be confident that there are 24 hours in a day and 7 days in a week. They should use their knowledge of hours and time to recognise and explain how they can find the time 24 hours on from a given time. Children should also be confident that the day starts at 12 o'clock midnight and ends at 12 o'clock midnight the day after.

ANSWERS

Question ❶: From 12 noon to 12 midnight there are 12 hours. There are 48 hours in 2 days. The hour hand goes around the clock 24 times in a full day.

Question ❷ a): 5 hours

Question ❷ b): 17 hours

Question ❸: Emma spends 19 hours sleeping, at school and playing football so has 5 hours left in the day to do other things.

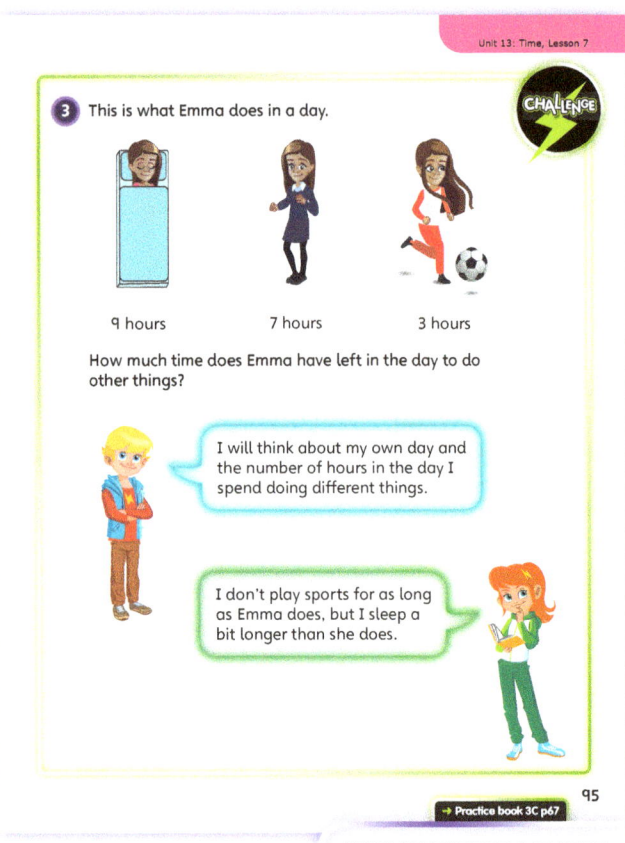

PUPIL TEXTBOOK 3C PAGE 94

PUPIL TEXTBOOK 3C PAGE 95

132

Unit 13: Time, Lesson 7

Practice

WAYS OF WORKING Independent thinking

IN FOCUS Question ❶ checks that children understand the basic facts of the make up of a day. Question ❷ gives children a pictorial representation of the time to scaffold their independent learning. Provide plastic clocks or laminated clock pictures to support children's work. Question ❸ revisits the potential misconception that the day lasts from 12 midnight until 12 noon.

In question ❹, children work out the number of hours from the given time to the end of the day. Discuss strategies they might use to find the answers more efficiently than counting every hour (for example, for 11 am to 12 midnight, children could work out 1 hour to 12 noon and then add 12).

STRENGTHEN Some children may find question ❸ challenging, and not be able to identify what they need to do. Discuss the question prior to the task. If any children are still unsure, offer them a timeline with the hours of the day recorded on it. For question ❹, children should use clocks and count the hours as they move the hands around the clock faces.

DEEPEN Question ❻ links fractions and time by using grids of 24. This question offers the opportunity for children to think of the day as a unit made of 24 equal parts. Ask children to think of different ways to represent a 24-hour day. Share different representations with the class. Ask: *How do they differ? What do they have in common?*

ASSESSMENT CHECKPOINT Children should be able to confidently explain that there are 24 hours in a day. They should be able to find the time 24 hours in the future and explain that the time stays the same while the day changes. Children should be able to solve simple problems confidently using their knowledge of the hours in a day and the days in a week.

ANSWERS Answers for the **Practice** part of the lesson can be found in the *Power Maths* online subscription.

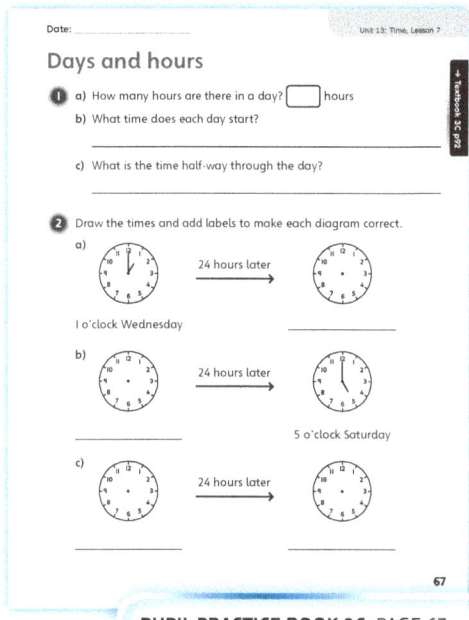

PUPIL PRACTICE BOOK 3C PAGE 67

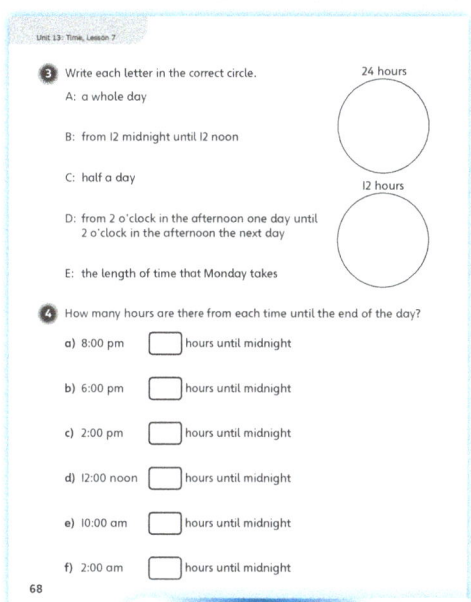

PUPIL PRACTICE BOOK 3C PAGE 68

Reflect

WAYS OF WORKING Pair work

IN FOCUS Children may need help finding the total number of hours they spend doing each activity. For example, they may know they eat three meals a day and have two snack times, but may struggle to estimate the total amount of time eating takes. For simplicity, estimate that each mealtime is perhaps half an hour, so total mealtimes plus snacks totals 2 hours. Try to round the time spent on each activity to the nearest hour. Children will look further at durations in Lessons 9 and 10 of this unit.

ASSESSMENT CHECKPOINT Children should know that there are 24 hours in a day, and be growing in confidence solving problems involving days and hours.

ANSWERS Answers for the **Reflect** part of the lesson can be found in the *Power Maths* online subscription.

After the lesson

- Are children confident that there are 24 hours in a day?
- How did you challenge children's assumptions about times they were not familiar with?
- Are children confident that the day starts at 12 o'clock midnight and ends at 12 o'clock midnight the day after?

PUPIL PRACTICE BOOK 3C PAGE 69

Unit 13: Time, Lesson 8

Hours and minutes – start and end times

Learning focus
In this lesson, children will learn to find start and end times to the minute for different events.

Before you teach
- Can children use the vocabulary of sequence?
- What real-life experiences could you provide?

NATIONAL CURRICULUM LINKS

Year 3 Measurement – time

Estimate and read time with increasing accuracy to the nearest minute; record and compare time in terms of seconds, minutes and hours; use vocabulary such as o'clock, am/pm, morning, afternoon, noon and midnight.

Compare durations of events [for example to calculate the time taken by particular events or tasks].

ASSESSING MASTERY

Children can find the start or end time to the nearest minute, when given a start time, an end time and a duration, including using digital notation.

COMMON MISCONCEPTIONS

Children may add the number of minutes each time. For example, if the start time is '12 minutes to 4' and the duration is 10 minutes, children add 10 to 12 and get the end time of '22 minutes to 4'. Ask:
- *Will the end time be before or after the start time?*

Children may be unsure how to find durations when they cross the hour boundary. Ask:
- *When do you arrive at school if you leave home at '12 minutes to 8' and the journey to school lasts 15 minutes?*

STRENGTHENING UNDERSTANDING

Before the lesson, discuss real-life examples of duration (for example, TV guides). Ask: *Why might it be useful to know the start time, end time or duration of an event?*

GOING DEEPER

Challenge children to look at a local bus timetable. Ask: *Where can you travel to if you leave now and need to arrive at another stop before x am?*

KEY LANGUAGE

In lesson: start time, end time, **duration**

Other language to be used by the teacher: amount, finish, forwards, backwards, time taken

STRUCTURES AND REPRESENTATIONS

Number line

RESOURCES

Mandatory: analogue clock manipulatives, laminated pictures of clock faces, digital clock

Optional: number lines

 In the eTextbook of this lesson, you will find interactive links to a selection of teaching tools.

Quick recap

Ask children simple addition and subtraction questions within 60. Check that children have a method to work out the answer, either using the column method or a number line.

Unit 13: Time, Lesson 8

Discover

WAYS OF WORKING Pair work

ASK

- Question 1 a): *What time is it at the moment? How do you know? Can you say the time another way? What would the time on a digital clock be? Is it am or pm? How do you know?*
- Question 1 b): *What does 'queuing time' mean? Is the time on the clock the start time or the end time?*

IN FOCUS Children have already explored finding durations, so they should be familiar with the idea of moving from a start time through a duration to an end time. Now that they are in Year 3, children will apply these concepts to times given to the nearest minute (when times are represented on both analogue and digital clocks).

PRACTICAL TIPS Provide children with real-life examples of durations, such as cooking instructions on food items. Ask children to consider what the end times might be, given a specific start time (or vice versa).

ANSWERS

Question 1 a): Max will get on the dodgems at 2:53 pm or 7 minutes to 3.

Question 1 b): Olivia should start queuing for the big wheel at 3:55 pm, or 5 minutes to 4.

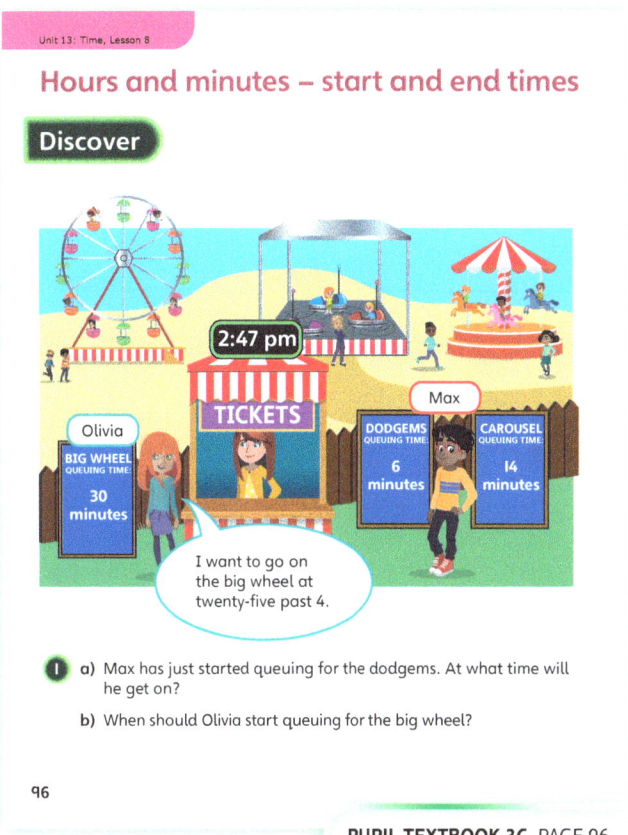

PUPIL TEXTBOOK 3C PAGE 96

Share

WAYS OF WORKING Whole class teacher led

ASK

- Question 1 a): *What time is it in the picture?*
- Question 1 a): *What will the clock look like when Max goes on the dodgems?*
- Question 1 a): *If you know the start time and the duration, how can you find the end time? Is it forwards or backwards in time?*
- Question 1 b): *If you know the end time and the duration, how can you find the start time? Is it forwards or backwards in time?*
- Question 1 b): *Why do you subtract 25 minutes first, then subtract 5 minutes to find the start time?*

IN FOCUS Ensure children are clear about what they are being asked to do in this task. They should understand the difference between start time, duration and end time, and how they link together. Provide real clocks to help children visualise the problem. Draw their attention to how this question is similar to those in previous lessons. In question 1 b), use the number line, alongside a clock, to explain what happens when children cross the hour.

STRENGTHEN Provide examples of problems where children have to find the start time, the duration or the end time. Encourage children to discuss what they need to find and the method they will use. Ask: *Do you add or subtract?* Ask children to work with durations that cross the hour. For example, ask: *Find the end time, if the start time is 4:48 and the duration is 25 minutes. Find the start time, if the end time is 3:16 and the duration is 55 minutes.*

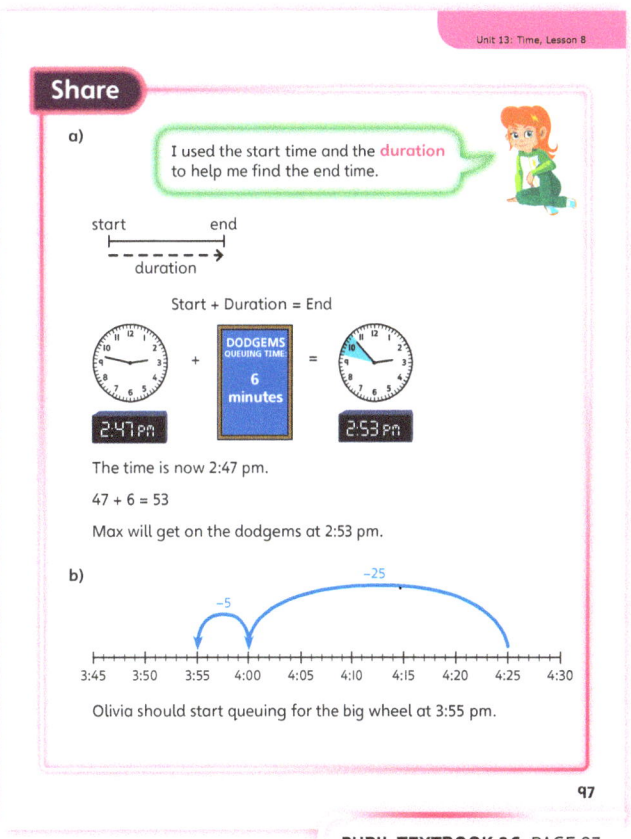

PUPIL TEXTBOOK 3C PAGE 97

135

Unit 13: Time, Lesson 8

Think together

WAYS OF WORKING Whole class teacher led (I do, We do, You do)

ASK

- Question ❶: *How will you use the start time and the duration to find the end time?*
- Question ❷: *How will you use the end time and the duration to find the start time?*
- Question ❸: *How can you use the number line to show how you found the end time?*

IN FOCUS Question ❶ scaffolds children's understanding of using the start time and duration to find the end time within an hour. Question ❷ asks children to find a start time, where the end time and the duration are known. They should use the model that they learnt in the lesson and subtract the time in two stages: first go back 9 minutes to 10 o'clock, then go back another 16 minutes. Make sure children understand how to cross the hour to find the start and end times. This will enable them to work more efficiently, using more easily recorded visual representations of time.

STRENGTHEN When children are working on questions ❶ and ❷, emphasise the link between the analogue clocks and a number line. In Year 2, children will have used a number line to find an end time. Strengthen children's understanding by using a number line to find the start and end times. This will prepare children for question ❸ and increase their familiarity with the different visual representations of time.

DEEPEN In question ❸, deepen children's ability to reason by asking: *What time will Bella go on the ride if the queue takes 22 minutes?* Make sure that children can model crossing the hour using the number line. Encourage children to use analogue clocks to answer the question as well. Ask: *Which method do you prefer? Why?*

ASSESSMENT CHECKPOINT At this point in the unit, children should be able to calculate start and end times, recognising the relationship that exists between start time, duration and end time. Children should be more confident in calculating the start or end time when crossing the hour.

ANSWERS

Question ❶: Max and Olivia can go on the carousel at 4:50 pm.

Question ❷: Luis started queuing at 9:44 am.

Question ❸ a): Bella will go on the helter-skelter at 3:04 pm or 4 minutes past 3.

Question ❸ b): The Khan family will get home at 6:34 pm. The most efficient method is to add an hour and take away 1 minute.

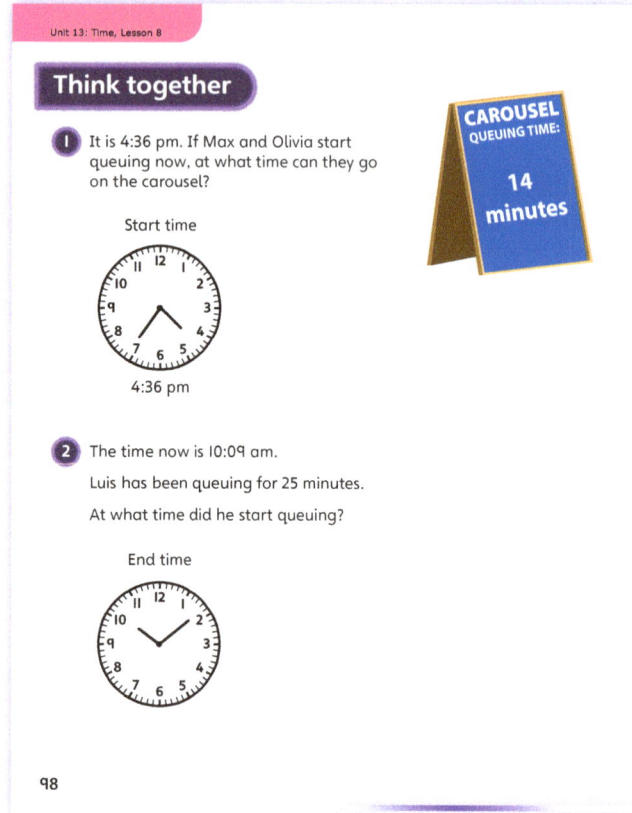

PUPIL TEXTBOOK 3C PAGE 98

PUPIL TEXTBOOK 3C PAGE 99

Unit 13: Time, Lesson 8

Practice

WAYS OF WORKING Independent thinking

IN FOCUS In question ❶ children find the end times given a range of start times and durations. Some of the end times require children to cross the hour. The use of clocks or number lines should help children to find the answers. In question ❷, children are given a duration and an end time and have to find the start time. This time children should realise that they need to subtract. Question ❸ addresses a common misconception. There are only 60 minutes in an hour, so when adding a duration to a start time, if the number of minutes is greater than 60, then the finish time should be expressed as the number of minutes past the next hour instead. Ask children to discuss with each other why the answer given cannot be correct.

STRENGTHEN To support children with questions ❶ and ❷, provide a number line. When children are working on question ❹, support them in identifying whether they have to move forward to find the end time or backwards to find the start time. Make sure children understand that the end time will always be after the start time, and use this fact to check and correct any errors that arise.

DEEPEN Ask further questions similar to the **Challenge**, in which children are asked to find the start or end times for durations longer than an hour (for example, 1 hour 15 minutes or 90 minutes).

ASSESSMENT CHECKPOINT Children should be confident in finding durations of time. They should be able to use various representations to find and demonstrate durations of time visually. Ask children to explain how they worked out the durations that cross the hour boundary. They should use this understanding to solve real-life problems independently.

ANSWERS Answers for the **Practice** part of the lesson can be found in the *Power Maths* online subscription.

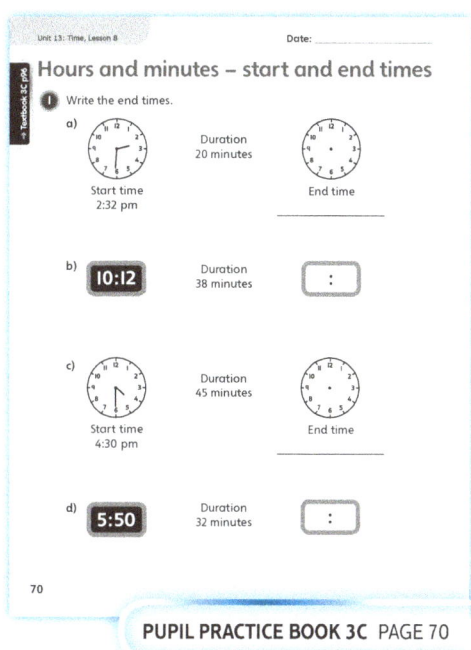

PUPIL PRACTICE BOOK 3C PAGE 70

PUPIL PRACTICE BOOK 3C PAGE 71

Reflect

WAYS OF WORKING Pair work

IN FOCUS Give children an opportunity to develop their reasoning independently. Ask children to share their answers as a pair and then to the wider group. Provide children with clocks and number lines. Ask: *Which method do you prefer to use? Why did you choose this method?*

ASSESSMENT CHECKPOINT Check that children can provide a method to find an end time given a duration. Make sure children are moving forward in time and they have not misunderstood the question. Check that children can cross the hour boundary by first working out how many minutes until the end of the hour.

ANSWERS Answers for the **Reflect** part of the lesson can be found in the *Power Maths* online subscription.

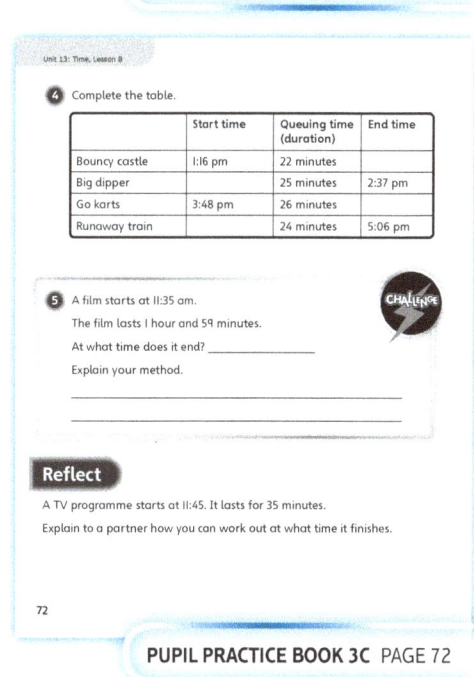

PUPIL PRACTICE BOOK 3C PAGE 72

After the lesson

- Are children confident in using the lesson's vocabulary?
- Can children confidently find a duration across the hour?
- Are children secure when choosing different types of representation to solve problems independently?

Unit 13: Time, Lesson 9

Hours and minutes – durations

Learning focus
In this lesson, children will learn to find a duration between two times, including using the 24-hour clock.

Before you teach
- Are children confident counting around the clock?
- Can children link a number line with a clock face?

NATIONAL CURRICULUM LINKS

Year 3 Measurement – time

Estimate and read time with increasing accuracy to the nearest minute; record and compare time in terms of seconds, minutes and hours; use vocabulary such as o'clock, am/pm, morning, afternoon, noon and midnight.

Compare durations of events [for example to calculate the time taken by particular events or tasks].

ASSESSING MASTERY

Children are able to confidently find the duration between two times, either by counting forwards or backwards from one time to the other. Children should find durations to the nearest minute, including 24-hour times.

COMMON MISCONCEPTIONS

Children may add the number of minutes each time (for example, if the start time is '12 minutes to 4' and the duration is 10 minutes, they add 10 to 12 and get an end time of '22 minutes to 4'). Ask:
- *Will the end time be before or after the start time?*

Children may be unsure how to find durations when they cross the hour boundary. Ask:
- *How many minutes to the next hour? How many more minutes after that?*

STRENGTHENING UNDERSTANDING

Give children a clock with moving hands, and cards with numbers of minutes (such as 5 minutes, 10 minutes, 15 minutes and 20 minutes). Children should work in pairs and take it in turns to pick a card. The card indicates the number of minutes that the minute hand has to travel. Children should show the new time on the clock and draw it. They score 3 points if they correctly say the original time, the new time and how many minutes the minute hand has travelled.

GOING DEEPER

Challenge children to look at a local bus timetable. Ask: *Where can you travel to in less than 20 minutes? In less than one hour?*

KEY LANGUAGE

In lesson: start time, end time, duration, finish

Other language to be used by the teacher: amount, time taken

STRUCTURES AND REPRESENTATIONS

Number line

RESOURCES

Mandatory: analogue clock manipulatives, digital clock

Optional: flashcards showing analogue clocks and written times

 In the eTextbook of this lesson, you will find interactive links to a selection of teaching tools.

Quick recap
Ask children to 'count on' to find the difference between two 2-digit numbers on a number line. Start with numbers such as 18 and 40 and then move to numbers such as 18 and 47. Encourage children to do this as efficiently as possible. They should record the jumps and the difference.

Unit 13: Time, Lesson 9

Discover

WAYS OF WORKING Pair work

ASK
- Question 1 a): *Where can you look to find how long the farmer takes to plough the field?*
- Question 1 a): *What clocks are shown in the pictures?*
- Question 1 a): *What time did the farmer start? What time did he finish ploughing?*
- Question 1 b): *How can you work out how long the farmer waits before he has a cup of tea?*

IN FOCUS Use the pictures to discuss the duration of some activities or clubs that children do. For example, ask: *When does choir practice start and finish? How long is it? What other activities could you find the duration of time for?*

PRACTICAL TIPS A number line and clock faces will help children to calculate the length of time. It will be useful to revisit how the number line can be used and to remind children that 1 hour = 60 minutes, not 100 minutes.

ANSWERS

Question 1 a): It takes the farmer 28 minutes to plough the field.

Question 1 b): Another 28 minutes go by before the farmer has a cup of tea.

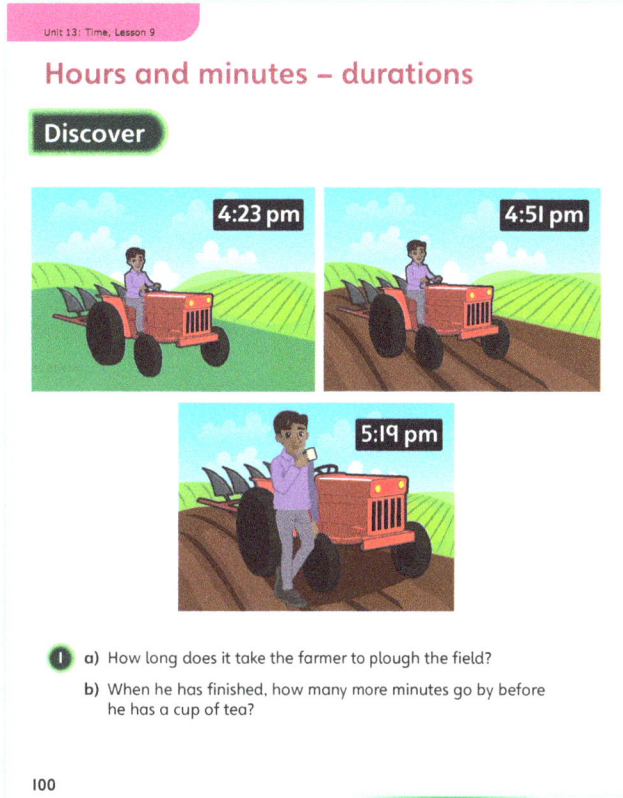

PUPIL TEXTBOOK 3C PAGE 100

Share

WAYS OF WORKING Whole class teacher led

ASK
- Question 1 a): *What does 'duration' mean?*
- Question 1 a): *To work out the duration of something, what information do you need to know?*

IN FOCUS Although children may have been introduced to duration in Year 2, this was non-statutory guidance, so do not assume they have done it or that they will remember it. It is important to scaffold children's use of the term 'duration' through your questioning and support. Providing concrete or pictorial representations of analogue and digital clocks will help children. Do not go into the full conversion of minutes and hours, as this is covered in Year 4.

STRENGTHEN Set up two clocks at the front of the class. Ask one child to make any time on the first clock, and another to make any time on the second clock. The class should then work out the duration between the two times.

Show children a TV guide where the start and end times of programmes are given. Ask children to devise their own duration questions based on what they see.

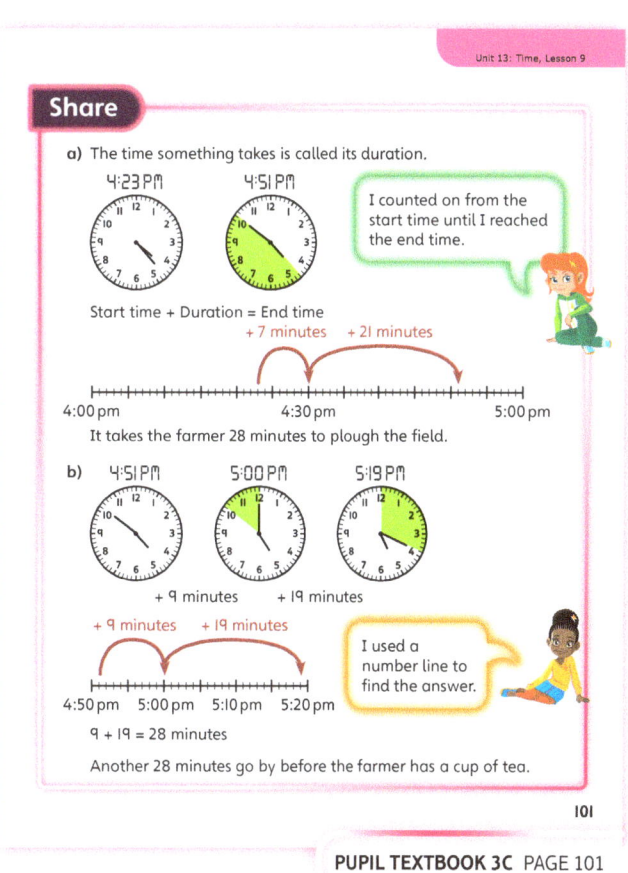

PUPIL TEXTBOOK 3C PAGE 101

139

Think together

WAYS OF WORKING Whole class teacher led (I do, We do, You do)

ASK
- Questions ❶ to ❸: *How will you count the duration of time between the start and finish times?*
- Questions ❶ to ❸: *How can you show the duration using a number line?*
- Questions ❶ to ❸: *How will you record how many minutes the duration is? Could you record it in a different way?*

IN FOCUS Question ❶ scaffolds children's ability to find the duration within an hour. Question ❷ asks children to find a duration less than 60 minutes long, but crossing the hour boundary. Question ❸ progresses to finding a duration that is greater than 60 minutes and crosses the hour.

STRENGTHEN If children are unsure how to complete questions ❶ and ❷, provide blank clock faces and ask them to mark the start and end times. They can then count the steps between the two times to measure the duration of time.

DEEPEN In question ❸, provide children with a number line showing two hours from 16:00 to 18:00, with each hour split into 60 minutes. Ask children to show how long the farmer took to collect the eggs on the number line.

ASSESSMENT CHECKPOINT At this point in the unit, children should be able to calculate 'duration' by calculating the length of time between two points. Children should also recognise how and why a measurement of duration can be different to the number of minutes mentioned in the question or those shown on the clock.

ANSWERS

Question ❶: It takes the farmer 44 minutes to milk the cows.

Question ❷: 25 + 22 = 47
The lorry driver was at the farm for 47 minutes.

Question ❸: It takes the farmer 1 hour 23 minutes or 83 minutes.

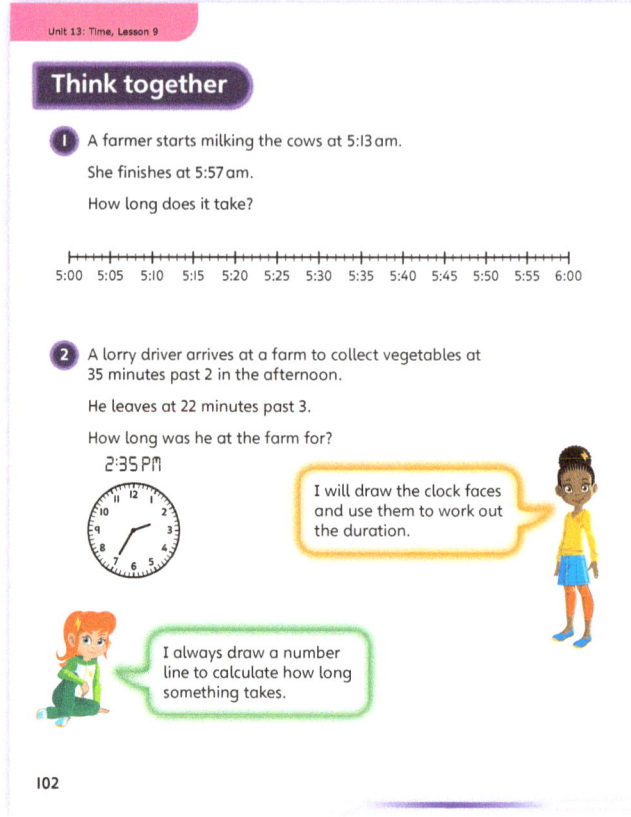

PUPIL TEXTBOOK 3C PAGE 102

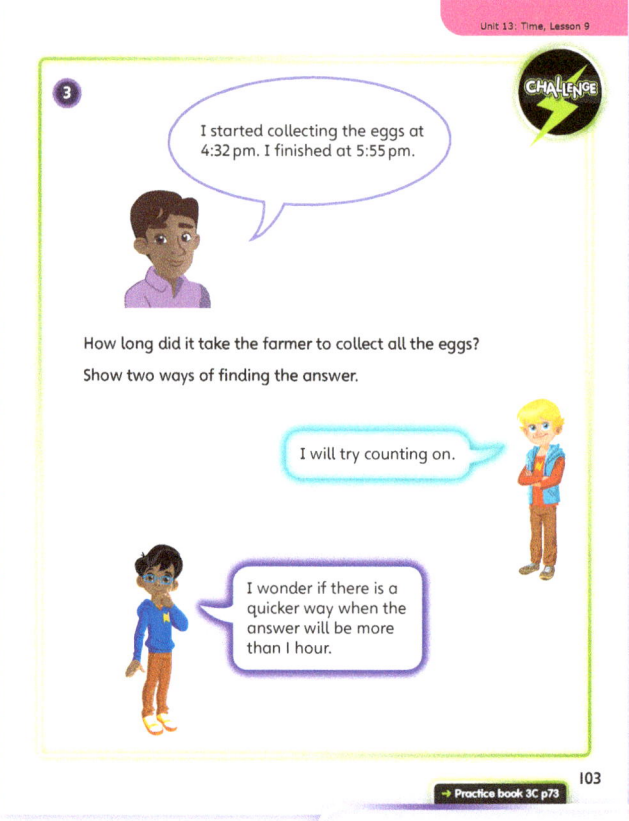

PUPIL TEXTBOOK 3C PAGE 103

140

Unit 13: Time, Lesson 9

Practice

WAYS OF WORKING Independent thinking

IN FOCUS In question ❶, children are given the start and end times and asked to find the duration. All the times are within the hour and clocks are provided for support. In question ❷, children find the duration again, but this time clocks cross the hour, meaning children have to first work out the difference to 60. In question ❸, the clocks have been stripped away, but children essentially follow the same process.

STRENGTHEN Children could use clocks and number lines to support them to find the duration. They should work out how many minutes they need to count on to get to the new time. This is the duration.

DEEPEN Give children more examples where the durations of time last longer than an hour. Ask: *How long is it between 1:50 am and 3:15 am?*

ASSESSMENT CHECKPOINT At this point in the unit, children should be confident in finding durations of time. They should be able to use multiple representations to find and visually demonstrate durations of time. Look for children explaining how they found durations that cross the hour boundary and how they can use this understanding to solve real-life problems involving time independently.

ANSWERS Answers for the **Practice** part of the lesson can be found in the *Power Maths* online subscription.

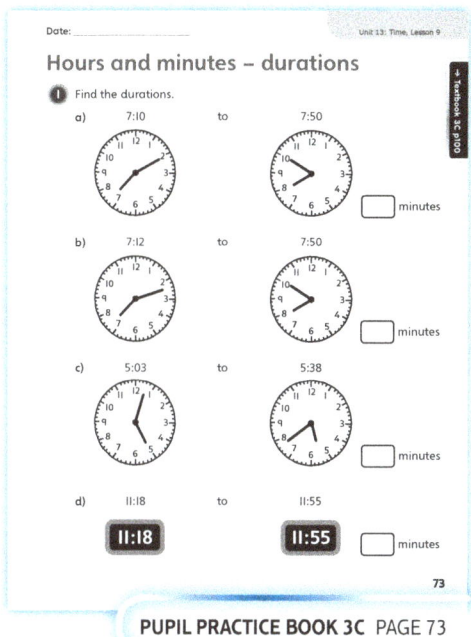

PUPIL PRACTICE BOOK 3C PAGE 73

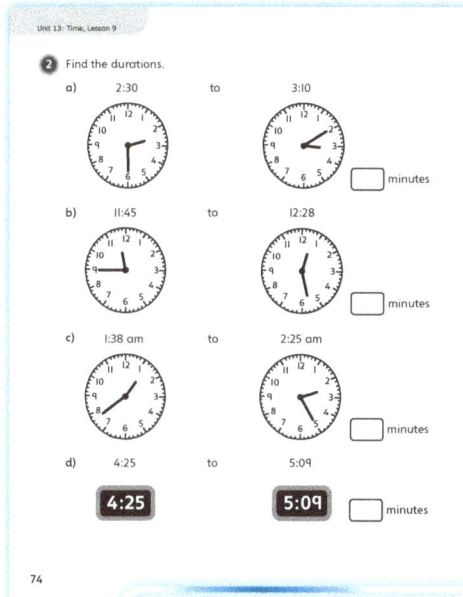

PUPIL PRACTICE BOOK 3C PAGE 74

Reflect

WAYS OF WORKING Independent thinking or pair work

IN FOCUS Give children an opportunity to develop their reasoning independently. Can they answer their own duration problems? What method do they prefer to use? Ask: *Why did you choose this method?* Encourage children to compare their method with a partner's. Ask: *Whose method is more efficient?*

ASSESSMENT CHECKPOINT Children should be able to identify a start and end time, and find the duration of the time between them. Children should be using different types of representation fluently. Do they use different representations when the duration is within an hour, or when the duration crosses an hour boundary? Listen to children's reasoning as to why they have chosen a particular representation.

ANSWERS Answers for the **Reflect** part of the lesson can be found in the *Power Maths* online subscription.

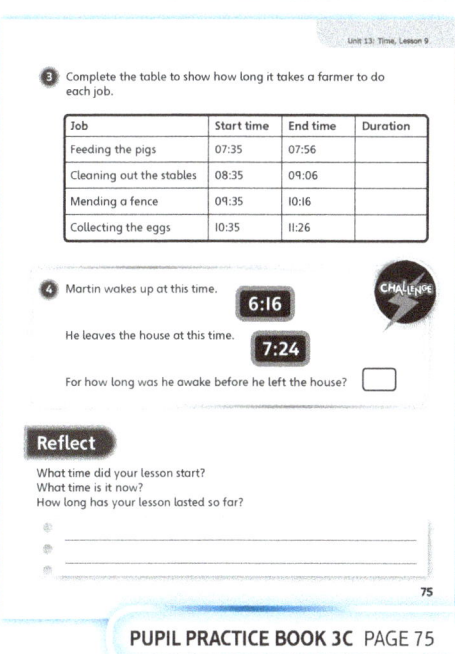

PUPIL PRACTICE BOOK 3C PAGE 75

After the lesson

- Are children secure in using the vocabulary they have learnt in the lesson?
- Can children find a duration that crosses the hour confidently?
- Can children choose different representations to solve problems independently?

Unit 13: Time, Lesson 10

Hours and minutes – compare durations

Learning focus
In this lesson, children will learn to compare durations of time.

Before you teach
- Are children confident in counting around a clock?
- Are children confident in using am and pm?
- Can children link a number line with a clock face?

NATIONAL CURRICULUM LINKS

Year 3 Measurement – time

Estimate and read time with increasing accuracy to the nearest minute; record and compare time in terms of seconds, minutes and hours; use vocabulary such as o'clock, am/pm, morning, afternoon, noon and midnight.

Compare durations of events [for example to calculate the time taken by particular events or tasks].

ASSESSING MASTERY

Children confidently compare given durations, and can also derive durations to compare. Children use their knowledge of place value and units of measurement to say whether a duration is a longer or shorter time. They can order durations, including working with digital times, 24-hour time and to the nearest minute.

COMMON MISCONCEPTIONS

Children may think that a duration that ends later, takes longer (when it may have started later and be shorter). Ask:
- *Would you rather play games between twenty-five to 4 and quarter past 4, or between five to 4 and twenty-five past 4? Why?*

STRENGTHENING UNDERSTANDING

Before the lesson, ask children to compare break time with lesson time. Ask: *Which is longer? Which is shorter?*

GOING DEEPER

Challenge children with a local bus timetable. Ask: *Where can you travel in less than 20 minutes? In one hour?*

KEY LANGUAGE

In lesson: duration, longer, longest, shortest, finish

Other language to be used by the teacher: amount, time taken, start time, end time, how long?, shorter, how long left?

STRUCTURES AND REPRESENTATIONS

Number line

RESOURCES

Mandatory: analogue clock manipulatives, digital clock

Optional: flashcards showing analogue clocks and written times

 In the eTextbook of this lesson, you will find interactive links to a selection of teaching tools.

Quick recap

Ask children to find the difference between two times using a number line – for example, 7:12 and 7:43. What jumps do they make? Ask: *How many minutes difference are there?* Repeat for a mixture of times.

Discover

WAYS OF WORKING Pair work

ASK

- Question 1 a): *What word can you use to describe the length of time that something takes?*
- Question 1 a): *What times are shown in the picture? How can you use the times to find the answer?*

IN FOCUS Use the picture to recap and briefly assess children's current understanding. Begin by discussing the duration of some of the activities in the school day. For example, does the morning assembly take more time or less time than registration?

PRACTICAL TIPS Displaying the schedule of a normal school day will help children to visualise how their day is split up and which activities take more time than the others. Using a number line may help children to see which number is larger or smaller and to compare the durations of activities.

ANSWERS

Question 1 a): Meet the Author:
9:06 am to 10:00 am = 54 minutes
Songs and Stories:
11:35 am to 12:20 pm = 45 minutes
Poetry Workshop:
1:40 pm to 2:40 pm = 60 minutes

Question 1 b): The Poetry Workshop activity lasts the longest.

PUPIL TEXTBOOK 3C PAGE 104

Share

WAYS OF WORKING Whole class teacher led

ASK

- Question 1 a): *What information do we need to work out the duration of an event?*
- Question 1 b): *How can you tell whether something takes a longer or shorter length of time than something else?*

IN FOCUS Use this opportunity to recap the previous lesson and to scaffold what duration actually means. Providing concrete or pictorial representations of analogue and digital clocks will help children to find a duration. Encourage children to be systematic in their approach. To compare durations, they first need to find each one. Give children the opportunity to show their understanding of times and what duration means. Encourage them to justify which activity takes longest, rather than guessing or shouting out.

STRENGTHEN Address the misconception that a duration that ends later always takes longer (when in fact it may have started later and be shorter). Set up two clocks at the front of the class. Ask one child to make the start time of the first library activity on the first clock, and the end time on the second clock. The class should then work out the duration between the two times. Repeat for all three activities. Set up a number line and ask children to place the duration of each activity on the number line. Ask: *Which number is largest? What does this mean?*

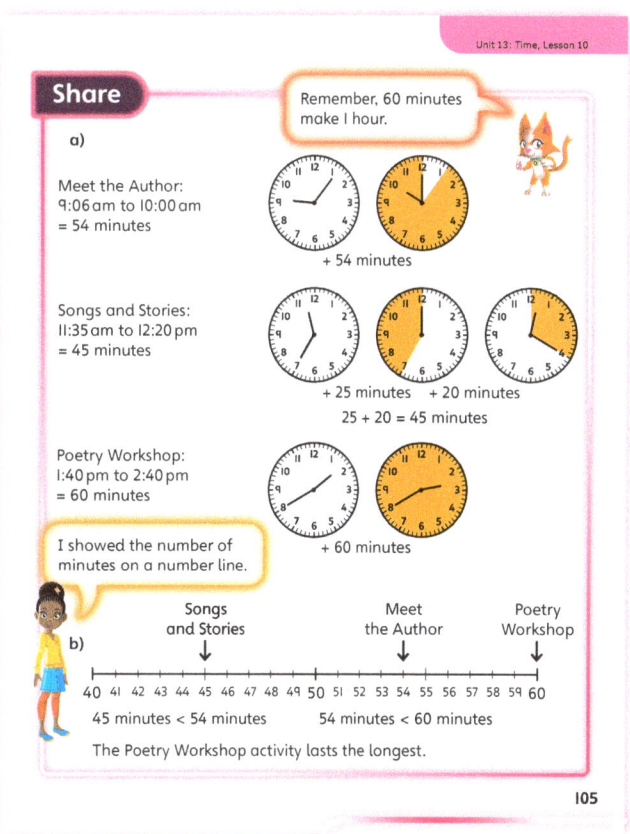

PUPIL TEXTBOOK 3C PAGE 105

Unit 13: Time, Lesson 10

Think together

WAYS OF WORKING Whole class teacher led (I do, We do, You do)

ASK
- Question ❶: *What are the start times? What are the end times? How will you count the duration of time between the start and end times?*
- Question ❷: *How can you tell whether something takes a longer or shorter length of time?*
- Question ❷: *How will you record how many minutes the durations are? Could you find and record the duration in a different way (perhaps using hours rather than minutes)?*
- Question ❸: *Do you always have to work out each duration to be able to compare them? What resources could you use to help compare the durations?*

IN FOCUS It is important for children to make the distinction between question ❶ and other questions that they have met so far. In question ❶, the times of both events are very similar but the second event starts 5 minutes later than the first one, and finishes 2 minutes later than the first. Ask children how they can use this information to work out which event is longer.

STRENGTHEN When working on question ❷, provide children with a number line that they can use to calculate the difference between the durations. In questions ❷ a) and b), children may subtract the end and start minutes from each other. In question ❷ c), remind children that there are 60 minutes in 1 hour (not 100). Provide laminated clock faces to draw on. Look for children who find the duration from 8:28 to 9:00, and 9:00 to 9:03 and add their answers.

DEEPEN Once children have demonstrated that they can solve question ❸, deepen their understanding of this kind of problem by challenging them to create a similar question for a partner. Ask: *What times would you use to make the question easier? How can you make the question harder? What resources could help to answer the question?*

ASSESSMENT CHECKPOINT At this point in the lesson, children should have demonstrated that they are able to confidently measure durations of time in minutes and hours. Children should be able to compare durations of time, explaining which is longer or shorter and giving their reasoning.

ANSWERS

Question ❶: Story Time = 29 minutes
Make a Book = 26 minutes
Story Time lasts longer.

Question ❷: A = 8:12 am until 8:48 am = 36 minutes
B = 8:43 am until 8:57 am = 14 minutes
C = 8:28 am until 9:03 am = 35 minutes
A (36 minutes) is the longest duration.

Question ❸: D (63 minutes), B (64 minutes), C (65 minutes), A (67 minutes)

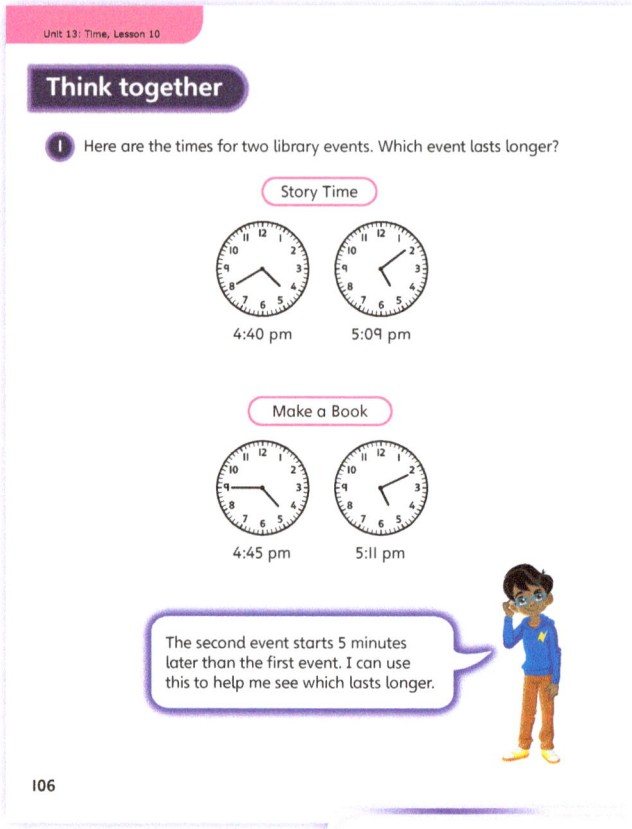

PUPIL TEXTBOOK 3C PAGE 106

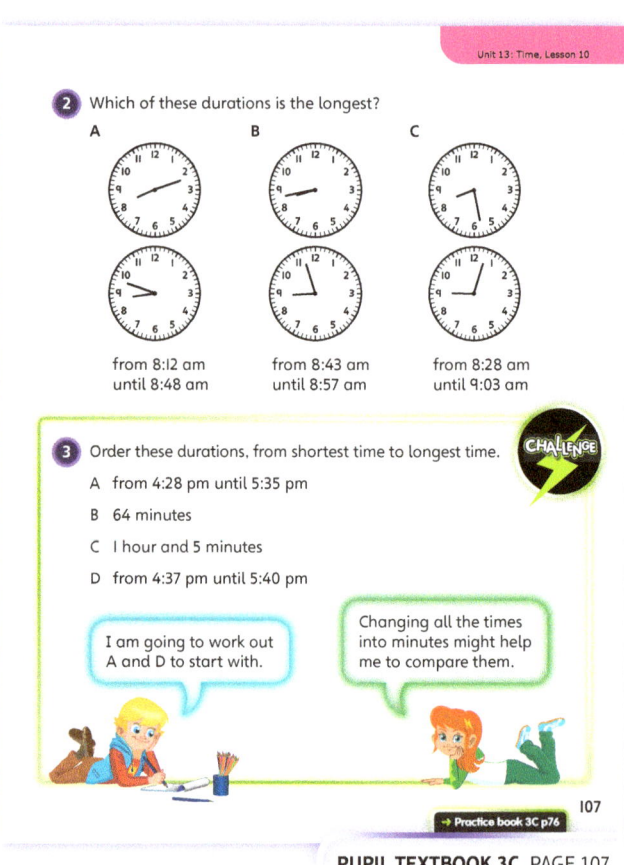

PUPIL TEXTBOOK 3C PAGE 107

144

Unit 13: Time, Lesson 10

Practice

WAYS OF WORKING Independent thinking

IN FOCUS In question ①, children could shade the sectors of time Alex spends practising to cement the process of counting a duration of time. They can then use a number line to help them find the duration. Questions ② and ③ encourage children to use their understanding of the duration of time to solve problems. In question ④, children work out durations that are longer than an hour and then compare them. Look out for children thinking the largest time is the fastest, this is often a misconception.

STRENGTHEN If children are struggling with question ②, encourage them to re-read the question. Ask: *Can you describe it? What is it about?* The time is represented in an abstract way. Ask: *How could you represent the times?* It is important for children to notice that the fee of £1 is for 65 minutes rather than 1 hour. Ask: *Think of car park fees that you may have seen. How much would it cost to leave the car there for 2 hours? 3 hours?*

DEEPEN Challenge children to work out start and end times if they are given a duration. For example, a film starts at 3:40 pm and lasts for 1 hr 30 minutes. What time does it end?

ASSESSMENT CHECKPOINT Children should be able to fluently measure durations of time in minutes and hours. They should be confident in comparing durations of time and using their understanding to solve mathematical problems.

ANSWERS Answers for the **Practice** part of the lesson can be found in the *Power Maths* online subscription.

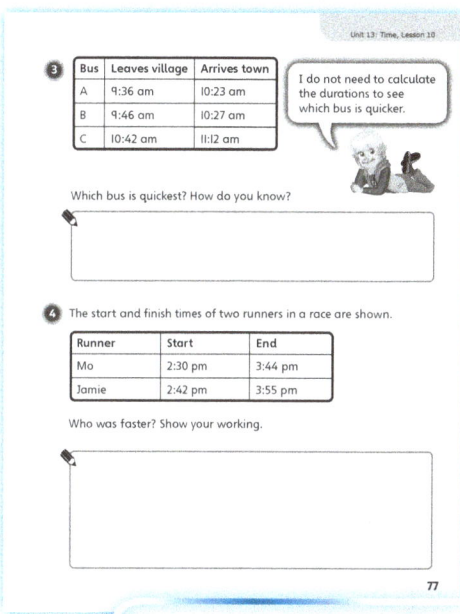

PUPIL PRACTICE BOOK 3C PAGE 76

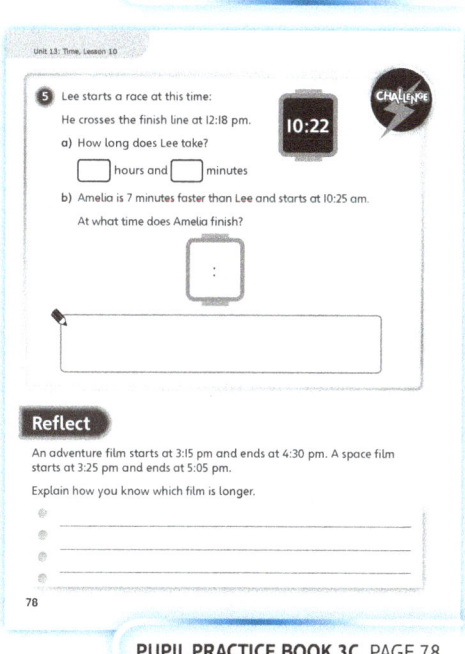

PUPIL PRACTICE BOOK 3C PAGE 77

Reflect

WAYS OF WORKING Independent thinking, pair work

IN FOCUS Give children the opportunity to record their methods. Ask: *What method do you prefer to use? Why?* Encourage children to compare their method with a partner's. Ask: *Whose method is more efficient?*

ASSESSMENT CHECKPOINT Look for clarity in children's explanations. They should be systematic in their approach and compare the lengths of time accurately. They should be able to use the appropriate vocabulary to explain how to compare durations of time.

ANSWERS Answers for the **Reflect** part of the lesson can be found in the *Power Maths* online subscription.

After the lesson

- Are children secure in using the vocabulary learnt in this lesson?
- Can children confidently find a duration that crosses an hour?
- Are children confident in choosing different representations to solve problems independently?

PUPIL PRACTICE BOOK 3C PAGE 78

Unit 13: Time, Lesson 11

Minutes and seconds

Learning focus
In this lesson, children will learn to measure events (such as a race) in seconds.

Before you teach
- Do children know that a minute is 60 seconds?
- Have children seen a stopwatch in real life?

NATIONAL CURRICULUM LINKS

Year 3 Measurement – time

Estimate and read time with increasing accuracy to the nearest minute; record and compare time in terms of seconds, minutes and hours; use vocabulary such as o'clock, am/pm, morning, afternoon, noon and midnight.

ASSESSING MASTERY

Children know how long a second is and how many seconds equal 1 minute. They can apply the information to estimate 1 minute, and time activities in seconds.

COMMON MISCONCEPTIONS

Children may think 1 second is any fast period of time, and that however quickly they count to 60, it is 1 minute. Ask:
- *How long does the second hand take to move once? Is it always the same?*

Children may mix up the three clock hands or the numbers on digital clocks, and so cannot identify seconds. Ask:
- *What part of the clock helps us count seconds? How can you tell?*

STRENGTHENING UNDERSTANDING

Give children a stopwatch and cards with numbers of seconds (such as 5 seconds, 15 seconds, 30 seconds, 45 seconds, 60 seconds). Children should work in pairs, taking turns to pick up a card. One child starts the stopwatch, the other says 'Stop' when they think the time has passed. The child who gets closest wins 1 point.

GOING DEEPER

Show 100 m race results from a school race, county race and the Olympics. Ask: *How many seconds did the winners take? Estimate how long you might take. Have a race. How do the results compare with the estimates?*

KEY LANGUAGE

In lesson: measure, seconds, minutes, stopwatch
Other language to be used by the teacher: amount, time taken

STRUCTURES AND REPRESENTATIONS

Bar model

RESOURCES

Mandatory: analogue and digital clocks that show hours, minutes and seconds
Optional: stopwatch

 In the eTextbook of this lesson, you will find interactive links to a selection of teaching tools.

Quick recap
Check simple addition and subtraction skills within 100 with children. Many of the questions in this lesson require children to find the sum or difference. Check they have a method to solve calculations like these.

Discover

WAYS OF WORKING Pair work

ASK
- Question 1 a): *Where might you use seconds every day [hint: timers]? Where can you see the number of seconds on Richard's clock? Where can you see the number of seconds on Amelia's clock?*
- Question 1 a): *Can you use a clock face to prove that there are 60 seconds in a minute? How?*
- Question 1 a): *There are 60 seconds in 1 minute. How many seconds remain until the end of a minute if 18 seconds have gone by?*

IN FOCUS Use the pictures to discuss some instances where a stopwatch may be used (for example, measuring the time in a race).

PRACTICAL TIPS Play a minute game in class. Say '*Go!*' and time 1 minute. Each child should close their eyes and silently use their own method to count in seconds; when they think 1 minute is up, they should put their hand up. Keep looking at the clock and make a note of the children whose estimates are closest to 1 minute. When everyone is done, ask them to open their eyes and to share their methods. Point out which children were closest to 1 minute. Ask: *Can you improve your estimate?*

ANSWERS

Question 1 a): Richard has been playing for 50 seconds. Amelia has been playing for 35 seconds.

Question 1 b): Lee could measure seconds using the clock on the wall, by counting the marks as the second hand moves.

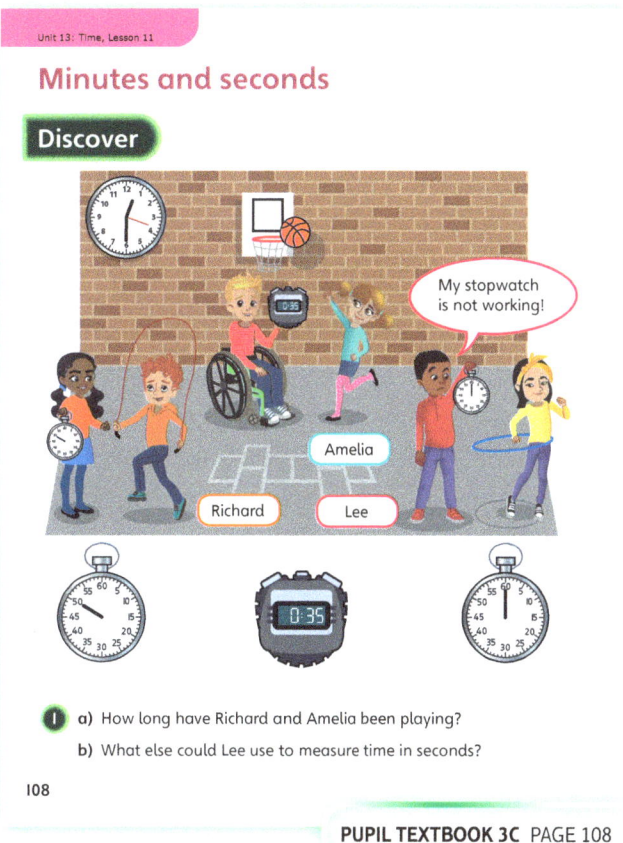

PUPIL TEXTBOOK 3C PAGE 108

Share

WAYS OF WORKING Whole class teacher led

ASK
- Question 1 a): *What are seconds used for?*
- Question 1 a): *How many seconds are there in 1 minute?*
- Question 1 b): *Why do we use a stopwatch? How could you measure seconds if you didn't have a stopwatch?*
- Question 1 b): *How do the two stopwatches in the picture differ?*
- Question 1 b): *Do all clocks have a second hand? Why? Why not?*

IN FOCUS Ensure children are confident reading the stopwatches in the picture. Make sure children understand that there are 60 seconds in 1 minute, and that they can show how this is measured on both a stopwatch and an analogue clock. Ask: *How do a stopwatch and an analogue clock differ?*

STRENGTHEN Some children may not have enough experience measuring in seconds. You could ask them to check their heart rate for 30 seconds. Can they predict what their heart rate will be in 1 minute? Ask children to think of activities that could happen in 1 minute and how they can measure them (for example, the number of steps they could walk or the number of times they can clap their hands together). Encourage children to set up their own tasks and record their findings.

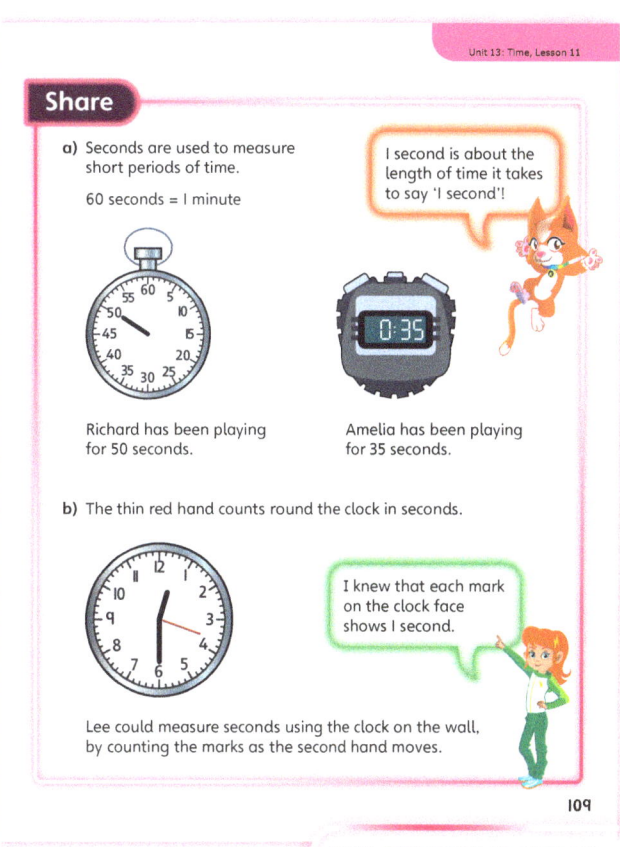

PUPIL TEXTBOOK 3C PAGE 109

Unit 13: Time, Lesson 11

Think together

WAYS OF WORKING Whole class teacher led (I do, We do, You do)

ASK
- Question ① : *What is the start time? What is the end time?*
- Question ① a): *What does the shaded face show?*
- Question ① a): *How will you count the duration of time between the start and end times?*
- Question ① b): *How can you calculate the duration?*
- Question ② : *How can you use the bar model to calculate the time left?*
- Question ② : *How can you calculate the difference?*

IN FOCUS Questions ① and ② give children pictorial representations of measuring seconds to scaffold their learning of start time, end time and duration. Provide plastic clocks or laminated stopwatch pictures to support their learning.

STRENGTHEN To strengthen understanding of question ①, ask children to mark the start times and the end times. Ensure children understand that each mark on the stopwatch in question ① a) shows one second. Ask children to think about how this question differs from other questions they have solved in previous lessons. Ask: *What do you know? What do you have to find out?*

DEEPEN When solving question ③, deepen children's understanding by asking them to show two different ways to work out the answers (for example, counting on in multiples of 30 seconds).

ASSESSMENT CHECKPOINT Children should be able to recognise that there are 60 seconds in 1 minute. They should be confident converting minutes into seconds by using different representations, including the bar model.

ANSWERS

Question ① a): Star jumps take 37 seconds.

Question ① b): Running takes 43 seconds.

Question ② : 60 − 48 = 12
There are 12 seconds left.

Question ③ a): 30 seconds

Question ③ b): 90 seconds

Question ③ c): 150 seconds

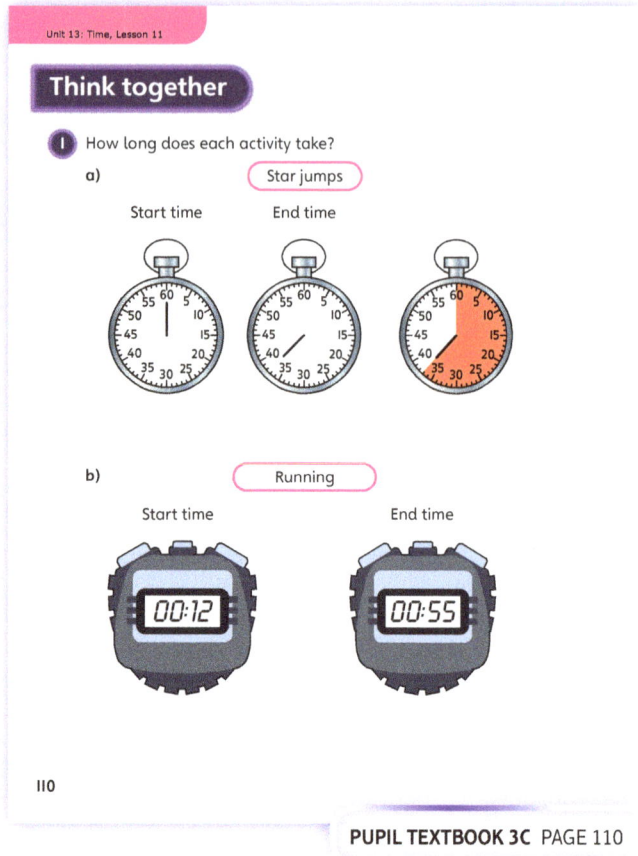

PUPIL TEXTBOOK 3C PAGE 110

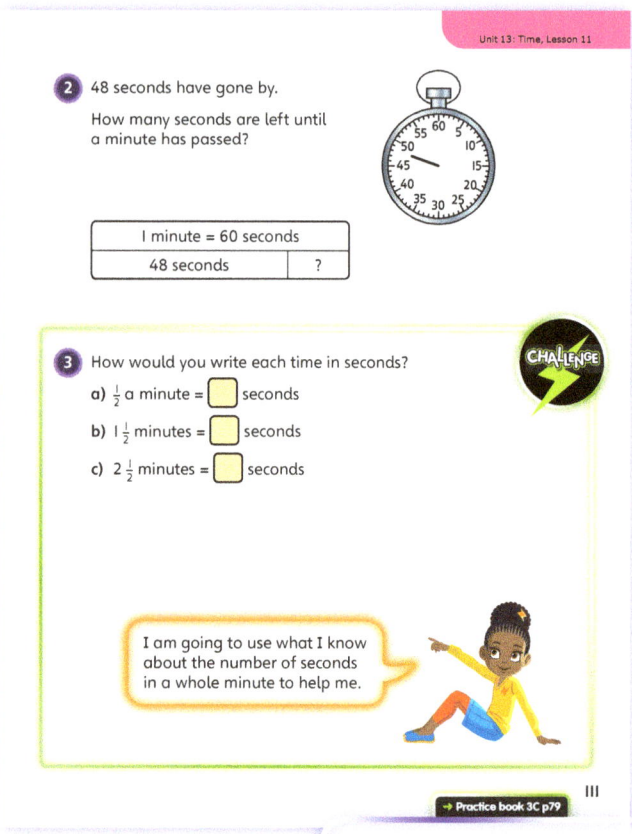

PUPIL TEXTBOOK 3C PAGE 111

Unit 13: Time, Lesson 11

Practice

WAYS OF WORKING Independent thinking

IN FOCUS Question ① gives children the opportunity to work with pictorial representations of time to develop their ability to measure time in seconds. Ask: *Which picture do you find the easiest to read? Which picture is the trickiest?* Children should be familiar with the different ways used to measure seconds.

STRENGTHEN Question ② offers concrete examples of measuring time in seconds. Children use their knowledge of fractions to convert between minutes and seconds. If children are struggling with the abstract representation of time, provide the concrete and pictorial representations that they have been using before.

DEEPEN Provide children with blank stopwatch faces and ask them to show each of the times included in question ④. Repeat with an analogue clock.

THINK DIFFERENTLY The wording of question ④ is slightly different to previous ones. Ask children what the word 'until' means in this context. Children may want to use a number line to count back.

ASSESSMENT CHECKPOINT At this point in the unit, children should be confident in estimating times in seconds and should know that there are 60 seconds in 1 minute. Children should be able to convert minutes to seconds and vice versa. Look for children using the concrete and pictorial representations they have worked with to support their learning.

ANSWERS Answers for the **Practice** part of the lesson can be found in the *Power Maths* online subscription.

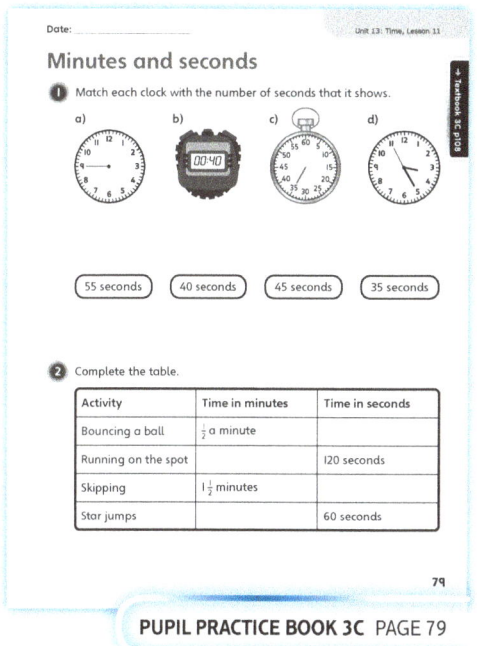

PUPIL PRACTICE BOOK 3C PAGE 79

PUPIL PRACTICE BOOK 3C PAGE 80

Reflect

WAYS OF WORKING Pair work

IN FOCUS The question revisits the misconception that counting to 60 takes 1 minute, regardless of how slow or fast you count. Give children time to discuss with a partner how they would help Bella overcome her misconception. Provide children with a stopwatch or analogue clock and ask them to use these to justify their answers.

ASSESSMENT CHECKPOINT Children should recognise that there are 60 seconds in 1 minute. They should be able to estimate with confidence a minute, and understand that 1 minute is made of 60 equal intervals, which are called seconds.

Children explore different ways of estimating 1 minute. Listen to children's reasoning as to why they chose the methods they did.

ANSWERS Answers for the **Reflect** part of the lesson can be found in the *Power Maths* online subscription.

After the lesson

- Are children confident about how many seconds are in 1 minute?
- Can children confidently estimate 1-minute intervals?
- Are children secure in converting from minutes to seconds and from seconds to minutes?

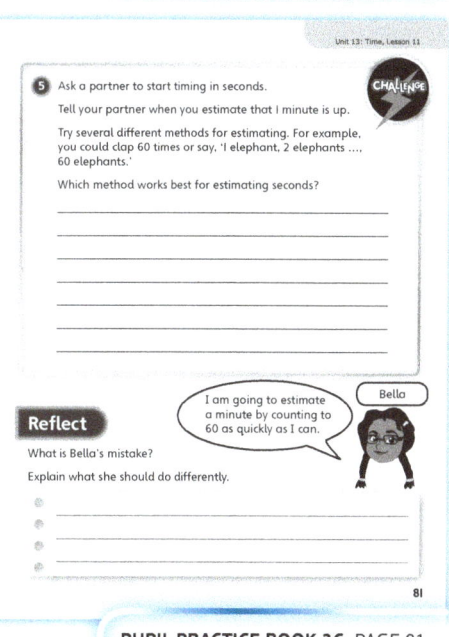

PUPIL PRACTICE BOOK 3C PAGE 81

Unit 13: Time, Lesson 12

Solve problems with time

Learning focus
In this lesson, children will choose the most appropriate unit of measure for different activities.

Before you teach
- Check that children know different units of measurement.
- Check children can convert minutes to seconds and hours to minutes.

NATIONAL CURRICULUM LINKS

Year 3 Measurement – time

Estimate and read time with increasing accuracy to the nearest minute; record and compare time in terms of seconds, minutes and hours; use vocabulary such as o'clock, am/pm, morning, afternoon, noon and midnight.

ASSESSING MASTERY

Children can choose the correct unit of measure for different activities. They should also be able to order different activities based on the units of measure.

COMMON MISCONCEPTIONS

Children sometimes think that a greater time means it is faster. For example, if Grace completes a race in 38 seconds and Mary takes 29 seconds, some children will think Grace is faster as the greater the number, the better. Discuss with children the context of a race or competition. Ask them to time two people racing and they should notice that the person who takes fewer seconds is the faster.

STRENGTHENING UNDERSTANDING

To support children choosing the correct unit of measurement, ask them to order different activities based on how long they might take. Ask them to think about whether each activity will take seconds, minutes, hours, days or years.

GOING DEEPER

Ask children to find an activity that somebody might use different units of time to measure (for example, someone going between two places might measure in minutes if they are driving or hours if they are walking). Ask children to discuss why there may be a difference.

KEY LANGUAGE

In lesson: seconds, minutes, hours, days, years, compare, order, unit of measure

Other language to be used by the teacher: faster, slower

RESOURCES

Optional: stopwatch

 In the eTextbook of this lesson, you will find interactive links to a selection of teaching tools.

Quick recap

Ask children to convert 1 minute 13 seconds into seconds. Then ask them to convert 2 hours and 7 minutes into minutes. Check that children know which methods to use to do the conversions.

150

Unit 13: Time, Lesson 12

Discover

WAYS OF WORKING Pair work

ASK

- Question 1 a): *What is the person doing in the image? What else can you see? Which will take longer to fill up, why?*
- Question 1 b): *How could you measure how long it would take to fill up the jug and the bath? What does it means to select the most appropriate unit?*

IN FOCUS Children should work in pairs to discuss the two questions. Make sure they have a good understanding of what the questions are asking. You may want to discuss each question in turn and discuss the children's thoughts before moving on to the next question. In question 1 a), children should reflect that the jug takes a shorter time to fill up as it has a smaller capacity (that is, the space inside the jug is less than the space inside the bath). In question 1 b), children should discuss using a stopwatch.

PRACTICAL TIPS Use a variety of different containers and fill them with water.

ANSWERS

Question 1 a): It will take longer to fill the bath than the jug as the bath has a greater capacity (that is, the space inside the bath is greater than the space inside the jug).

Question 1 b): I would use seconds to measure how long it would take to fill the jug.
I would use minutes and seconds to measure how long it would take to fill the bath.

Share

WAYS OF WORKING Whole class teacher led

ASK

- Question 1 a): *Why will the bath take longer to fill up?*
- Question 1 b): *Why do you think it only takes seconds to fill up the jug? What units do you think you can measure the time it takes to fill the bath in?*

IN FOCUS In question 1 a), discuss with children that it will take a longer time to fill up the bath as it holds more water. Some children may start to say it depends how fast the water is coming out of the tap, but explain that we will assume the water comes out at the same rate. In question 1 b), fill a jug with water and have children time it. They will see it takes seconds to fill. Discuss whether it makes sense to measure filling the bath in seconds too. Children should see that it could be measured in minutes and seconds, rather than just seconds.

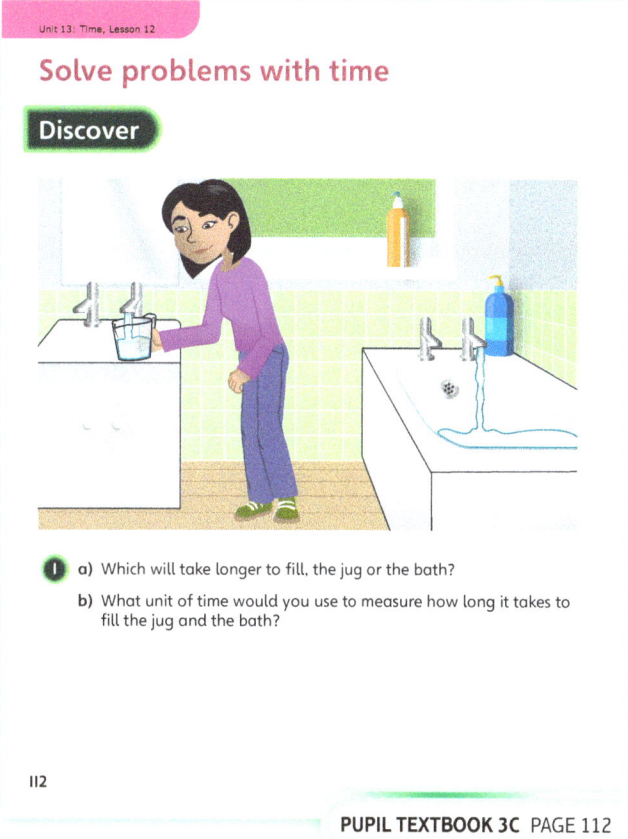

PUPIL TEXTBOOK 3C PAGE 112

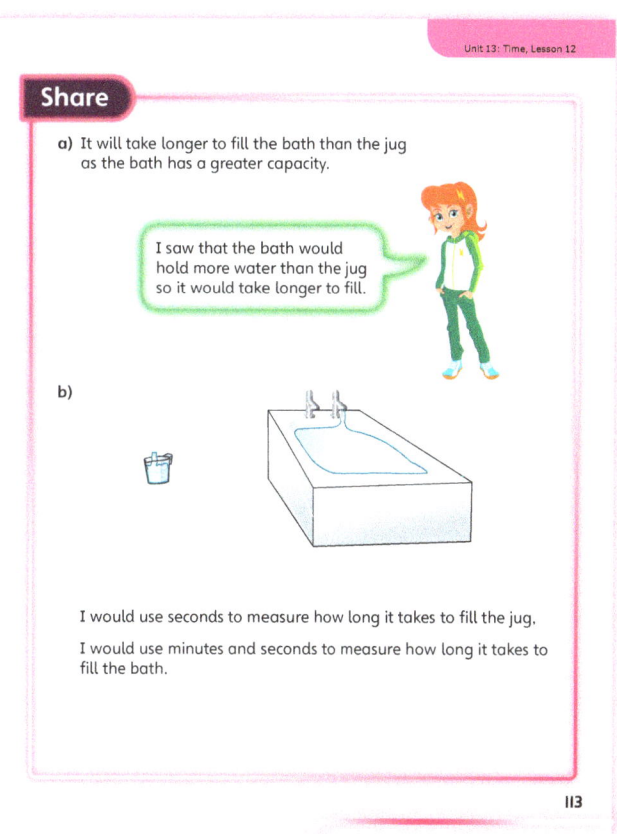

PUPIL TEXTBOOK 3C PAGE 113

Unit 13: Time, Lesson 12

Think together

WAYS OF WORKING Whole class teacher led (I do, We do, You do)

ASK
- Question ①: *What activities are there? Do you think the activities take a long time? What units of measure fit each one?*
- Question ②: *Who took the greater number of seconds? Does this mean Aki is faster or slower? Why?*
- Question ③: *Why do you think it might take Emma only minutes to get the seaside? Why might it take Max hours?*

IN FOCUS In question ①, discuss the different units of time and the activities. Ensure that children understand the activities. Discuss with children how long they think the activities will take. You might want to compare the length of one activity to the others. Question ②, focuses on the misconception that the greater the time, the faster a child has run. In question ③, children discuss and consider why Emma and Max could both be right by considering the distance they are away from the seaside. Children should also consider the method of transport.

STRENGTHEN For question ①, discuss which of the activities will take the shortest time and which will take the longest. This may help children decide which unit of time best matches each activity. For question ②, ask children to think about doing an activity in the classroom and timing it. Note that the children who finish first have a smaller time.

DEEPEN Ask children to come up with activities that could be done in seconds, minutes, hours, days and years. Ask them to share with a partner to check.

ASSESSMENT CHECKPOINT Use the questions to check that children understand what units of time different activities could be measured in.

ANSWERS

Question ①: Playing a football game – minutes.
Blowing up a balloon – seconds.
Flying to the moon – days.
Tree growing tall – years.

Question ②: Aki is incorrect because the race should be run as quickly as possible.
Ambika won the race as she took 18 seconds and Aki took 25 seconds.

Question ③: It depends on how far away they each live. It could take one minutes and the other hours.

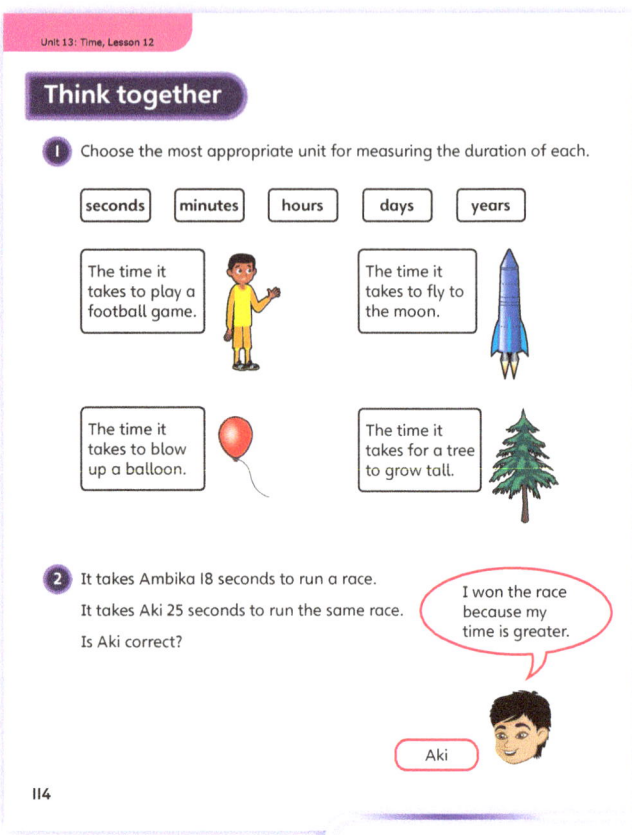

PUPIL TEXTBOOK 3C PAGE 114

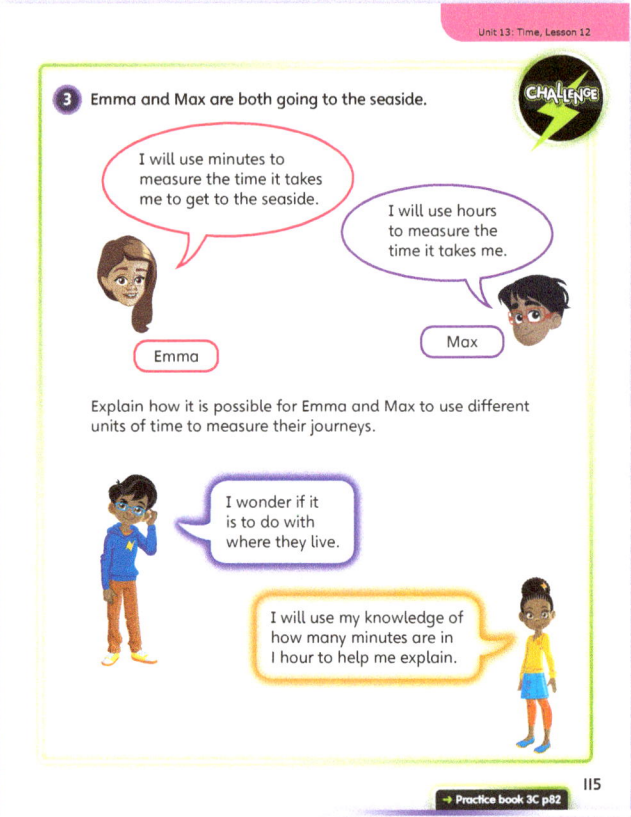

PUPIL TEXTBOOK 3C PAGE 115

Unit 13: Time, Lesson 12

Practice

WAYS OF WORKING Independent thinking

IN FOCUS Question ❶ is similar to the question asked in **Discover** in the main lesson. Question ❷ asks children to match up activities to the unit of time they are measured in. In question ❸, children choose the correct units of time to complete the sentences. Question ❹ checks that children can compare times that are given in mixed units of time and work out who is slowest. They should realise that a time of 11 minutes and any number of seconds (less than 60) will be shorter than a time of 12 minutes and a number of seconds. In question ❺, children have to convert times to the same unit in order to work out how much time is left.

STRENGTHEN For questions ❶ to ❸, ask children to think about each activity and have them consider how long it might take. You might want children to compare the activities with each other to work out which is longest and shortest. This may help them to choose the correct unit of time.

DEEPEN Ask children to come up with activities that could be done in seconds, minutes, hours, days and years. Ask them to share their ideas with a partner to check.

ASSESSMENT CHECKPOINT Use questions ❶ to ❸ to check that children can select the correct unit of measure for different activities, and questions ❹ and ❺ to assess how children can order durations that are given in different units of measure.

ANSWERS Answers for the **Practice** part of the lesson can be found in the *Power Maths* online subscription.

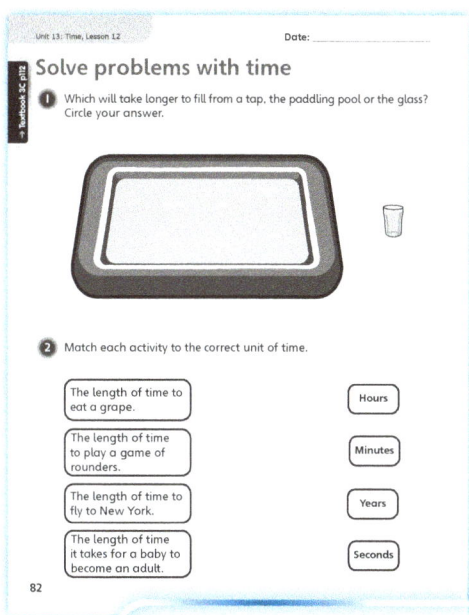

PUPIL PRACTICE BOOK 3C PAGE 82

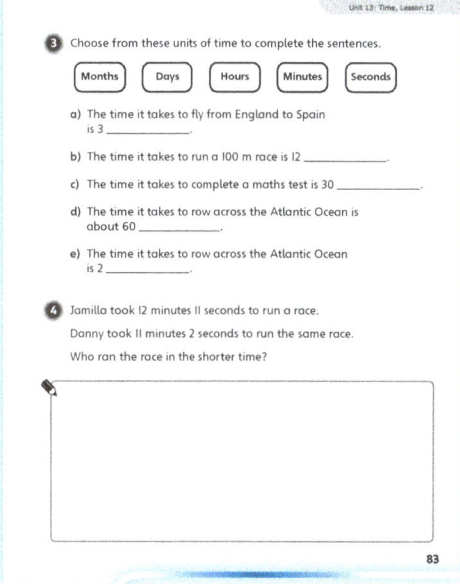

PUPIL PRACTICE BOOK 3C PAGE 83

Reflect

WAYS OF WORKING Whole class

IN FOCUS Ask children to think independently about activities that take the lengths of time suggested. Share these as a class and create a list of the different activities. Where children disagree, discuss if the answer could be both (for example, it might take someone seconds to get to school because they live next door, whereas someone else might take minutes).

ASSESSMENT CHECKPOINT Check that children have an understanding of how long different activities take and what they would measure the length of time of each in.

ANSWERS Answers for the **Reflect** part of the lesson can be found in the *Power Maths* online subscription.

PUPIL PRACTICE BOOK 3C PAGE 84

After the lesson

- Can children choose the correct units of time for different activities?
- Do children know that the *shorter* the time something takes the *faster* it is?

Unit 13: Time

End of unit check

Don't forget the unit assessment grid in your *Power Maths* online subscription.

WAYS OF WORKING Group work adult led

IN FOCUS

- Question ① assesses children's ability to recognise the months that have 30 days. Ask: *How do you remember which months have 30 days and which months have 31 days?*
- Question ② assesses children's ability to recognise the number of months in a year, the number of days in a year and the number of hours in a day. Ask: *What is a leap year?*
- Question ③ assesses children's ability to tell the time in different ways such as 'am' and 'pm', minutes past, minutes to, 12-hour and 24-hour time. It also provides an opportunity to discuss similarities and differences between times such as '4:58 am', '4.58 pm' and '2 minutes to 5 am'. Ask: *How are these times the same and different?*
- Question ④ assesses children's ability to find the end time when given the start time and duration, and to use different representations to record the time. Ask: *What is the start time/end time? What does duration mean?*
- Question ⑤ assesses children's ability to recognise that there are 60 seconds in 1 minute, and to convert minutes into seconds. Ask: *How many seconds are there in 1 minute?*
- Question ⑥ assesses children's ability to find a duration in hours and minutes when given the start and end times.

ANSWERS AND COMMENTARY

Children will demonstrate mastery by knowing how many days there are in each month. Children should be able to use different ways to record the time, including using the 12-hour and 24-hour clocks, digital and analogue clocks and am and pm.

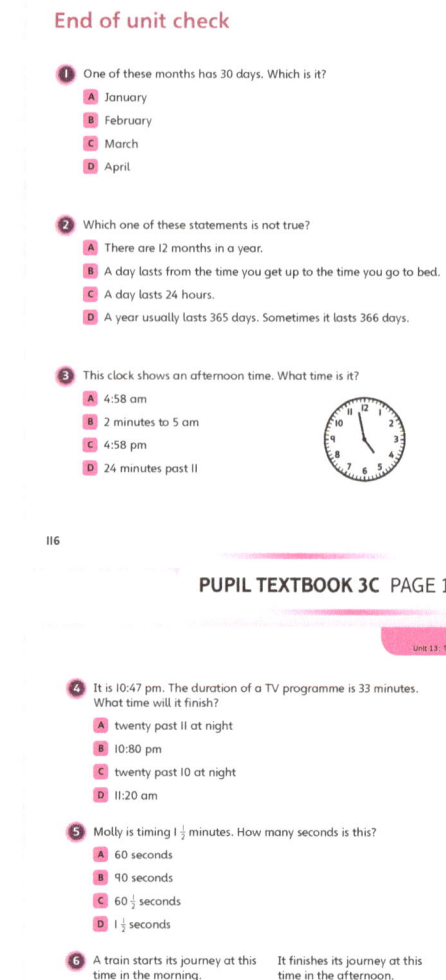

PUPIL TEXTBOOK 3C PAGE 116

PUPIL TEXTBOOK 3C PAGE 117

Q	A	WRONG ANSWERS AND MISCONCEPTIONS	STRENGTHENING UNDERSTANDING
1	D	Choosing A, B or C may indicate a lack of understanding about the number of days in each month.	To help children gain fluency in their understanding and use of time: • Make sure there are calendars in the classroom showing all the months of the year. • Clearly label all representations of analogue and digital clocks around the classroom. • Give access to 12-hour and 24-hour clocks throughout the day. • Provide train and bus timetables and TV guides. Use these to: find start and end times or durations of TV programmes; convert 12-hour time to 24-hour time and vice versa; and find the durations of journeys.
2	B	Choosing A, C or D may indicate a lack of understanding about time.	
3	C	Choosing A or B may indicate a lack of understanding as to what am means. Choosing D may indicate that children are not able to read time to the nearest minute.	
4	A	Choosing B or C may indicate that children lack experience in working with durations across an hour.	
5	B	Choosing A may indicate children have misread the question. C and D indicate a lack of understanding of how to find half of 1 minute in seconds.	
6	2 h 55 m	Other answers may indicate that children do not understand durations crossing one or more hours – for example, they may have added the start and end times hours and/or minutes.	

154

Unit 13: Time

My journal

WAYS OF WORKING Independent thinking

ANSWERS AND COMMENTARY

Question **1** : For each time, children may record answers such as those shown below:

a) I know that the time is 25 minutes to 3 because:
  • the hour hand has passed the 2 and is travelling towards the 3
  • the minute hand is in the 'to' section of the clock
  • the minute hand is pointing at 7 which means 35 minutes past
  • I can count 25 minutes to 3.

b) I know that the time is 17 minutes past 8 because:
  • the minute hand is pointing 2 minutes away from 15 minutes past, so it is 17 minutes past 8
  • the minute hand is in the 'past' section of the clock.

c) I know that the time is 9 minutes to 5 because:
  • the minute hand is in the 'to' section of the clock
  • the minute hand is pointing at 1 minute away from '10 minutes to'.

Encourage children to give reasons. Ask:
• Where is the minute hand pointing to? Can you count the minutes?
• What part of the clock is the minute hand in, 'to' or 'past'? How do you know?
• What do the hour hands tell you on each clock face?

Question **2** : Answers should demonstrate that children can draw times on analogue clock faces and write the corresponding times in words.

Power check

WAYS OF WORKING Independent thinking

ASK
• What did you know about reading and writing times before this unit?
• What new ideas and words have you learnt?
• How confident do you feel about reading and recording the time?
• Do you think you could look at a TV schedule at home and tell how long a programme lasts? What time does it start? What time does it end?

Power play

WAYS OF WORKING Independent thinking or pair work

IN FOCUS This game will assess children's ability to recognise times presented in different ways, and to identify the start and end times of 23-minute durations. It may help them to have the representations used in this unit at hand.

ANSWERS AND COMMENTARY If children are unable to follow the route accurately, give them opportunities to practise matching written times with pictorial and concrete representations of analogue and digital clocks.

After the unit

• How will you encourage children to measure time beyond this unit? For example, using time in role-play scenarios.
• Is your school a 'time-rich' environment? Is there a calendar, a digital clock and an analogue clock in each room? How could resources be improved to deepen children's fluency and use of time?

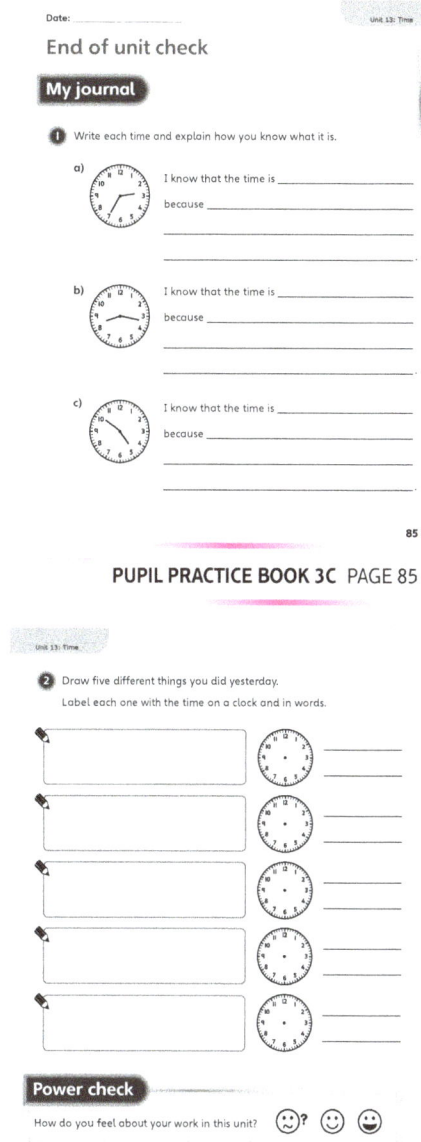

PUPIL PRACTICE BOOK 3C PAGE 85

PUPIL PRACTICE BOOK 3C PAGE 86

PUPIL PRACTICE BOOK 3C PAGE 87

Strengthen and **Deepen** activities for this unit can be found in the *Power Maths* online subscription.

155

Unit 14
Angles and properties of shapes

Mastery Expert tip! 'When I taught this unit, I used examples I could refer to within the school environment – square or rectangular tiles, vertical and horizontal lines around the classroom, parallel and perpendicular lines marked on the school playground. We turned quarter, half, three-quarter and full turns, varying the starting positions.'

Don't forget to watch the Unit 14 video!

WHY THIS UNIT IS IMPORTANT

This unit explores the concept of right angles. Right angles are linked to the concepts of parallel, perpendicular, vertical and horizontal lines and are linked with the angle properties of 2D shapes. Vertical and horizontal lines of symmetry are also explored and, finally, children describe and construct 3D shapes.

Angles are introduced to children as a measure of a turn, establishing that a right angle is a quarter turn, two right angles (quarter turns) make a half turn, three right angles make a three-quarter turn and four right angles make a full turn. Children will learn that angles less than a right angle are called acute angles and angles greater than a right angle (but less than two right angles) are called obtuse angles. Children will revise the names of 2D shapes. These include triangles (right-angled and isosceles), quadrilaterals (square, rectangle, rhombus, trapezium, parallelogram and kite), pentagons, hexagons and octagons. Children will revise the names of 3D shapes: cube, cuboid, pyramid, prism, cylinder, sphere and cone. This unit will provide an important foundation for further development of the concept of geometry in later years, such as measuring and drawing angles accurately using a protractor and describing the properties of 2D shapes in more detail.

WHERE THIS UNIT FITS

→ Unit 13: Time
→ **Unit 14: Angles and properties of shapes**
→ Unit 15: Statistics

This unit builds on children's understanding of the names and some of the properties of 2D and 3D shapes. It extends children's basic comprehension of these shapes with an emphasis on identifying right angles, lines of symmetry, vertical and horizontal lines and parallel and perpendicular lines and edges.

Before they start this unit, it is expected that children:
- understand what is meant by a 2D shape and are able to recognise and name most of them
- understand what is meant by a 3D shape and are able to recognise and name most of them.

ASSESSING MASTERY

Children who have mastered this unit will understand that angles are a measure of a turn and will recognise acute and obtuse angles. Children will recognise and identify vertical and horizontal lines in diagrams and 2D shapes. Children will identify pairs of parallel or perpendicular lines in diagrams and 2D shapes. Children will begin to describe 2D and 3D shapes in terms of the properties of right angles, parallel and perpendicular edges and lines of symmetry.

COMMON MISCONCEPTIONS	STRENGTHENING UNDERSTANDING	GOING DEEPER
Children may not recognise right angles, symmetry or parallel lines when shapes are in different orientations.	Mark 2D shapes with their right angles and parallel/perpendicular lines and then turn them 45 degrees. Record both on squared paper.	Encourage children to consider 'always, sometimes, never' relating to right angles, and parallel and perpendicular lines in 2D shapes, especially in quadrilaterals.
Children may mix up the terms parallel and perpendicular and horizontal and vertical.	Give plenty of opportunities to explore all these concepts in and around school.	Children explore changing a pair of parallel lines into a pair of perpendicular lines and vice versa by turning one line a quarter turn.

Unit 14: Angles and properties of shapes

UNIT STARTER PAGES

Use these pages with the whole class to revise the names and some of the properties of 2D and 3D shapes introduced in KS1. Use the actual shapes rather than just the images on the page. Do children recognise any of the key language? Discuss where they may have heard some of these words previously.

STRUCTURES AND REPRESENTATIONS

2D shapes:

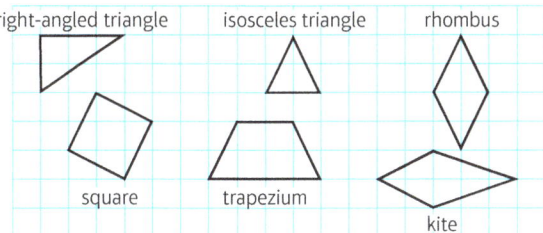

3D shapes:

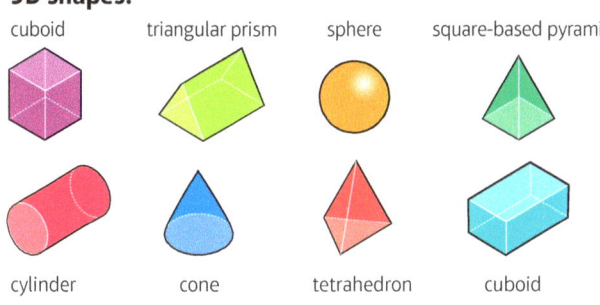

Lines:

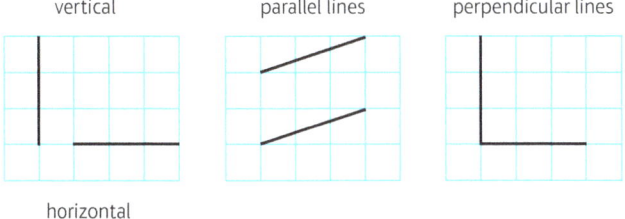

KEY LANGUAGE

There is some key language that children will need to know as part of the learning in this unit:

- right angle, quarter turn, half turn, acute angle, obtuse angle, clockwise, anticlockwise
- vertical, horizontal, parallel, perpendicular
- triangle, quadrilateral, square, rectangle, parallelogram, trapezium, rhombus, kite, pentagon, hexagon
- cube, cuboid, sphere, pyramid, prism, cylinder, cone, triangular prism, square-based pyramid, tetrahedron
- describe, property, 2D, 3D, edges, faces, vertices, draw accurately

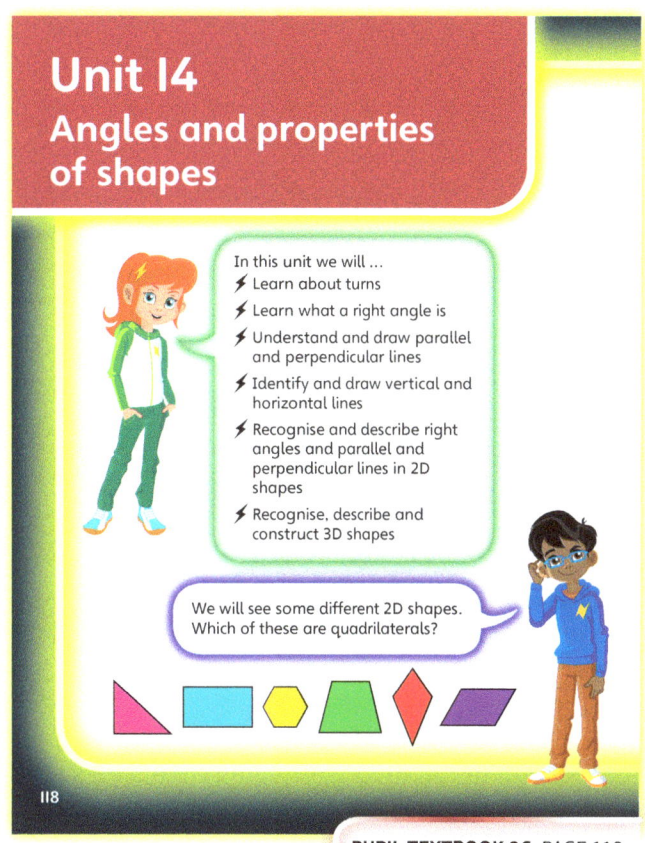

PUPIL TEXTBOOK 3C PAGE 118

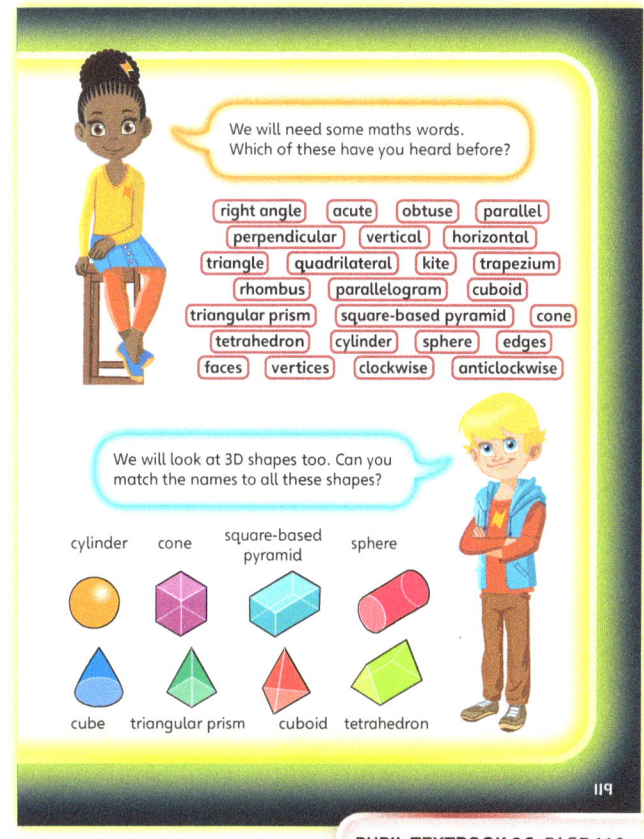

PUPIL TEXTBOOK 3C PAGE 119

Unit 14: Angles and properties of shapes, Lesson 1

Turns and angles

Learning focus

In this lesson, children understand angles as a measure of turn. Children learn that a right angle is a quarter turn, two right angles make a half turn, and four right angles make a whole turn.

Before you teach

- Do children know the difference between clockwise and anticlockwise?
- Can children describe a half turn and a quarter turn?

NATIONAL CURRICULUM LINKS

Year 3 Geometry – properties of shapes

Recognise angles as a property of shape or a description of a turn.

Identify right angles, recognise that two right angles make a half turn, three make three quarters of a turn and four a complete turn; identify whether angles are greater than or less than a right angle.

ASSESSING MASTERY

Children can follow rotations in steps of a quarter turn, both clockwise and anticlockwise. Children can explain the effect of turning two quarter turns, and how many right-angle turns are equivalent to a full turn.

COMMON MISCONCEPTIONS

Children may not recognise turns if they are presented in different orientations. Ask:
- *Face the window. Face this corner. Where will you be facing after a quarter turn? After a half turn?*

Children may struggle to sense which way is clockwise and which is anticlockwise, especially when given different starting positions. Ask:
- *If you turn clockwise, is that to the left or the right? What do you see first when you turn anticlockwise?*

STRENGTHENING UNDERSTANDING

Give children the opportunity to turn themselves, physically. Practise as a class: start facing the walls but then also try turning from a diagonal position, facing a corner. Practise turning clockwise and listing the things children see in order. Repeat with a turn anticlockwise. Give children a clock face with movable hands to place on the floor in front of them. Then use toy figures on a simple map to enact the turns.

GOING DEEPER

Use a simple treasure map on a grid (or ask children to draw one with four or five features marked). Ask children to write instructions to get from one feature to another, using the language of turns.

KEY LANGUAGE

In lesson: angle, **right angle**, turn, direction, quarter turn, half turn, complete turn, clockwise, anticlockwise

Other language used by the teacher: three-quarter turn, right-angle turn, whole turn, north, south, east, west

RESOURCES

Optional: toy figures, diagrams of eight-point compass (or chalks so this can be drawn on the playground), an object to represent the rover

 In the eTextbook of this lesson, you will find interactive links to a selection of teaching tools.

Quick recap

In pairs, ask children to write down all of the words they know about angles and turns. They should then try to describe to each other two of the words that they have written down.

Unit 14: Angles and properties of shapes, Lesson 1

Discover

WAYS OF WORKING Pair work

ASK
- Question ❶ a): *What do you think the arrows on the buttons mean? What do you think each button does?*
- Question ❶ b): *Where is the rover facing now? Where does the rover need to face? Which button would you press to turn the rover in the direction it needs to face?*
- Question ❶ b): *Is there another way you could get the rover to face the canyon?*

IN FOCUS The purpose of this activity is for children to develop an understanding of quarter turns and how they can be different depending on which way children turn. Encourage children to use words such as clockwise and anticlockwise rather than just left and right. For those children confident with simple turns, you could ask them to find at least one more way that the rover can face the canyon by pressing multiple buttons. Check also that children know what turns they need to make from the starting position to face the other landmarks.

PRACTICAL TIPS This could be enacted with a small toy vehicle or toy robot. Alternatively, a child could take the place of the rover in a role play. There is an opportunity for taking this task into the hall or the playground. Parts of the room could be labelled according to the rover's environment; or, for the rover, children could draw a 'map' in chalk on the playground. Children should experience for themselves turning two half turns in either direction so that they can see the end result is the same.

ANSWERS

Question ❶ a): The buttons are for making quarter turns in either a clockwise or an anticlockwise direction.

Question ❶ b): To face the canyon, the rover needs to make a quarter turn in the anticlockwise direction.

Share

WAYS OF WORKING Whole class teacher led

ASK
- Question ❶ a): *Which way is the arrow on this left-hand button facing? And on the right-hand button?*
- Question ❶ a): *What do you think this left-hand button does to the rover? And what about this right-hand button?*
- Question ❶ b): *Where has the rover turned to face?*
- Question ❶ b): *Could you have done it another way?*

IN FOCUS Work through the **Share** section. You might want to ask children to face the front and pretend they are the rover. This will help them see what a quarter turn means and which way they turn. Take time to make sure children are using the correct language, highlighting it throughout. For question ❶ b), you could discuss how children could have used three quarter turns clockwise to face the canyon. Again, act this out in the classroom with children all pretending to be the rover.

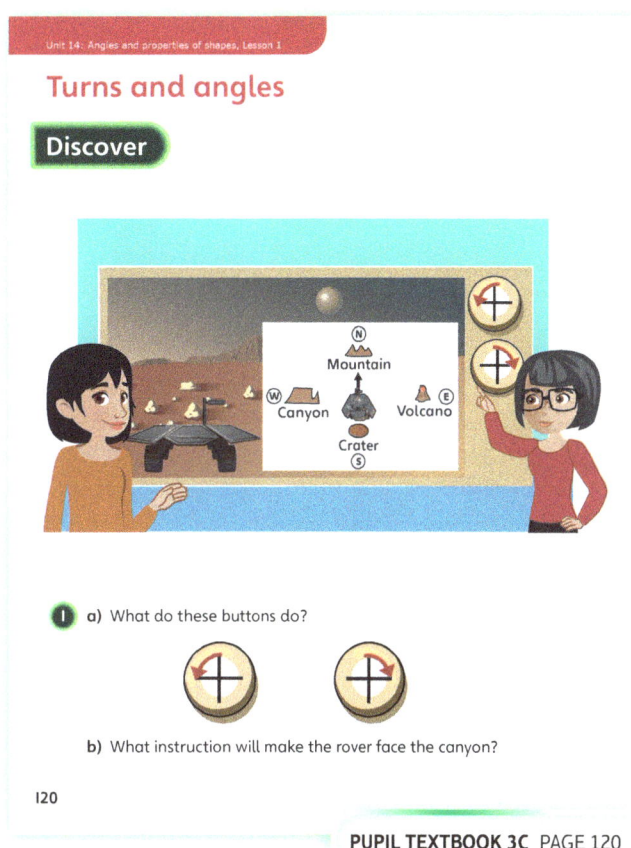

PUPIL TEXTBOOK 3C PAGE 120

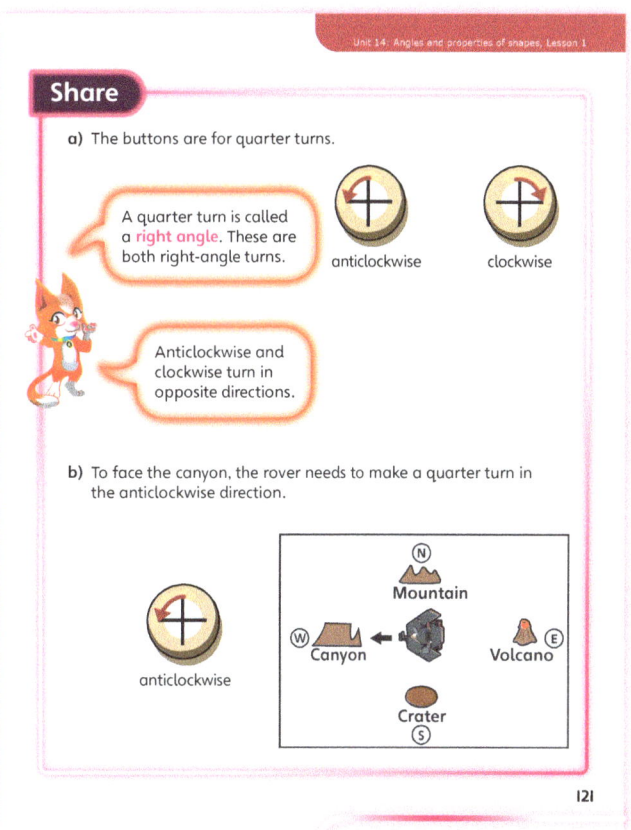

PUPIL TEXTBOOK 3C PAGE 121

Unit 14: Angles and properties of shapes, Lesson 1

Think together

WAYS OF WORKING Whole class teacher led (I do, We do, You do)

ASK
- Question ① a): *Imagine a clock face on top of this diagram. The volcano is at 12 o'clock. Which way will the minute hand turn? What will the rover face after a quarter turn clockwise?*
- Question ① b): *Why are the answers to questions ① a) and b) different if they are both making quarter turns?*
- Question ②: *Why is making a quarter or right-angled turn trickier on this map?*

IN FOCUS These questions focus mainly on children making quarter turns. The language of a right-angle turn is also introduced, and you may need to discuss this with children. Throughout, you should encourage children to act out the turns as if they were the rover. Question ① presents children with the difference between a clockwise and anticlockwise turn. For question ③, children should make a physical copy of the triangle on a piece of paper that they can turn. Having actually performed those turns, rather than just trying to picture them in their heads, will help children to answer each part of the question.

STRENGTHEN Children could use a toy or make the turns themselves. They should practise making quarter turns, then two quarter turns as a sequence, and then three and four turns as a sequence.

Children may struggle to remember and have a sense of which way a clockwise and an anticlockwise turn takes them. This should be practised from different starting positions. To help, you could try placing a clock face at their feet.

DEEPEN Challenge children to explain the relationship between a three-quarter turn in one direction and a one-quarter turn in the opposite direction. Ask them to explain how to reverse the turn instruction to get back to the start position. Use question ③ to get children to describe half and full turns in terms of quarter turns. Ask: *What do you notice?*

ASSESSMENT CHECKPOINT Children should now understand the language of quarter, half and whole turns, realising that a quarter turn clockwise will have a different result from a quarter turn anticlockwise. Children will begin to understand how instructions can be reversed to get back to the original position.

ANSWERS

Question ① a): It would face the crater.

Question ① b): It would face the mountain.

Question ② a): It would face the weather station.

Question ② b): It would face the unidentified object.

Question ③ a):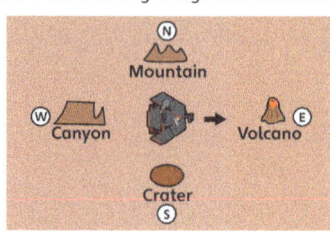

Question ③ b):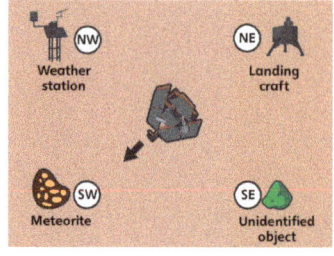

Question ③ c):

Question ③ d):

The shape will return to its starting position.

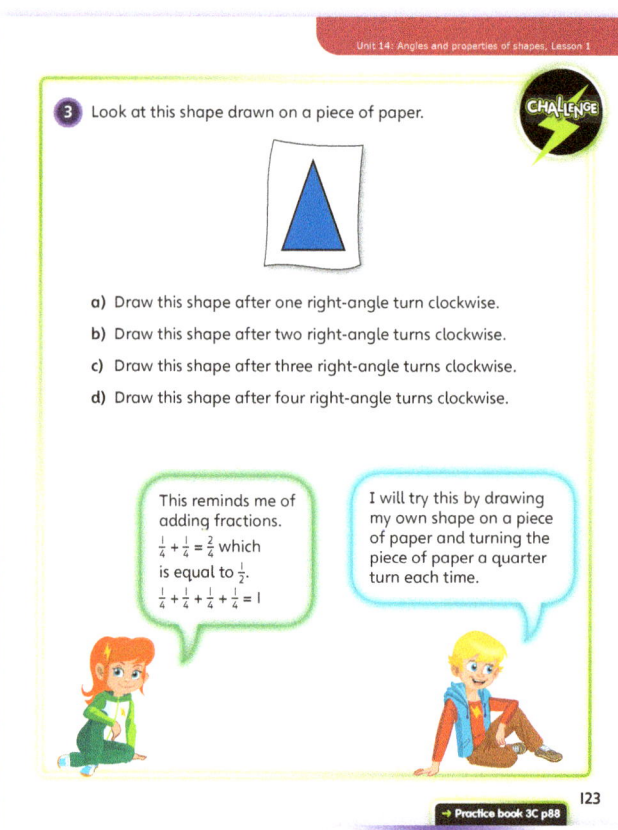

PUPIL TEXTBOOK 3C PAGE 122

PUPIL TEXTBOOK 3C PAGE 123

Unit 14: Angles and properties of shapes, Lesson 1

Practice

WAYS OF WORKING Independent thinking

IN FOCUS Questions ① and ② focus on quarter turns in different contexts.

Question ③ demonstrates that a quarter turn clockwise is the same as a three-quarter turn anticlockwise.

STRENGTHEN Allow children to use a toy figure to move according to each question and act out the turns.

DEEPEN Ask children to discuss and explain to each other what types of turn need to include clockwise and anticlockwise and when it does not matter. To explore this further, children can work in pairs: one child gives an instruction and the other reverses it.

Question ⑤ challenges children to apply their understanding of turns to the rotation of simple shapes.

THINK DIFFERENTLY Question ④, with some compass points being on a 45 degree angle, tests understanding of right-angle turns from non-horizontal or non-vertical lines. It is usual to describe quarter, half or three-quarter turns as one instruction, rather than a half turn followed by a quarter turn, but accept these suggestions.

ASSESSMENT CHECKPOINT Children should now understand the language of quarter, half and whole turns, realising that a quarter turn clockwise will have a different result from a quarter turn anticlockwise. Children will be beginning to understand how instructions can be reversed to get back to the original position.

ANSWERS Answers to the **Practice** part of the lesson can be found in the *Power Maths* online subscription.

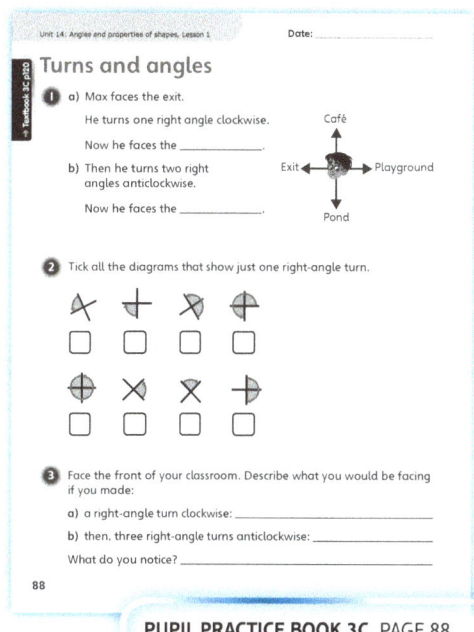

PUPIL PRACTICE BOOK 3C PAGE 88

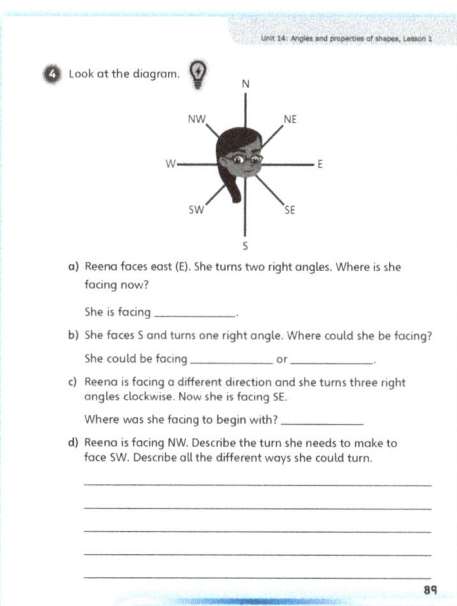

PUPIL PRACTICE BOOK 3C PAGE 89

Reflect

WAYS OF WORKING Pair work

IN FOCUS Ask children to make a quarter turn clockwise. Alternate between this and the language of right-angle turns. Then ask children to make two quarter turns from the starting position. Ask: *What do you notice?* Continue, each time returning to the start and discussing what happens when you make an additional quarter turn each time.

ASSESSMENT CHECKPOINT Ask children to discuss with each other the language around the turn they have just made. Ensure children use the correct language and can associate it with the appropriate turn.

ANSWERS Answers to the **Reflect** part of the lesson can be found in the *Power Maths* online subscription.

After the lesson ⏸

- How many right-angle turns are required for a whole turn?
- Do you have any advice for how to remember the difference between clockwise and anticlockwise?

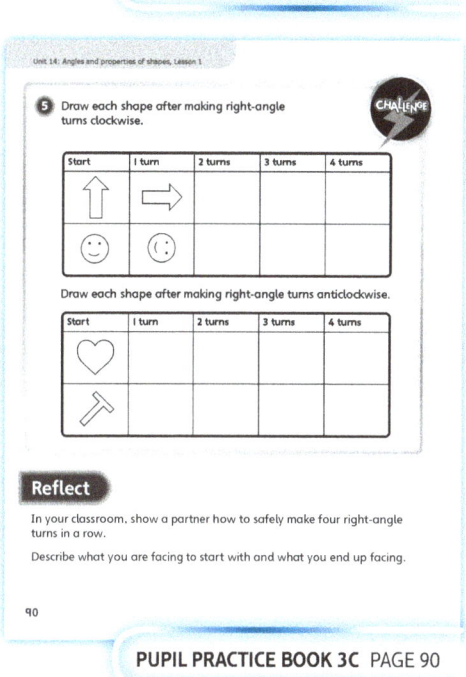

PUPIL PRACTICE BOOK 3C PAGE 90

Unit 14: Angles and properties of shapes, Lesson 2

Right angles in shapes

Learning focus
In this lesson, children develop their understanding of right angles to include the measure of an angle in a shape. They learn the symbol that indicates a right angle and are introduced to the idea of perpendicular lines.

Before you teach
- Can children find the angles in a shape?
- Do they know what right angle means?

NATIONAL CURRICULUM LINKS

Year 3 Geometry – properties of shapes

Recognise angles as a property of shape or a description of a turn.

Identify right angles, recognise that two right angles make a half turn, three make three quarters of a turn and four a complete turn; identify whether angles are greater than or less than a right angle.

ASSESSING MASTERY

Children can recognise right angles in shapes in different orientations, and can draw a pair of perpendicular lines to create a right angle.

COMMON MISCONCEPTIONS

Children may struggle to recognise right angles that are oriented diagonally – where neither line is horizontal or vertical. Ask:
- *How can you check whether this is a right angle or not? Does it help to turn the paper round?*

Children may not understand how the measure of a turn is linked to the measure of an angle in a shape. Ask:
- *How can you check whether this angle is the same as, greater than or less than a quarter turn?*

STRENGTHENING UNDERSTANDING

Encourage children to use their reasoning skills to predict if an angle is a right angle, greater than a right angle or less than a right angle; then support children to make right-angle measures, or to use the corner of a 2D square to check. Help children to understand how to orient and read the measure in order to check the angle.

GOING DEEPER

Challenge children to use the reasoning of a grid to justify whether or not an angle is a right angle, using the properties of the grid to support their reasoning, especially when two diagonal lines form the angle.

KEY LANGUAGE

In lesson: right angle, angle, angle measurer, curved, straight, shape

Other language used by the teacher: square, rectangle, arrow, right-angled triangle, reflex angle, greater than (>), less than (<), predict, prediction, diagonally, measure of turn, quarter turn, half turn

STRUCTURES AND REPRESENTATIONS

2D shapes

RESOURCES

Mandatory: right-angle measurer (folded paper, ruler or 2D square or rectangle)

Optional: square pieces of paper or card

 In the eTextbook of this lesson, you will find interactive links to a selection of teaching tools.

Quick recap
Ask children to face the front and make one right-angled turn clockwise. Ask: *What do you notice about the turn?* Then ask children to make other turns through a given number of right angles in different directions.

Discover

WAYS OF WORKING Pair work

ASK

- Question 1 a): *How can you accurately fold a piece of paper to create perfect halves?*
- Question 1 b): *How can you use your right-angle measurer to see where the right angles in the shapes are?*

IN FOCUS This is the first time that children are asked to recognise right angles in a shape. Allow them to make a right-angle measurer by folding paper or thin card and show them how to use it to test which angles are right angles. Discuss together how children can tell that some of the angles are *not* right angles or that a shape will *not* contain right angles. Curved lines cannot make right angles.

PRACTICAL TIPS Provide children with square pieces of paper or thin card for them to fold. Check that their folds are correct.

ANSWERS

Question 1 a): Children should accurately make and mark their right-angle measurers.

Question 1 b): The blue rectangle has four right angles – one at each vertex.
The purple trapezium has two right angles – both on the right-hand side of the shape.

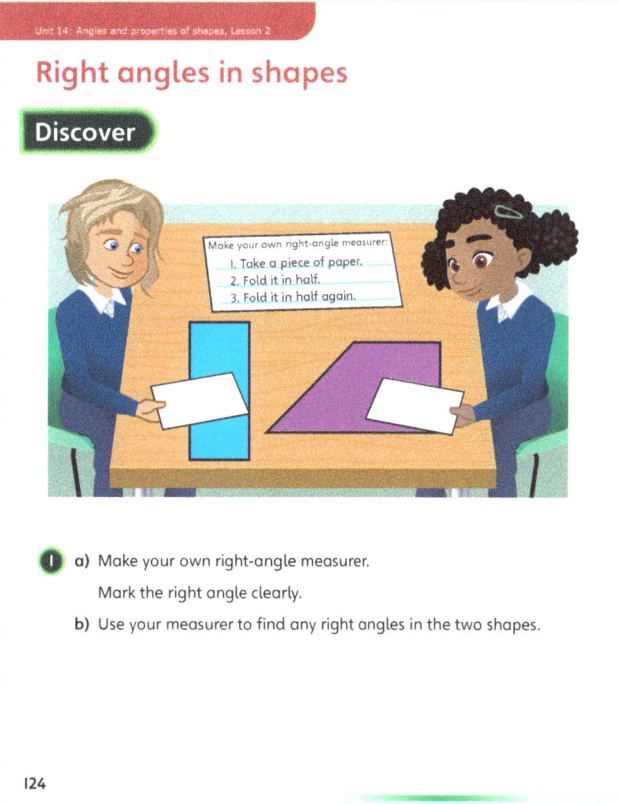

PUPIL TEXTBOOK 3C PAGE 124

Share

WAYS OF WORKING Whole class teacher led

ASK

- Question 1 a): *How does the right-angle measurer work?*
- Question 1 b): *How can you use the measurer to check if the angles in this rectangle/trapezium are right angles? How do you line up the measurer to check an angle accurately?*
- Question 1 b): *Can you predict whether any of these angles are not right angles before you check?*
- Question 1 b): *How many right angles are there in each shape?*

IN FOCUS Question 1 a) shows children how to use an angle measurer. Work through this together with children again. In question 1 b), use the angle measurer to find the right angles. Show children how they can use their measurer to check whether an angle is a right angle, greater than a right angle or less than a right angle. Also show children how to mark right angles using conventional notation. Discuss with children whether the angles in each of the shapes are less than, equal to or greater than a right angle.

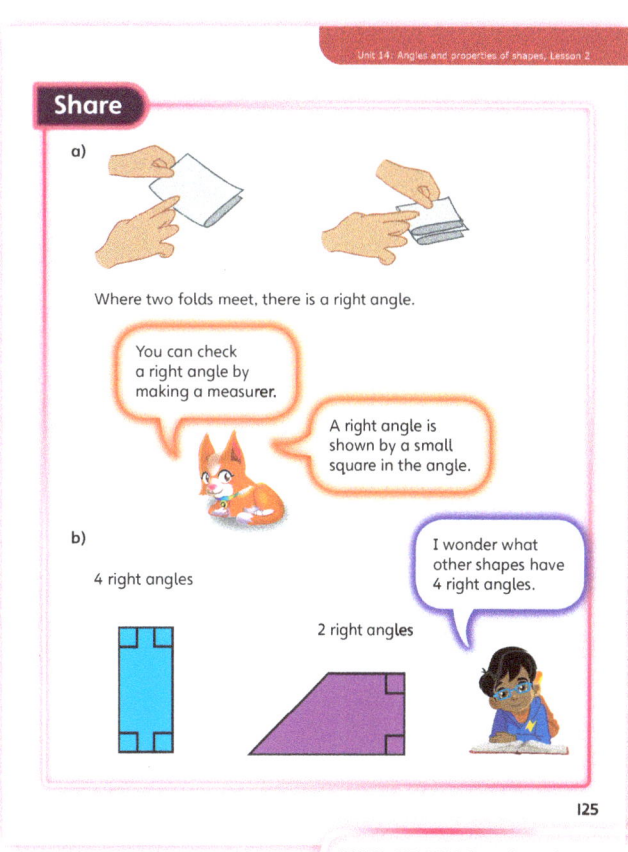

PUPIL TEXTBOOK 3C PAGE 125

163

Unit 14: Angles and properties of shapes, Lesson 2

Think together

WAYS OF WORKING Whole class teacher led (I do, We do, You do)

ASK
- Question ①: *Can you measure the angle in a curve?*
- Question ②: *Should you measure the angles inside or outside the shape?*
- Question ④: *How could you use the pattern of the grid to decide where to draw the lines?*

IN FOCUS Questions ① and ② develop an understanding of recognising right angles in different orientations, and recognising non-examples (for example, curved lines).

STRENGTHEN Encourage children to practise using a measurer and also to make predictions before they measure as a way of practising their visual reasoning. This will mean that the measurer can become a means of checking. Use large scale 2D shapes to make the task easier.

DEEPEN Challenge children to explain how to use a grid to find right angles in different orientations (diagonals). Question ④ challenges children to use the grid to decide where to draw right angles. The lines do not need to cross, and can be any length, but all should be drawn carefully, joining the dots not going between them.

ASSESSMENT CHECKPOINT Children should be able to use an informal paper angle measurer to check if an angle is a right angle, to identify angles that are clearly not right angles, and to explain that curved lines cannot show right angles.

ANSWERS

Question ①: The sports field has eight right angles: one at each vertex of the main outline and then four where the vertical line crosses the centre of the field.

Question ②: A = 2, B = 4, C = 0, D = 3

Question ③: The shape has five right angles and one reflex angle of 270°.
So Dexter is correct in thinking that one of the angles is a three-quarter turn and not a quarter turn.

Question ④: Lines should be added to each diagram that are perpendicular to the existing lines, for example:

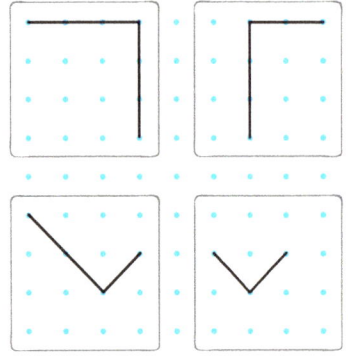

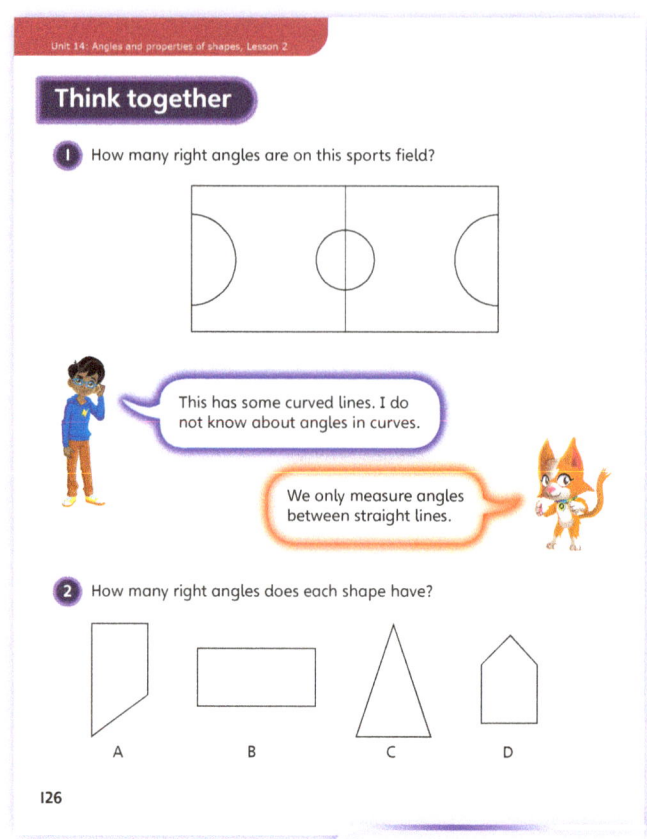

PUPIL TEXTBOOK 3C PAGE 126

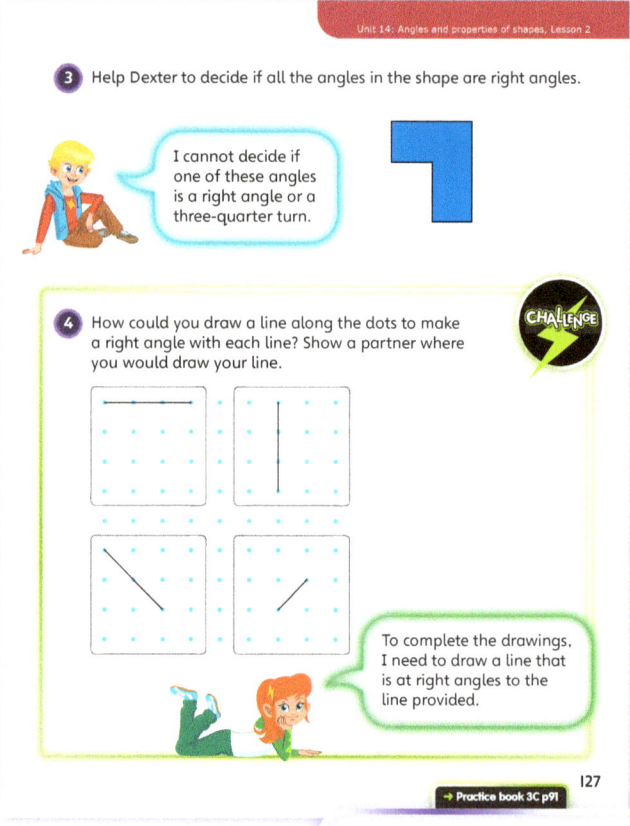

PUPIL TEXTBOOK 3C PAGE 127

Unit 14: Angles and properties of shapes, Lesson 2

Practice

WAYS OF WORKING Independent thinking

IN FOCUS Question ❶ tests recognition of right angles, and the understanding that they cannot be formed by a curved line.

Question ❷ focuses on right angles in 2D shapes that are oriented in different ways. Can children recognise a right angle, even when the lines that form the angle are not horizontal and vertical?

Question ❹ focuses on forming right angles on a dotted grid when one line is given. Children can use the dots to form many different right angles: they could draw a perpendicular on either side of each line, and at different distances along the line. The bottom two grids give examples of slanted lines. Ensure that children can identify which dots are diagonally opposite the dots on their line, in order to draw a right angle.

Question ❻ is a logic puzzle with clues involving recognition of right angles. Encourage children to work systematically. For example, after reading clue 1, they can put a 'cross' in the box containing the rhombus (top, centre). They can continue working through the clues, putting a tick or cross in each box.

STRENGTHEN Encourage children to use visual reasoning and justify predictions before checking with an angle measurer. Examples on a slant could be turned to vertical/horizontal. Provide a variety of physical 2D shapes for children to identify the right angles in each. Ask: *Is this angle less than a right angle? Is it greater than a right angle?*

DEEPEN There are several solutions to question ❹. Challenge children to find all the different solutions to each. If you insist the lines join the dots, there will be a limited number.

THINK DIFFERENTLY Question ❸ challenges children to identify plausible misconceptions about right angles.

ASSESSMENT CHECKPOINT Children should be able to identify correctly the internal right angles of a variety of 2D shapes and be able to show pairs of perpendicular lines on a grid, including those on a diagonal. They will be beginning to identify angles within shapes that are greater than or less than a right angle and will know that right angles cannot be formed on curved lines.

ANSWERS Answers to the **Practice** part of the lesson can be found in the *Power Maths* online subscription.

Reflect

WAYS OF WORKING Independent thinking

IN FOCUS This is an open opportunity for children to demonstrate their understanding. They should use the right-angle notation to identify the three right angles. Once complete, children should share their shape with others.

ASSESSMENT CHECKPOINT Children accurately show a shape with exactly three right angles and use the correct notation to denote them.

ANSWERS Answers to the **Reflect** part of the lesson can be found in the *Power Maths* online subscription.

After the lesson

- Can children recognise right angles in different orientations?
- Can children use an angle measurer to check predictions made based on visual reasoning on grids?

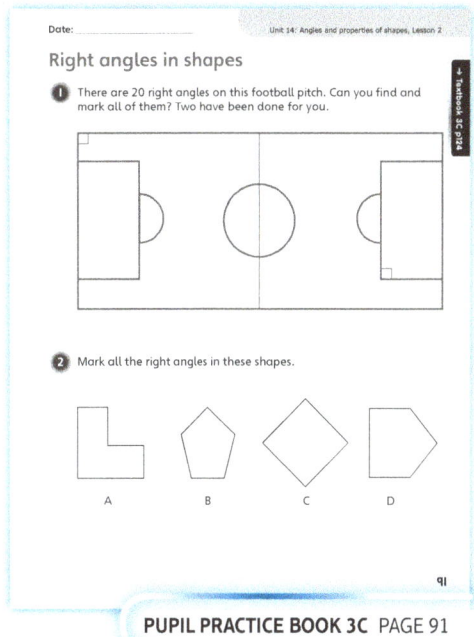

PUPIL PRACTICE BOOK 3C PAGE 91

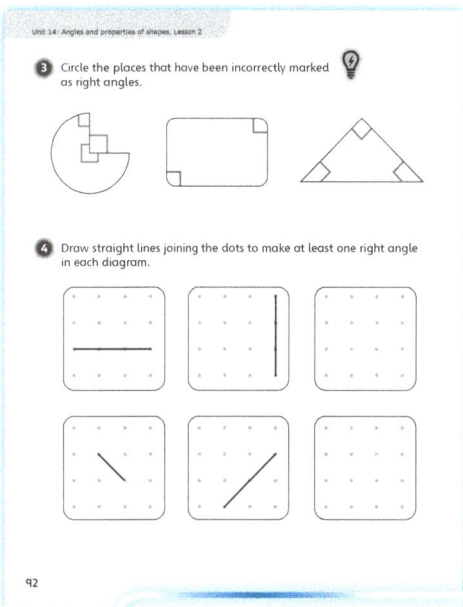

PUPIL PRACTICE BOOK 3C PAGE 92

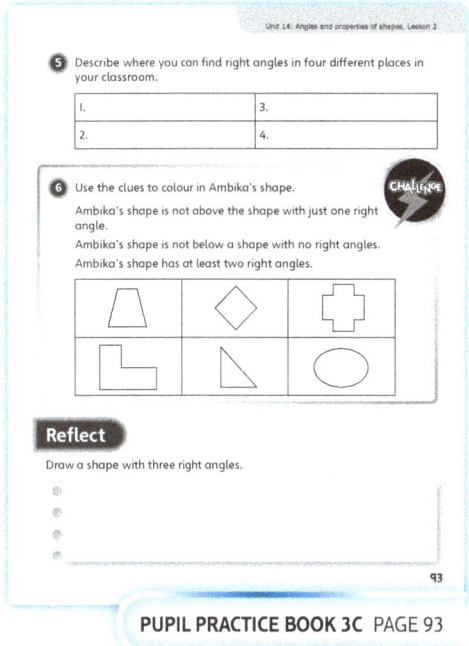

PUPIL PRACTICE BOOK 3C PAGE 93

165

Unit 14: Angles and properties of shapes, Lesson 3

Compare angles

Learning focus
In this lesson, children learn how to recognise angles that are greater than, equal to or less than a right angle. They are introduced to the terms acute and obtuse.

Before you teach
- Can children show you a right angle with their arms or a folded piece of paper?
- Do children know that a right angle is a quarter turn?

NATIONAL CURRICULUM LINKS

Year 3 Geometry – properties of shapes

Identify right angles, recognise that two right angles make a half turn, three make three quarters of a turn and four a complete turn; identify whether angles are greater than or less than a right angle.

Recognise angles as a property of shape or a description of a turn.

ASSESSING MASTERY

Children can identify acute and obtuse angles in relation to a right angle in different orientations, using visual reasoning and checking with a right-angle measurer. They can use reasoning to predict or justify whether an angle is greater or less than a right angle.

COMMON MISCONCEPTIONS

Children may think that an angle is a measure of the size of the space between two lines, and so may think that a given angle is larger if the lines extend further. Ask:
- *How could you compare the size of these two angles to see which is smaller/larger?*

STRENGTHENING UNDERSTANDING

Children could use two rods hinged together at a point, or a folded piece of card, to create angles greater than, equal to or less than a right angle in different orientations. Children should start with a right angle, and close the rods (or card) in towards each other for an acute angle and open them further out for an obtuse angle. They could also demonstrate the angles by creating an angle between their arms, or they could move the hands on a clock face to form right angles first, then obtuse or acute angles.

GOING DEEPER

Challenge children to make reasoned predictions before checking with a measurer. They could use the properties of a given background (such as squared paper or a clock face) to justify their reasoning.

KEY LANGUAGE

In lesson: compare, angle, right angle, **acute**, **obtuse**, greater than (>), less than (<), equal to (=), turn, measure

Other language to be used by the teacher: predict, space, size, straight line, half turn, quarter turn, complete turn

STRUCTURES AND REPRESENTATIONS

Geoboard

RESOURCES

Mandatory: geoboards and bands or square dotted paper to represent geoboards

Optional: clock faces with movable hands, a pair of hinged rods, pipe cleaners, two rulers or folded card to make angles, squared paper

 In the eTextbook of this lesson, you will find interactive links to a selection of teaching tools.

Quick recap
Ask children to make or draw their own right angle and then share with the rest of the class. Compare all the right angles and ask: *Which look accurate? How could you make them more accurate?* (For example, by folding paper or using a ruler.)

166

Unit 14: Angles and properties of shapes, Lesson 3

Discover

WAYS OF WORKING Pair work

ASK

- Questions ❶ a) and b): *Which angle do you need to measure in order to answer the question?*
- Questions ❶ a) and b): *Are you looking for roof angles less than or greater than a right angle?*
- Questions ❶ a) and b): *How can you measure each angle?*
- Questions ❶ a) and b): *Can you see any right angles?*

IN FOCUS Here, children's attention is drawn to the angle formed at the peak of the roof, and they consider how this angle relates to the pitch of the roof. Children may want to discuss the physical/engineering reasons that underlie the context, but should return to a comparison of the angle at the apex of each roof.

PRACTICAL TIPS Children could form the different angles of each roof by opening or leaning books to an approximation of the given angles. Alternatively, children could construct them from folded card, and place them above a box. This could form the basis of an interesting experiment in pouring water from a watering can to represent the rain.

ANSWERS

Question ❶ a): House C would be good in a snowy country. A steeper roof means the snow can fall off more easily.

Question ❶ b): Houses B and D would suit a dry country as the angle at the top of the roof is greater than a right angle and therefore the roofs are less steep.

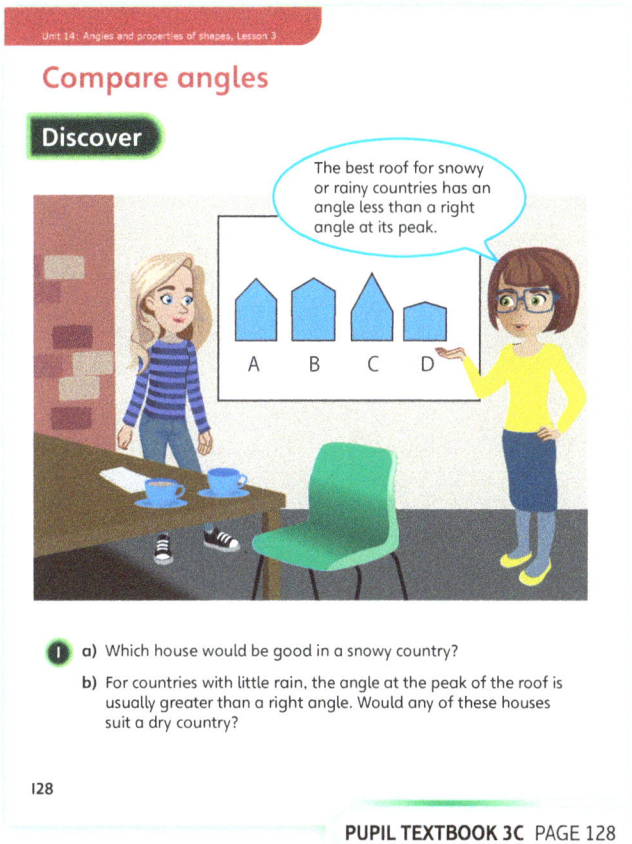

PUPIL TEXTBOOK 3C PAGE 128

Share

WAYS OF WORKING Whole class teacher led

ASK

- Question ❶ a): *Can you predict which roof angle is less than a right angle just by looking?*
- Question ❶ a): *How should you place a measurer to check your predictions?*
- Question ❶ b): *Are you looking for roof angles less than or greater than a right angle for this part of the question?*

IN FOCUS Question ❶ a) focuses on identifying the acute angle but without introducing this language.

Question ❶ b) focuses on identifying the two obtuse angles but, once again, the language will be formally introduced by Sparks in **Think together**. Informally, you could explain to children that the angles smaller than a right angle are the more 'pointy' angles, whereas the angles greater than a right angle are the 'flatter' angles. However, emphasize that when angles are close to a right angle, children should always check their size using a right angle measurer and not guess.

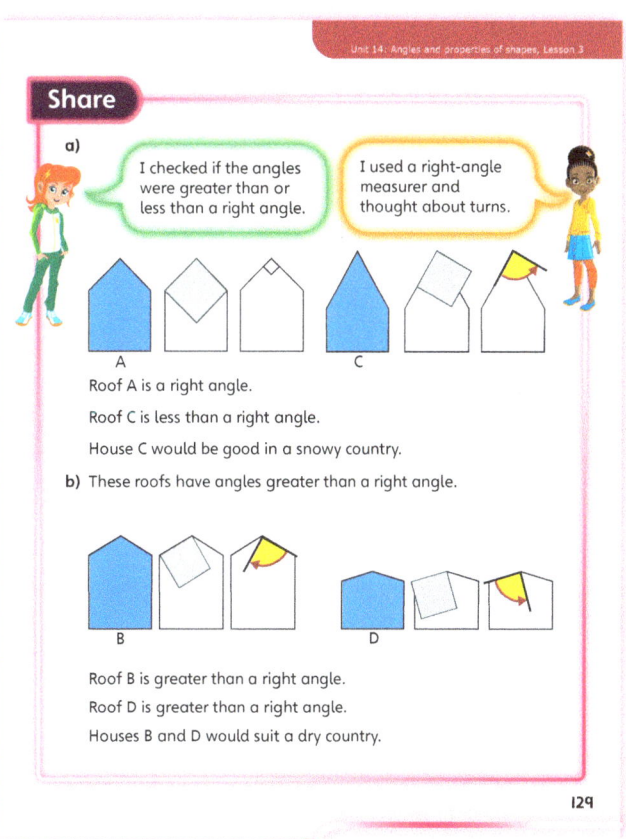

PUPIL TEXTBOOK 3C PAGE 129

167

Unit 14: Angles and properties of shapes, Lesson 3

Think together

WAYS OF WORKING Whole class teacher led (I do, We do, You do)

ASK
- Question ①: *Can you tell just by looking whether each angle is greater than or less than a right angle?*
- Question ①: *Is an acute angle greater than or less than a right angle?*
- Question ②: *Which numbers on a clock are at right angles to each other?*
- Question ②: *Is the angle between two numbers next to each other on the clock less or more than a right angle?*

IN FOCUS Question ② uses the angles between numbers on a clock face to identify acute and obtuse angles.

Question ③ asks children to show or make acute and obtuse angles on a geoboard. It is very helpful if children can have access to geoboards to try this activity out themselves, as it can be difficult to visualise whether or not an angle is different to another or whether it is the same angle in a different orientation.

STRENGTHEN In question ①, ask children to make predictions and check them using a right-angle measurer to compare with the given angle. Children should also practise forming different angles by opening the covers of a book or pivoting their arms about a point. They could use geoboards to explore question ②: it is helpful for children to see that the angles between 12 and 3, 3 and 6, 6 and 9, and 9 and 12 are all right angles. Link this with quarter turns and the fact that the angle between 12 and 6 is a half turn – as is the angle between all pairs of opposite numbers.

DEEPEN In question ③, challenge children to justify their reasoning about different or similar acute and obtuse angles represented on the geoboards in different orientations. Can children recognise when two angles are the same (rotations or reflections of one another)? Extend question ③ by asking children to start with the right angle in a diagonal orientation on the board.

ASSESSMENT CHECKPOINT Children should be able to recognise, form or draw angles that are greater than, less than or equal to a right angle and explain the terms acute and obtuse in relation to a right angle.

ANSWERS

Question ①: A = less than, B = greater than, C = equal to, D = greater than, E = equal to, F = less than

Question ②: A = equal to, B = less than, C = less than, D = greater than, E = equal to, F = greater than

Question ③: Isla can make many different acute and obtuse angles by starting at different pins on the boards. Children may notice that the length to which the elastic is pulled does not matter (that is, the length of the arms does not affect the size of the angle).

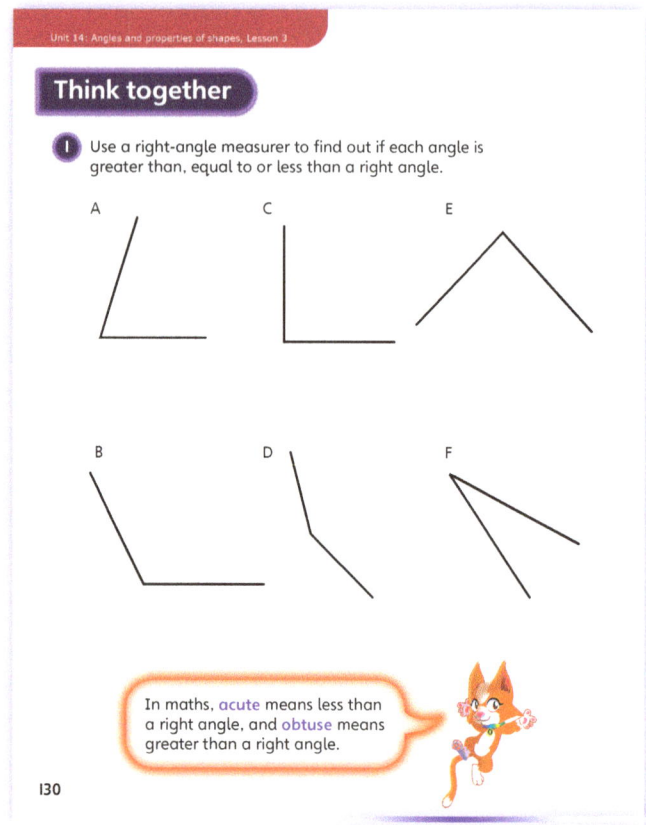

PUPIL TEXTBOOK 3C PAGE 130

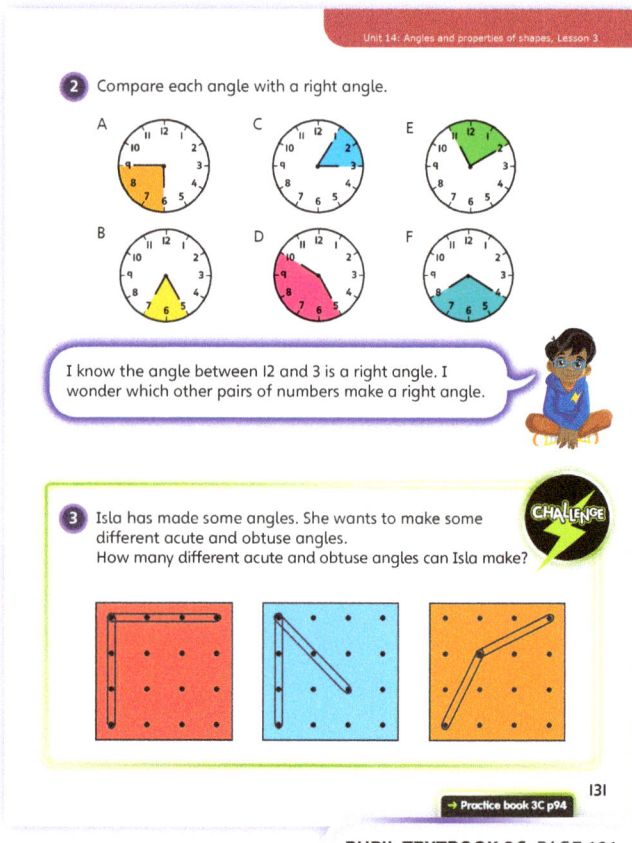

PUPIL TEXTBOOK 3C PAGE 131

Unit 14: Angles and properties of shapes, Lesson 3

Practice

WAYS OF WORKING Independent thinking

IN FOCUS The focus in questions ① to ④ is on recognising and drawing angles that are less than, greater than or equal to a right angle.

Question ⑤ asks children to make predictions about acute and obtuse angles and to justify their reasoning. You might want to start by asking children to count all the angles they can see inside the rectangle, and making sure they can see them all. There are 26 in total.

STRENGTHEN Encourage children to demonstrate acute and obtuse angles as turns, using pipe cleaners bent into angle shapes or clock faces with movable hands.

DEEPEN Challenge children to make generalisations about the angles between numbers on a clock face. For example: between consecutive/alternate numbers the angles are acute; between three numbers (1 and 4, 2 and 5 and so on) the angle is a right angle; between more than three numbers the angle is obtuse; opposite numbers are on a straight line (or a half turn).

THINK DIFFERENTLY Question ④ will require children to form angles in various orientations to ensure they draw three different angles of each type that are not simply rotations or reflections.

ASSESSMENT CHECKPOINT Children should be able to recognise, form or draw angles that are greater than, less than or equal to a right angle and explain the terms acute and obtuse in relation to a right angle.

ANSWERS Answers to the **Practice** part of the lesson can be found in the *Power Maths* online subscription.

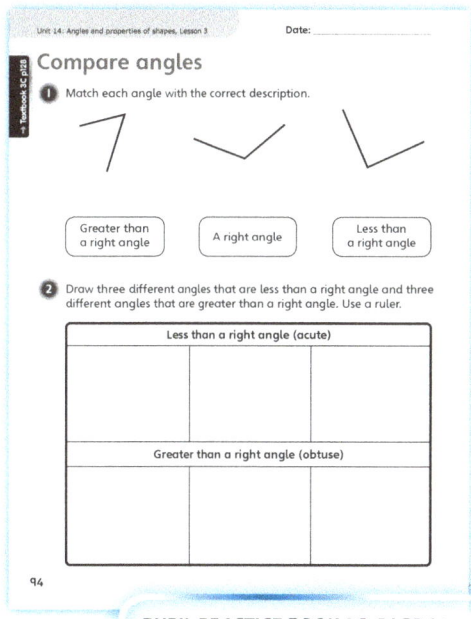

PUPIL PRACTICE BOOK 3C PAGE 94

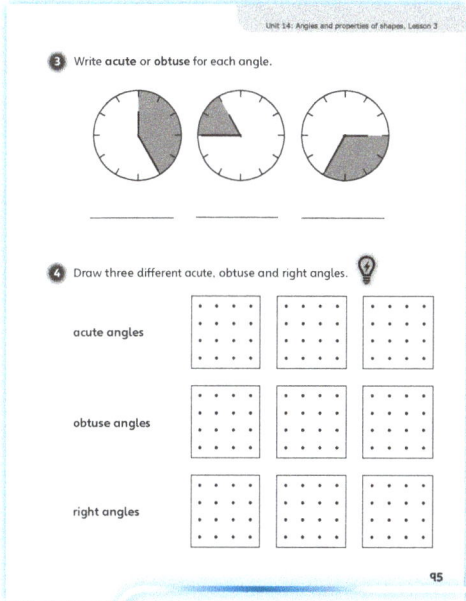

PUPIL PRACTICE BOOK 3C PAGE 95

Reflect

WAYS OF WORKING Independent thinking

IN FOCUS This section asks children to explore their understanding of angles in the school environment. Some children may look around their classroom, while others may visualise angles that are in another part of the school.

ASSESSMENT CHECKPOINT Can children justify their suggestions by comparing the angles with a right angle?

ANSWERS Answers to the **Reflect** part of the lesson can be found in the *Power Maths* online subscription.

After the lesson ⏸

- Can children recognise angles that are greater than, less than or equal to a right angle in different orientations?
- Do children understand the terms acute and obtuse, explaining them in comparison to a right angle?

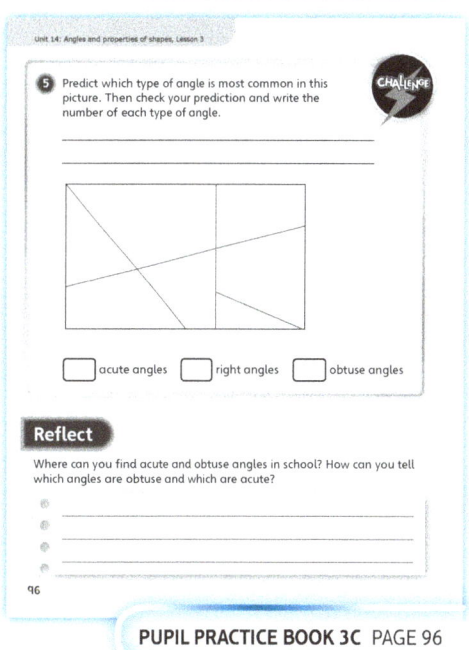

PUPIL PRACTICE BOOK 3C PAGE 96

Unit 14: Angles and properties of shapes, Lesson 4

Measure and draw accurately

Learning focus
In this lesson, children develop their ability to draw and measure accurately in centimetres and millimetres, and apply this to reasoning about 2D shapes.

Before you teach
- Do children know any common mistakes made when measuring with a ruler?
- Can children explain the difference between centimetres and millimetres?
- Can children draw a line to a specific length – for example, 3 cm 5 mm long?

NATIONAL CURRICULUM LINKS

Year 3 Geometry – properties of shapes

Draw 2D shapes and make 3D shapes using modelling materials; recognise 3D shapes in different orientations and describe them.

Identify horizontal and vertical lines and pairs of perpendicular and parallel lines.

ASSESSING MASTERY

Children can measure and draw accurately in centimetres and millimetres and use this skill to form and measure 2D shapes of given dimensions.

COMMON MISCONCEPTIONS

Children may not appreciate the importance of accuracy when drawing and measuring to form specific shapes. Ask:
- *Why is it important to measure the lines accurately when drawing a square?*

Children may not have mastered the ruler skills required. Ask:
- *Are you holding the ruler correctly to measure this line? Can you explain the mistake?*

STRENGTHENING UNDERSTANDING

Help children to measure single lines, and to mark accurately, by supporting their motor skills where necessary. For example, they need to hold the ruler still, once they have positioned it correctly.

GOING DEEPER

Challenge children to draw squares and rectangles of any given dimensions given as whole centimetres and millimetres.

KEY LANGUAGE

In lesson: measure, measurement, accurately, ruler, centimetre (cm), millimetre (mm), wide, width, length, predict, check, square, diagonal, opposite, corner

Other language to be used by the teacher: wide, long, rectangle, horizontal, vertical

STRUCTURES AND REPRESENTATIONS

2D shapes

RESOURCES

Mandatory: ruler, scissors, landscape A4 paper cut into 10 cm strips, squared paper

Optional: pre-cut 10 cm squares, plastic/wooden squares and rectangles

 In the eTextbook of this lesson, you will find interactive links to a selection of teaching tools.

Quick recap
Ask children to draw and measure lines using a ruler. Alternatively, they could measure the length of a pencil or book instead of drawing lines. Check that children can use a ruler correctly and line up the start of the line or object with 0.

Unit 14: Angles and properties of shapes, Lesson 4

Discover

WAYS OF WORKING Pair work

ASK
- Question 1 a): *Where should you place the ruler to start the measurement?*
- Question 1 a): *How do you measure exactly 10 cm?*
- Question 1 a): *How many whole 10 cm squares can be made?*

IN FOCUS Question 1 a) further develops the skill of measuring accurately (from Unit 7). Here, children are challenged to follow instructions to form a square, something that requires accurate measurement.

PRACTICAL TIPS The dimensions of the paper strips are designed to match the width of an A4 piece of paper presented horizontally. Show a piece of A4 paper held horizontally and explain that what is shown in the Textbook is a 10 cm strip cut from this size paper. Cut out an approximate strip of depth 10 cm to demonstrate. Then give each child an accurately cut 10 cm deep strip of A4 paper. If the mixed units of 29 cm 7 mm is a distraction for children, then the strips could be pre-cut to a width of 25 cm, for example, and the instructions adapted accordingly.

ANSWERS

Question 1 a): Two 10 cm squares can be cut from the piece of paper.

Question 1 b): The piece of paper left over will be a rectangle because it is 10 cm tall and 9 cm 7 mm wide.

PUPIL TEXTBOOK 3C PAGE 132

Share

WAYS OF WORKING Whole class teacher led

ASK
- Question 1 a): *Why do you need to measure and make a mark at the top and the bottom?*
- Question 1 a): *How can you make sure your square is 10 cm along each side?*
- Question 1 b): *How will you find out how wide the left-over piece is?*

IN FOCUS This shows children a method for measuring accurately: make two opposing marks and then join them. This skill is transferable to other craft or design activities.

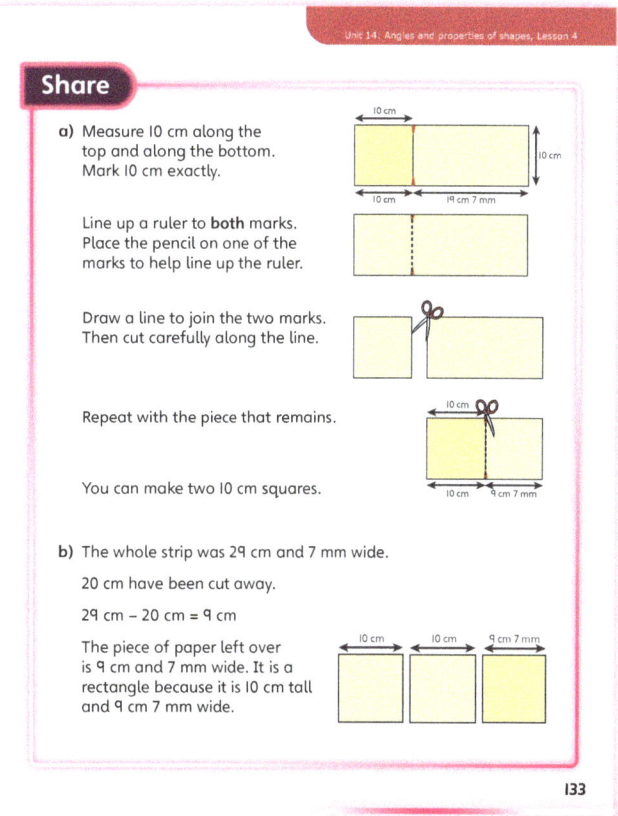

PUPIL TEXTBOOK 3C PAGE 133

Unit 14: Angles and properties of shapes, Lesson 4

Think together

WAYS OF WORKING Whole class teacher led (I do, We do, You do)

ASK
- Question 1 a): *Which line will you measure and draw first?*
- Question 1 b): *Which line will you cut first?*
- Question 1 b): *Do you need to cut out each shape separately?*
- Question 3: *Is Astrid correct?*

IN FOCUS Question 1 focuses on accurate measuring. Both parts use one of the 10 cm squares that were made previously, so it is important that these are accurate. If necessary, provide pre-cut 10 cm squares.

Question 3 provides an opportunity for reasoning about the relative lengths of diagonals of squares.

STRENGTHEN Support children with their motor skills where necessary. Children should focus on the skill of measuring accurately, but may need someone to hold the ruler while they draw lines to join points. When using squared paper, encourage children to keep to the vertical and horizontal grid lines for relevant sides.

DEEPEN Challenge children to draw squares on squared paper to explore whether both the diagonals of a square are the same length or different lengths and whether the diagonals are always longer than the sides. Children could also explore diagonals in rectangles.

ASSESSMENT CHECKPOINT Children should realise that accurate drawing and measuring is an important skill. They should be able to use a ruler correctly to make the necessary markers to draw and measure the length of the sides of squares and rectangles accurately.

ANSWERS

Question 1 a): Children should draw a 3 cm × 3 cm square on their 10 cm square and then cut out the smaller square and the rectangles that it leaves.

Question 1 b): The side lengths of the smaller square are both 3 cm; the side lengths of the larger square are both 7 cm; the side lengths of the two rectangles are 3 cm and 7 cm in each case.

Question 2: A = 3 cm × 4 cm × 5 cm
The third side measures 5 cm.
B = 3 cm × 2 cm × 3·6 cm (or 36 mm)
The third side measures 3·6 cm or 36 mm.

Question 3: Check children's predictions.
Actual diagonal lengths are:
A = approximately 42 mm
B = approximately 99 mm

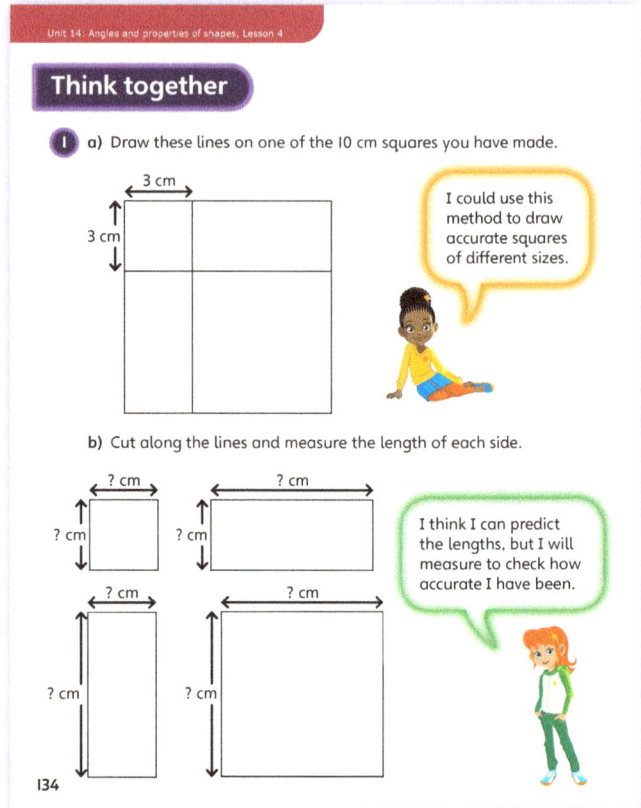

PUPIL TEXTBOOK 3C PAGE 134

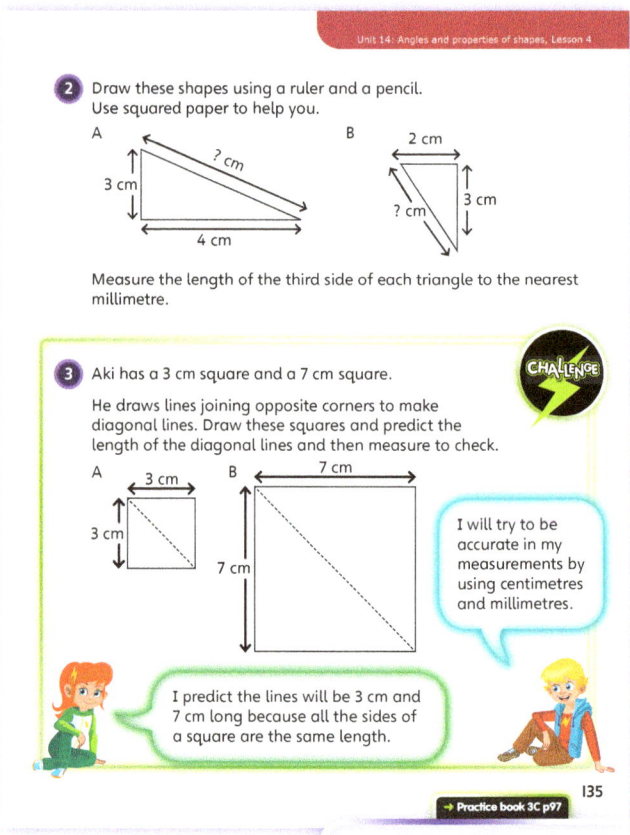

PUPIL TEXTBOOK 3C PAGE 135

Unit 14: Angles and properties of shapes, Lesson 4

Practice

WAYS OF WORKING Independent thinking

IN FOCUS Questions 1 to 3 focus on measuring and drawing in centimetres and millimetres.

STRENGTHEN Build children's confidence by asking them to measure the side lengths of shapes whose sides are a whole number of centimetres. To measure the side lengths of a particular shape, you could give children a stencil of the shape and ask them to draw around it. Suggest children mark a dot at each corner of the shape, to prevent them measuring from the wrong mark. They align the zero point of their ruler next to one dot, and read the length of the side by looking at where the other dot meets their ruler. Children could measure the lengths of wooden or plastic squares and rectangles for further practice.

DEEPEN Question 3 challenges children to enhance their accuracy in measuring. Question 4 challenges children to form smaller shapes within a larger square, and then to measure them accurately.

For further practice, ask children to measure 2D squares and rectangles and then make accurate drawings of them. They could also explore right-angled triangles. Can children explain why they do not need to measure the diagonal side before they draw it?

THINK DIFFERENTLY In question 3, children will need to think about where to start for these diagrams, because they will find that they cannot always begin with the vertical/horizontal from the left. If the copies are not accurate, encourage children to explain why they think this has happened.

ASSESSMENT CHECKPOINT Children should realise that accurate drawing and measuring is an important skill. They are able to use a ruler correctly to make the necessary markers to draw and measure the length of the sides of squares and rectangles accurately.

ANSWERS Answers to the **Practice** part of the lesson can be found in the *Power Maths* online subscription.

Reflect

WAYS OF WORKING Independent thinking

IN FOCUS Children need to use the correct words to break the skill down into its key steps.

ASSESSMENT CHECKPOINT Do children's explanations take into account common errors, such as letting the ruler slip or measuring from the wrong mark?

ANSWERS Answers to the **Reflect** part of the lesson can be found in the *Power Maths* online subscription.

After the lesson

- Can children measure and join two opposing marks to form a given line?
- Are children able to apply their understanding of millimetres and centimetres to drawing 2D shapes?
- Where children are unsuccessful, what is causing the inaccuracies (measuring, using a ruler or a different reason)?

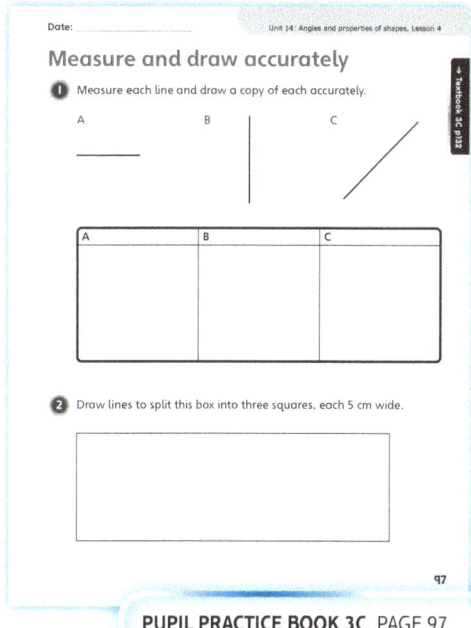

PUPIL PRACTICE BOOK 3C PAGE 97

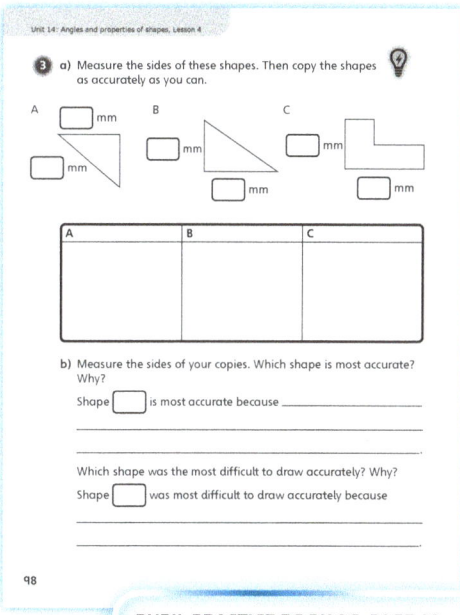

PUPIL PRACTICE BOOK 3C PAGE 98

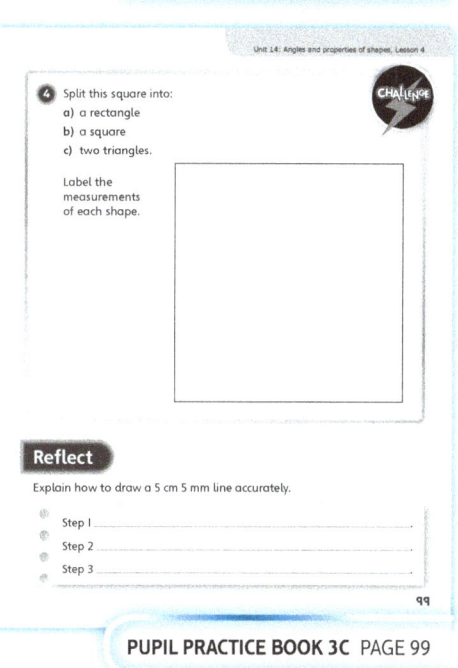

PUPIL PRACTICE BOOK 3C PAGE 99

Unit 14: Angles and properties of shapes, Lesson 5

Horizontal and vertical

Learning focus
In this lesson, children learn to identify and draw horizontal and vertical lines.

Before you teach
- Do children know what it means to sit up straight?
- Do children know what the horizon is?

NATIONAL CURRICULUM LINKS

Year 3 Geometry – properties of shapes

Identify horizontal and vertical lines and pairs of perpendicular and parallel lines.

ASSESSING MASTERY

Children can identify horizontal and vertical lines, construct horizontal and vertical lines, and recognise where lines are neither vertical nor horizontal. They are able to explain the relevance of horizontal and vertical lines in examples relating to their environment.

COMMON MISCONCEPTIONS

Children may find it difficult to remember which term relates to which property. **Horizon**tal is flat, like the **horizon**; **vert**ical stands up straight like **vert**ebrae in your back. Children may use the terms 'flat' or 'straight' instead. Ask:
- *How could you describe the two types of line without using the words 'horizontal' or 'vertical'?*
- *Can a straight line be neither horizontal nor vertical?*

STRENGTHENING UNDERSTANDING

Children could use a range of equipment, such as string, metre sticks or PE benches, to explore the concepts practically. They could try to balance a ball or marble on a table or a plank to keep it still (either horizontally or vertically). Discuss words which have the same roots.

GOING DEEPER

Provide a set of horizontal and non-horizontal lines that have a base line (similar to the ground in the **Discover** picture). Ask children to label those that are horizontal; and then to measure using a ruler those that are not horizontal to see how far one end needs to be moved in order to make it horizontal.

KEY LANGUAGE

In lesson: horizontal, vertical, straight, lower, higher, height, level, raised, lowered, right angle, line of symmetry, mirror line

Other language to be used by the teacher: plumb line, balanced, metre (m), centimetre (cm)

STRUCTURES AND REPRESENTATIONS

2D shapes

RESOURCES

Mandatory: squared paper

Optional: plumb line, PE equipment, marbles, mirrors, clear pictures of a flat and straight horizon

 In the eTextbook of this lesson, you will find interactive links to a selection of teaching tools.

Quick recap

Ask children to draw five different lines, each 10 cm long. Ask them to draw some on an angle, one that goes straight across the page and one that goes straight up.

Discover

WAYS OF WORKING Pair work

ASK

• Question 1 a): *What is the same and what is different about the shelves?*
• Question 1 a): *Why are shelves designed to be flat (that is, horizontal)?*
• Question 1 b): *How could the shelf be adjusted to make it work better?*

IN FOCUS Children will need to compare the shelves in the picture, and begin to search for an accurate way to describe the type of line they will come to know as horizontal.

PRACTICAL TIPS This context could be adapted in a number of ways. Children could explore horizontal surfaces by looking at when a ball balances motionless on a plank, or they could try walking around with a tennis ball balanced on a racket.

Alternatively, children could look at shelves in different parts of the school and judge why they are all level. Show children pictures of the horizon where there is a clear straight and level line between sea and sky, or land and sky; explain that this is called the horizon, which can help you remember that the word horizontal means level and flat, left to right (or right to left).

ANSWERS

Question 1 a): One of the shelves is not horizontal. It is not the same height all the way along.

Question 1 b): The shelf can be fixed by lowering the right-hand side to 1 m 50 cm.

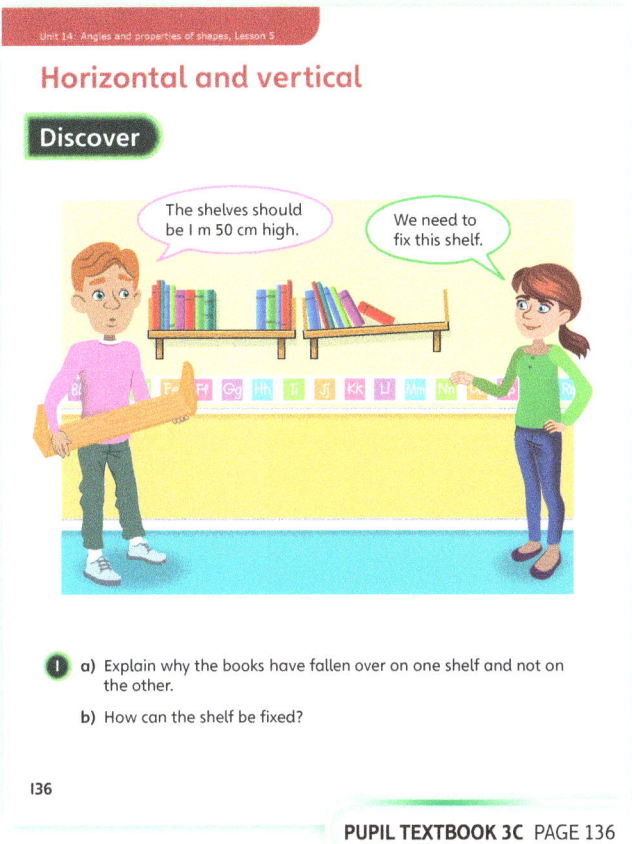

PUPIL TEXTBOOK 3C PAGE 136

Share

WAYS OF WORKING Whole class teacher led

ASK

• Question 1 a): *How do you know that one shelf is not horizontal?*
• Question 1 b): *In how many different ways could the shelf be fixed?*

IN FOCUS This might be the first time children have used the word horizontal. They may need to explore its meaning in terms of their experience of other situations relating to how level something is (for example, balancing on a beam, climbing a hill, rolling down a hill or riding a bike). In question 1 b), you might want to explore the two ways of fixing the crooked shelf: either lowering the right-hand end or raising the left-hand end.

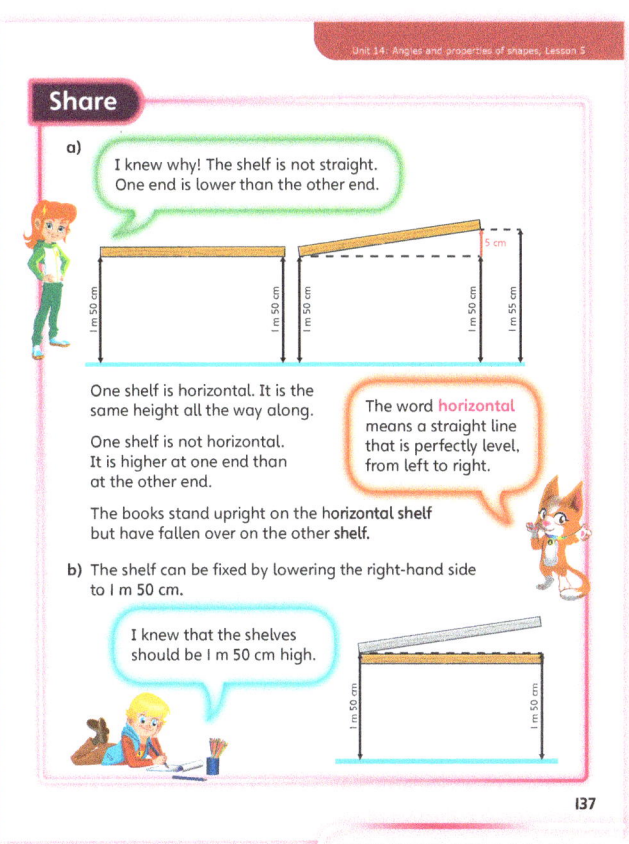

PUPIL TEXTBOOK 3C PAGE 137

175

Unit 14: Angles and properties of shapes, Lesson 5

Think together

WAYS OF WORKING Whole class teacher led (I do, We do, You do)

ASK
- Question ❶: *What do horizontal and vertical mean?*
- Question ❶: *Which are the horizontal parts of each fence?*
- Question ❶: *Which are the vertical parts of each fence?*
- Question ❷: *What causes a plumb line to hang vertically?*
- Question ❸: *Are lines of symmetry always vertical?*

IN FOCUS Question ❷ demonstrates a method for testing for vertical lines in the environment and deepens children's understanding of vertical as straight down due to gravity.

Question ❸ applies the concept of vertical and horizontal lines to the context of symmetry in simple polygons. Note that if shapes were turned a quarter turn, then the vertical symmetry lines would become horizontal, and vice versa. This is picked up in **Practice** question ❺.

STRENGTHEN Explore the language of straight across and straight down in physical terms, by dropping or rolling a ball and by balancing objects.

DEEPEN Challenge children to explore the relationship between gravity and our understanding of horizontal and vertical lines. Ask children to find examples around school where objects that should be vertical or horizontal are not (for example, a fence falling down, a bent netball post, a picture hung crookedly) and where objects are deliberately *not* vertical or horizontal for a particular purpose (ramps, for example).

ASSESSMENT CHECKPOINT If children are able to recognise both horizontal and vertical lines of symmetry in question ❸ and choose the correct term (vertical or horizontal) to describe the line of symmetry, then they will show good understanding.

ANSWERS

Question ❶: Fences A and B both have vertical posts, but fence B does not have horizontal posts. Fences A and C both have horizontal posts but fence C does not have vertical posts. A is the only fence with both horizontal and vertical posts.

Question ❷: Children should correctly use a plumb line to check whether something in the classroom is vertical.

Question ❸ a): A is symmetrical. It has a horizontal line of symmetry.
B is symmetrical. It has a horizontal and a vertical line of symmetry.
C is not symmetrical.
D is symmetrical. It has a horizontal and a vertical line of symmetry.
E is symmetrical. It has a horizontal line of symmetry.
F is symmetrical. It has a horizontal line of symmetry.

Question ❸ b): Children should accurately draw some more shapes with a horizontal line of symmetry.

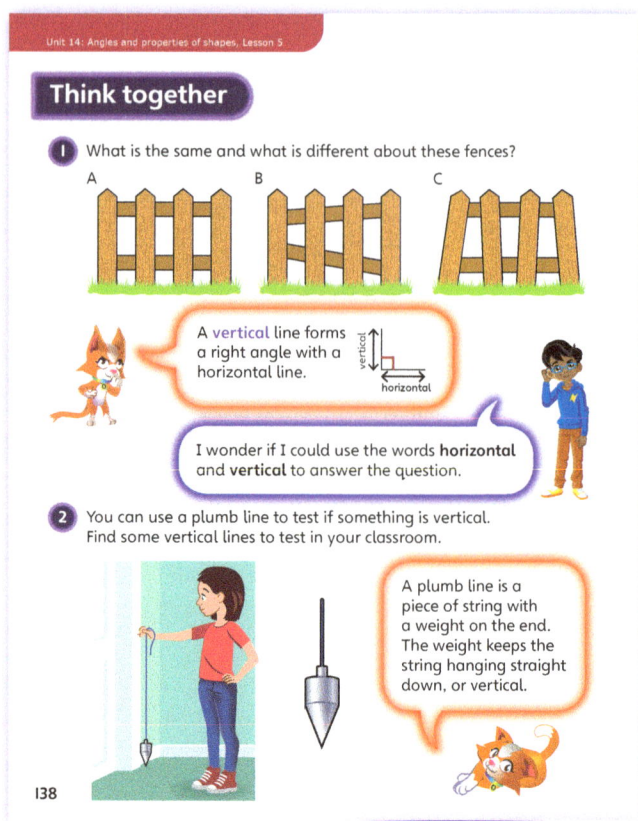

PUPIL TEXTBOOK 3C PAGE 138

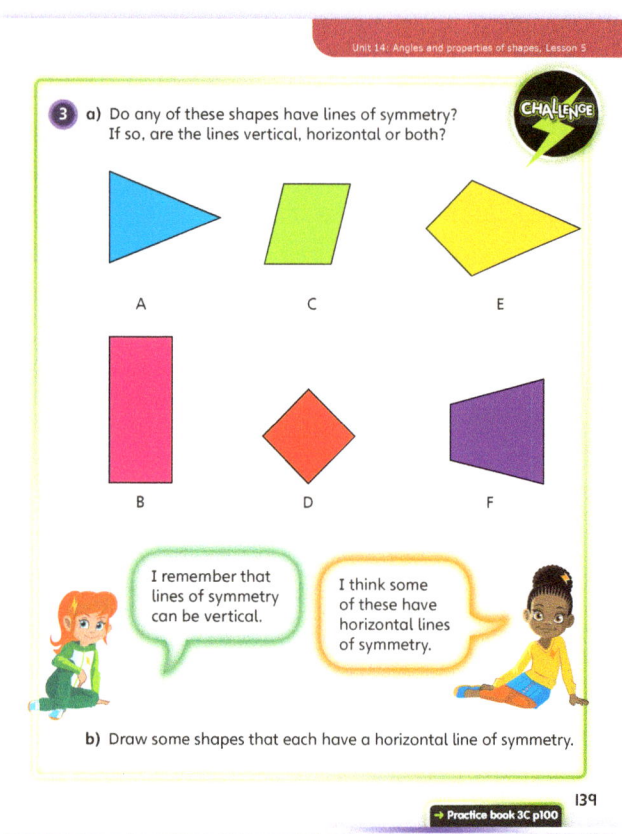

PUPIL TEXTBOOK 3C PAGE 139

Unit 14: Angles and properties of shapes, Lesson 5

Practice

WAYS OF WORKING Independent thinking

IN FOCUS Questions ❶ and ❷ are about recognising horizontal and vertical lines, and identifying lines that are neither.

Question ❸ requires children to draw the given lines.

Question ❹ looks for vertical and horizontal lines of symmetry.

Question ❻ challenges children to use their measuring skills to show whether or not the lines are vertical or horizontal. Children could predict and then use measuring to test their predictions.

STRENGTHEN Explore horizontal and vertical lines with practical equipment such as string metre rules – or by aligning PE benches.

DEEPEN Challenge children to apply their measuring skills to prove which lines are, or are not, vertical or horizontal by measuring from the top and bottom and from the left and right. Ask children to design artwork with vertical and horizontal lines and some lines that are neither horizontal nor vertical.

THINK DIFFERENTLY Question ❺ challenges children to think back to question ❸ in **Think together** on right-angle turns to change the mirror line from a horizontal to a vertical, or vertical to horizontal, by turning the shape a right-angle quarter turn clockwise or anticlockwise.

ASSESSMENT CHECKPOINT Children should now be able to draw and identify lines that are vertical, horizontal or neither and use this skill to describe lines of symmetry in simple polygons.

ANSWERS Answers to the **Practice** part of the lesson can be found in the *Power Maths* online subscription.

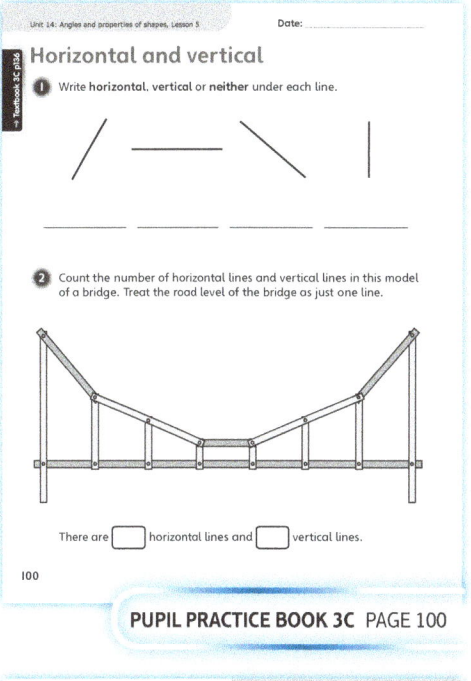

PUPIL PRACTICE BOOK 3C PAGE 100

PUPIL PRACTICE BOOK 3C PAGE 101

Reflect

WAYS OF WORKING Independent thinking

IN FOCUS Children should discuss and justify their ideas, based on reasoning about the physical properties of horizontal and vertical lines and surfaces. They could consider where it is necessary, helpful or not important for items to be vertical or horizontal.

ASSESSMENT CHECKPOINT Do children understand what the words 'horizontal' and 'vertical' mean, and can they give examples of horizontal and vertical lines that they see in everyday life? For example, horizontal lines include: surface of a table, book shelf, towel rail; vertical lines include: lamp post, railings, wall of a house.

ANSWERS Answers to the **Reflect** part of the lesson can be found in the *Power Maths* online subscription.

PUPIL PRACTICE BOOK 3C PAGE 102

After the lesson

- Have children developed a good understanding of the two types of line in relation to everyday experience?
- Can children apply their measuring skills to construct and identify horizontal and vertical lines?

177

Unit 14: Angles and properties of shapes, Lesson 6

Parallel and perpendicular

Learning focus
In this lesson, children learn to identify and construct parallel and perpendicular lines.

Before you teach
- Can children use their hands or arms to show a right angle?
- Do children know how to measure the distance between two lines accurately?

NATIONAL CURRICULUM LINKS

Year 3 Geometry – properties of shapes

Identify horizontal and vertical lines and pairs of perpendicular and parallel lines.

ASSESSING MASTERY

Children can explain that parallel lines are a constant distance apart and, even if the lines continued indefinitely, they would never cross. Children can construct parallel lines that meet this property. They also understand that perpendicular lines intersect at right angles and can construct a range of lines perpendicular to another line.

COMMON MISCONCEPTIONS

Children may assume that parallel lines must be of identical length. Ask:
- *How can you tell if these two lines are parallel?*

Children may assume that any two lines which do not intersect are parallel. Ask:
- *Would these two lines cross if they were continued?*

Children may confuse the two terms. Explain:
- *The word parallel has the letter 'l' twice in the middle, which itself shows a pair of parallel lines.*

STRENGTHENING UNDERSTANDING

Explore parallel and perpendicular lines through drawing, observing the environment around school, art activities such as weaving, and looking at pieces of modern art (for example, the work of Piet Mondrian).

GOING DEEPER

Can children identify parallel lines in a variety of contexts? For example, they may reason that because opposite sides of a rectangle are the same length, a rectangle has two pairs of parallel lines. Alternatively, children may be able to use dots on grid paper to identify any number of parallel lines (including non-horizontal and non-vertical parallel lines).

KEY LANGUAGE

In lesson: parallel, perpendicular, right angle, distance, ruler, concertina, angle, sign, describe, diagram

Other language to be used by the teacher: measure, identical length, constant distance, construct, intersect, extend, extended

RESOURCES

Mandatory: ruler, paper for folding, square paper, square dotted paper

Optional: examples of modern art that contain parallel and perpendicular lines (Piet Mondrian), circles (see Practice Book, question ⑤) with 3–10 dots

 In the eTextbook of this lesson, you will find interactive links to a selection of teaching tools.

Quick recap

Ask children to draw a horizontal line 15 cm long and a vertical line 10 cm long. Check children understand the words vertical and horizontal and can use a ruler accurately.

Unit 14: Angles and properties of shapes, Lesson 6

Discover

WAYS OF WORKING Pair work

ASK
- Question ❶: *How can the folds be made accurately?*
- Question ❷: *What sorts of lines will the creases make?*

IN FOCUS Children will need to visualise what the creases will look like, before folding and checking for themselves. There may be a need for a discussion around why there could be differences – for example, if some people fold more accurately than others.

PRACTICAL TIPS Children could fold their own sheets of scrap paper, trying to match the concertina Isla is folding. Challenge children to line up the bottom edge with the first fold so that the lines are equally spaced.

Alternatively, children could explore the situation by lining up pencils so they are all facing the same direction, or placing cones in a line so that they form stripes that would never cross.

ANSWERS

Question ❶ a): Isla's piece of paper will have parallel lines made by the folds.

Question ❶ b): Max's piece of paper will have the same parallel lines as Isla's but also a perpendicular line where he folded his paper in half.

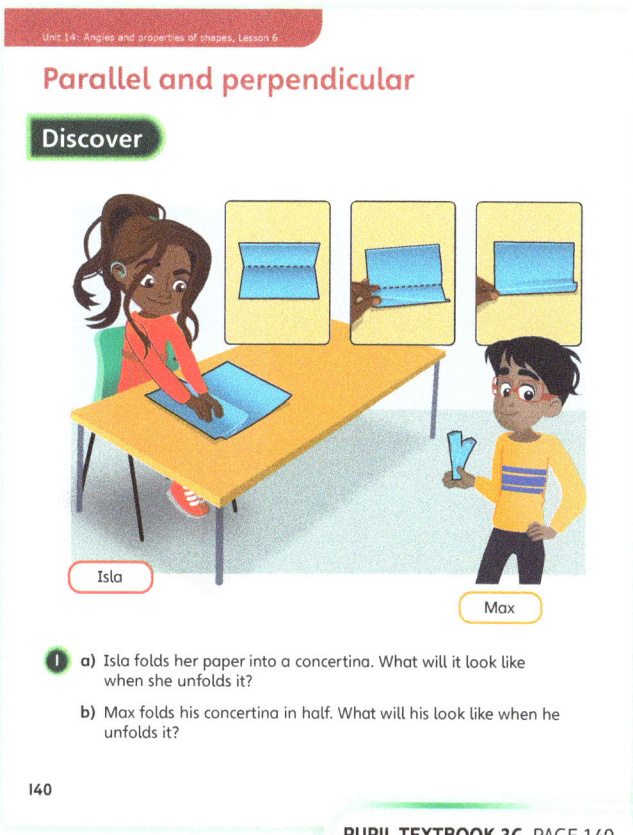

PUPIL TEXTBOOK 3C PAGE 140

Share

WAYS OF WORKING Whole class teacher led

ASK
- Question ❶ a): *How could you use measuring to check if your creases are parallel?*
- Question ❶ a): *Would these creases still be parallel if you turned the page a quarter turn?*
- Question ❶ b): *How many horizontal/vertical lines will there be when the paper is opened up?*
- Question ❶ b): *At what angle do the lines on Max's paper cross?*

IN FOCUS Children will need to understand that parallel lines are a constant distance apart and would never intersect (cross over), even if extended indefinitely.

Children will also need to understand that perpendicular lines are related to right angles. They either *meet* at a right angle (more often found in 2D shapes) or *cross over* each other at a right angle.

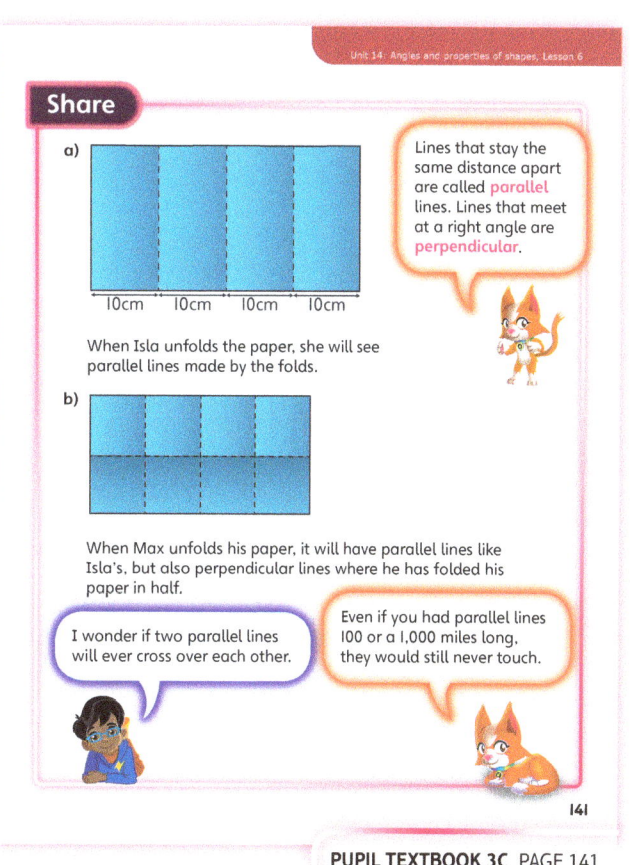

PUPIL TEXTBOOK 3C PAGE 141

179

Think together

WAYS OF WORKING Whole class teacher led (I do, We do, You do)

ASK
- Question ❶: *What types of line can you see in this picture?*
- Question ❶: *How do you think the artist created this picture?*
- Question ❸: *How can you tell if lines are parallel or perpendicular?*

IN FOCUS Question ❶ shows vertical and horizontal parallel lines, which then form many pairs of perpendicular lines. By discussing this picture, children will learn that: pairs of parallel lines can be at different distances apart from other pairs; parallel lines can be vertical or horizontal; perpendicular lines can join in the middle of a line, not just at the ends; and perpendicular lines can cross.

Question ❸ a) addresses several misconceptions: sometimes children think that: parallel lines must be the same length; if lines do not actually cross then they must be parallel; and lines actually have to cross to be perpendicular.

STRENGTHEN Explore parallel lines in different orientations and of different lengths through drawing or lining up strips of paper or straight objects. In question ❸ a), ask children to copy the lines onto squared paper, or use a pair of rulers or rods on the page to continue the lines as far as is necessary, to see if the pairs are parallel, perpendicular or neither.

DEEPEN Question ❸ b) challenges children to use the grid to justify how to draw lines parallel to diagonal lines. Can children spot how to do this by either using the diagonals of the squares or by using the ratio of squares up to squares across?

ASSESSMENT CHECKPOINT Children should be able to use the images in this lesson to explain what they understand about parallel and perpendicular lines and know which word describes each pair of lines. Children should explain that lines can be parallel or perpendicular even when the lines are not vertical or horizontal.

ANSWERS

Question ❶: Children should correctly point to the parallel and perpendicular lines in the picture.

Question ❷: Children should accurately draw parallel lines by drawing along each side of a ruler. These lines are parallel because they are the same distance apart all the way along.

Question ❸ a): Children should disagree. The pairs of lines in A and B are parallel because they are the same distance apart all the way along. The lines in C and D are not parallel because, if they were extended, they would meet.

Question ❸ b): Children should use the grid lines to make sure the lines they draw are always the same distance from the other line. These will then be parallel.

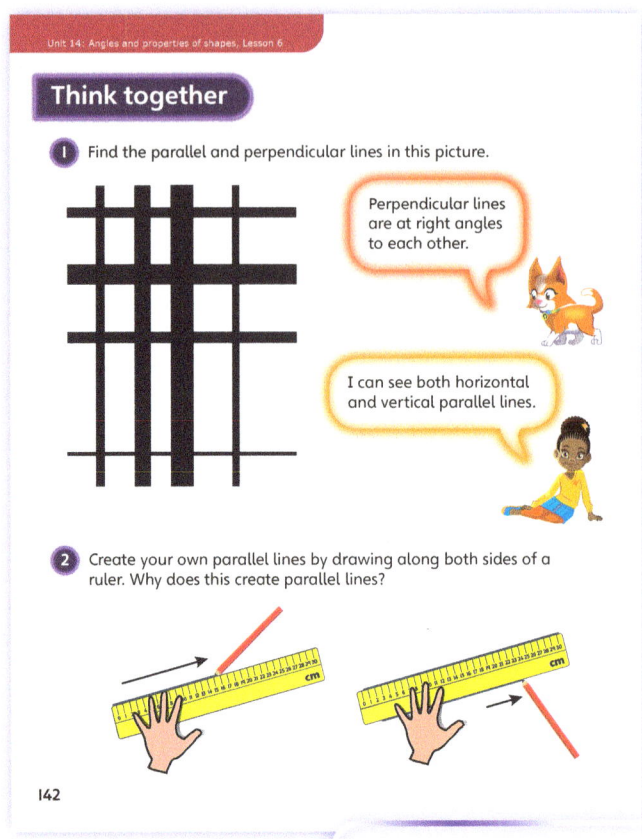

PUPIL TEXTBOOK 3C PAGE 142

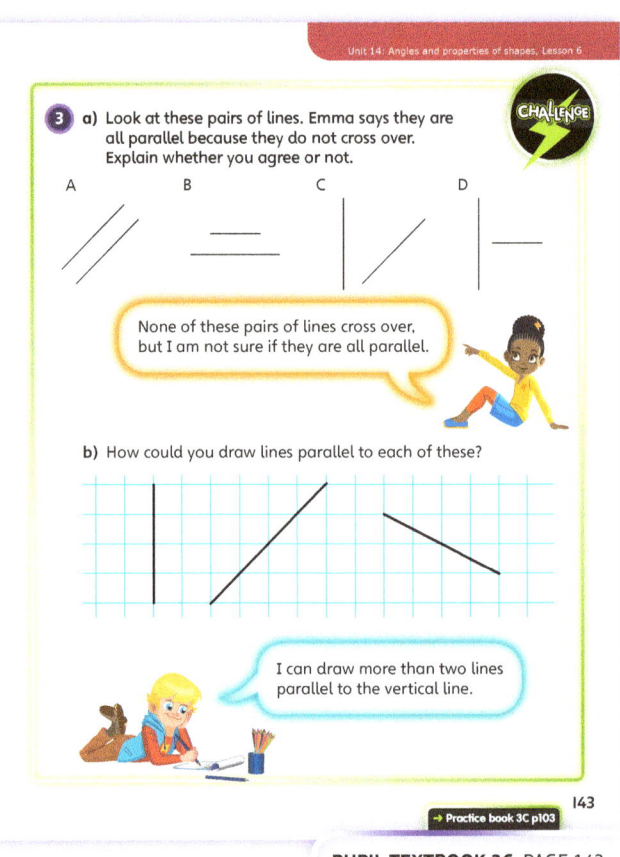

PUPIL TEXTBOOK 3C PAGE 143

Unit 14: Angles and properties of shapes, Lesson 6

Practice

WAYS OF WORKING Independent thinking

IN FOCUS Question ❶ asks children to demonstrate their understanding of parallel lines and in question ❷, they have to draw perpendicular lines. Children should use a ruler to support them. Look for the different methods they may use.

Question ❸ focuses on recognising parallel lines. In question ❹, in each part, children are asked to draw another line to make a pair of parallel lines.

STRENGTHEN Explore parallel and perpendicular lines by using physical equipment (such as rods or rulers of the same and of different lengths), or walking along the lines of a football pitch or a netball court, or lying down to make a pair of parallel or perpendicular lines with a partner.

DEEPEN Question ❺ challenges children to explore parallel and perpendicular lines by joining dots around a circle. Challenge children to find as many possibilities as they can. Children could also explore circles with three, four, five, seven, nine and ten dots.

ASSESSMENT CHECKPOINT Use questions ❶, ❸ and ❹ to assess whether children are confident to identify and draw parallel lines, including non-horizontal and non-vertical parallel lines. They will need to use a ruler to draw the first line, and mark at least three points that are equidistant in order to draw the second line. Use question ❷ to assess whether children are able to draw a pair of perpendicular lines using a right-angle measurer, and mark the right angle on them.

ANSWERS Answers to the **Practice** part of the lesson can be found in the *Power Maths* online subscription.

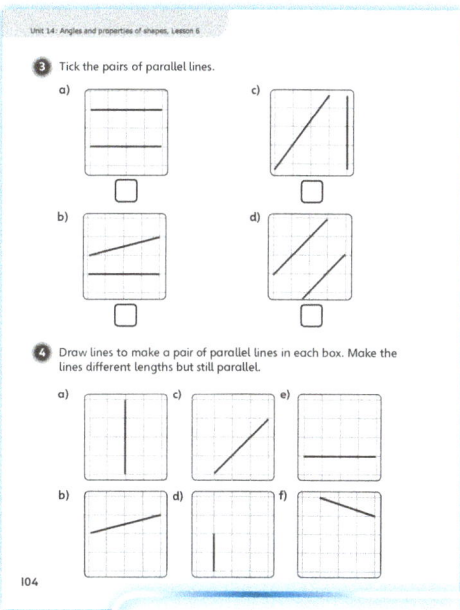

PUPIL PRACTICE BOOK 3C PAGE 103

PUPIL PRACTICE BOOK 3C PAGE 104

Reflect

WAYS OF WORKING Pair work

IN FOCUS Children should write their own responses and then compare these with those of their partner. Children could then try to justify their comments, or adopt their partner's reasoning if convinced by it.

ASSESSMENT CHECKPOINT Do children's responses demonstrate a clear understanding of the fundamental properties of both types of line and address some of the misconceptions?

ANSWERS Answers to the **Reflect** part of the lesson can be found in the *Power Maths* online subscription.

After the lesson ⏸

- Are children able to explain clearly the difference between parallel and perpendicular lines?
- Can children construct both kinds of line with reasonable accuracy, and justify their method?
- What further opportunities will you give children to reinforce the language covered in this lesson?

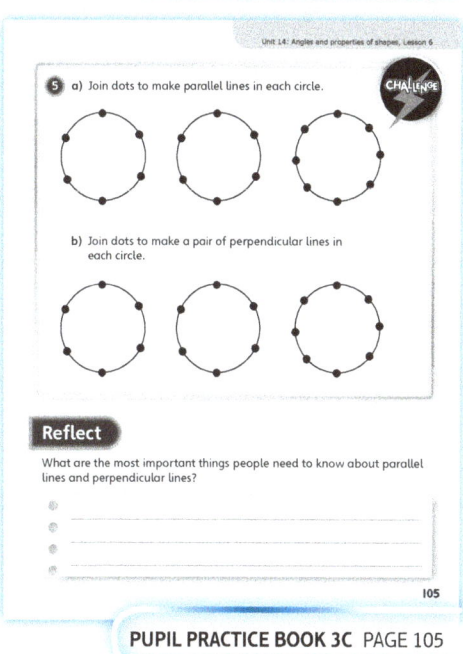

PUPIL PRACTICE BOOK 3C PAGE 105

181

Unit 14: Angles and properties of shapes, Lesson 7

Recognise, draw and describe 2D shapes

Learning focus
In this lesson, children apply their understanding of types of line and angle to the properties of 2D shapes.

Before you teach
- Can children draw four different triangles?
- Do children know what is the same and what is different about squares, rectangles and pentagons?

NATIONAL CURRICULUM LINKS

Year 3 Geometry – properties of shapes

Draw 2D shapes and make 3D shapes using modelling materials; recognise 3D shapes in different orientations and describe them.

ASSESSING MASTERY

Children can describe common 2D shapes using their angle and line properties, including an understanding of vertical and horizontal symmetry.

COMMON MISCONCEPTIONS

Children may rely on recognising common representations of shapes and may not recognise a shape in a different orientation, or shapes that do not look like the regular versions they are used to. Ask:
- *What is true of all rectangles? Or hexagons? Or quadrilaterals? How can you check if a shape is a square?*

Children may think that if a shape has two pairs of parallel sides, then it must be a square or a rectangle. Ask:
- *Does this shape* [show a rhombus or a parallelogram] *have two pairs of parallel sides? How can you tell that it is not a rectangle or a square?*

STRENGTHENING UNDERSTANDING

Explore polygons using constructions such as geoboards, stencils or tangible representations of the shape. Trace around 2D shapes by placing them in different orientations, marking the right angles in each and identifying the pairs of parallel or perpendicular sides.

GOING DEEPER

Challenge children to explore the properties of parallel and perpendicular lines in different polygons. Ask leading questions such as: *Do all quadrilaterals have a pair of parallel sides? Can you draw a trapezium with perpendicular sides? Can a triangle have parallel sides? Could the equal angles in a kite be right angles?*

KEY LANGUAGE

In lesson: 2D shapes, quadrilateral, pentagon, parallel, perpendicular, acute angle, line of symmetry

Other language to be used by the teacher: kite, parallelogram, trapezium, rhombus, symmetrical, obtuse angle, right angle, hexagon, polygon

STRUCTURES AND REPRESENTATIONS

2D shapes

RESOURCES

Mandatory: sticks or pencils of equal length

Optional: matchsticks, lolly sticks, base 10 equipment or rectangular shapes, mini whiteboards, plastic or wooden 2D shapes, sorting table

 In the eTextbook of this lesson, you will find interactive links to a selection of teaching tools.

Quick recap

Ask: *What 2D shapes do you know?* Ask children to draw and name as many as they can on their mini whiteboards with the aim of creating a class list of 10 different 2D shapes. Create a board with the shapes on and have 10 children stand at the front of the class, with each person holding their drawn shape on their mini whiteboard.

Unit 14: Angles and properties of shapes, Lesson 7

Discover

WAYS OF WORKING Pair work

ASK
- Question 1 a): *What is a quadrilateral?*
- Question 1 a): *What are parallel lines?*
- Question 1 a): *How can you check whether any of the lines are parallel?*
- Question 1 b): *What could you do to check if this is true for all sizes of this shape?*

IN FOCUS Children are searching the picture for as many properties as they can, based on their understanding of types of 2D shape, types of line and types of angle. Recap on the meaning of the terms parallel, perpendicular, vertical and horizontal, and the names of quadrilaterals. Which properties prove that the shape being studied is a rectangle?

Discuss what is the same and what is different about a range of rectangles to ascertain that all rectangles, including squares, always have two pairs of parallel and equal length sides, and that adjacent sides are perpendicular.

PRACTICAL TIPS This activity could be re-created as part of a PE lesson. Alternatively, the shapes could be modelled using matchsticks, lolly sticks or base 10 equipment or rectangular shapes placed end to end.

ANSWERS

Question 1 a): The children have made a rectangle. It has two pairs of parallel sides.

Question 1 b): All rectangles, including squares, have two pairs of parallel lines.

Share

WAYS OF WORKING Whole class teacher led

ASK
- Question 1 a): *Can you explain why a rectangle is a type of quadrilateral?*
- Question 1 b): *Would the opposite sides still be parallel if you turned the rectangle?*
- Question 1 b): *Would the opposite sides still be parallel if the rectangles were smaller or larger?*

IN FOCUS Children are exploring the idea that a rectangle must always have two pairs of parallel lines (sides). The justification is that opposite sides are of equal length, so the lines joining them must be a constant distance apart. It may also be worth discussing that, as all the angles are right angles, the adjacent sides are perpendicular.

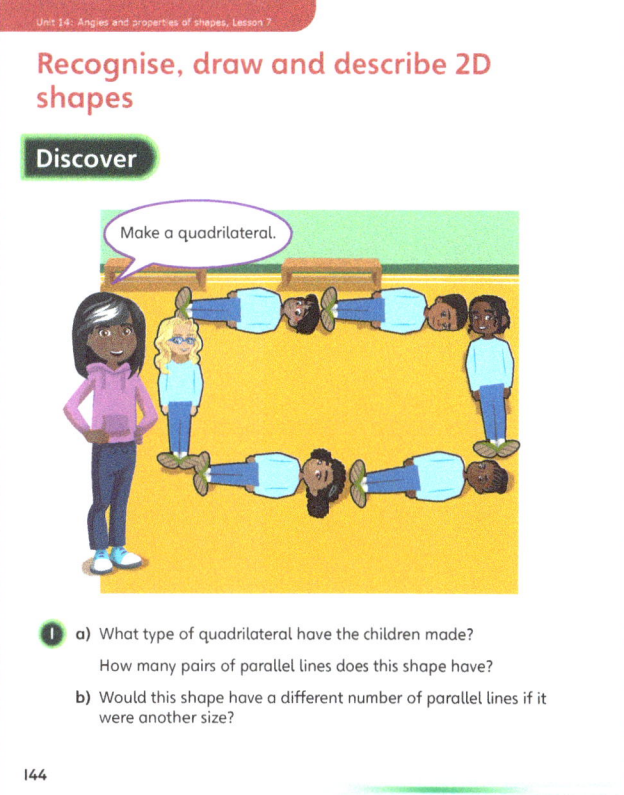

PUPIL TEXTBOOK 3C PAGE 144

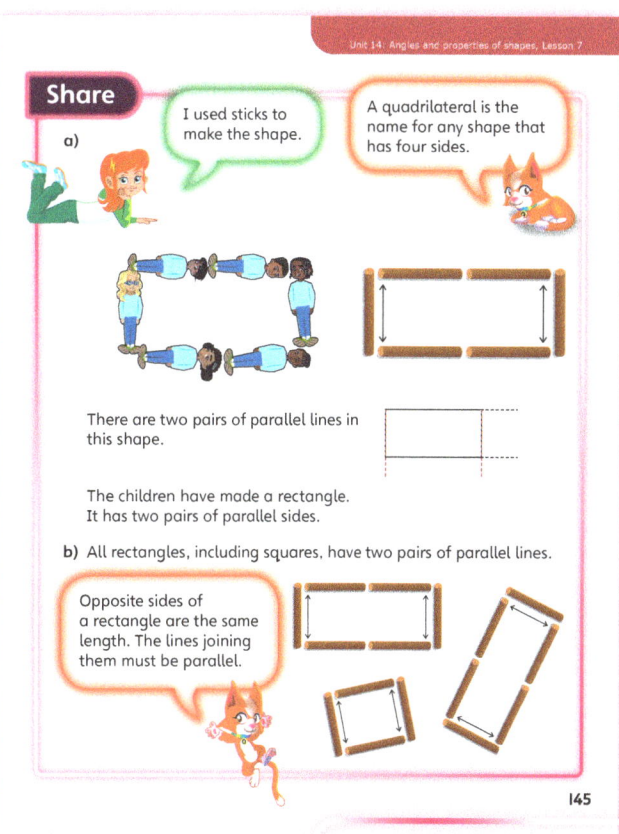

PUPIL TEXTBOOK 3C PAGE 145

183

Unit 14: Angles and properties of shapes, Lesson 7

Think together

WAYS OF WORKING Whole class teacher led (I do, We do, You do)

ASK

• Question ①: *What shapes can you see? How many angles does each shape have? How many sides does each shape have?*
• Question ①: *Which shape has right angles? Can you point to the right angles?*
• Question ②: *How many sides does each shape have?*
• Question ②: *How many sides does a hexagon have? How many sides does a pentagon have?*
• Question ③: *What do the words parallel and perpendicular mean?*

IN FOCUS In question ①, children are shown three simple shapes and they work out which has right angles. Take the opportunity to discuss the different shapes. Ask: *What do you see? How many sides and how many angles does each shape have? What are the names of the shapes?*

In question ②, children are presented with shapes that are less regular and which may not be so familiar to them. Ask children to describe the shapes before answering the questions. As you work through each part you may want to show a regular example of the shape and compare it to those shown in question ②.

In question ③, children draw their own shapes and work with a partner to describe the shape using a variety of key words. Use Dexter's example as a worked example to start them off.

STRENGTHEN Encourage children to explore making the different shapes using sticks. Remind children that they can make shapes with acute or obtuse angles. Provide 2D shapes for them to use as aids.

DEEPEN Ask children to draw their own shapes that have certain given properties. For example, ask: *Can you draw a hexagon with two pairs of parallel sides and two right angles?*

ASSESSMENT CHECKPOINT Use questions ① and ② to assess whether children are secure in their knowledge of the names and properties of 2D shapes.

ANSWERS

Question ①: Shape B is the only shape with a right angle. It has 4 right angles.

Question ② a): A and C

Question ② b): D

Question ② c): B

Question ② d): A, B and D

Question ② e): B and D

Question ③: Children should accurately copy and complete Dexter's square on squared paper.
They should then draw different 2D shapes on squared or dotted paper, and talk about the features of the shapes using language such as: perpendicular sides, quadrilateral, pentagon, right angle and parallel sides.

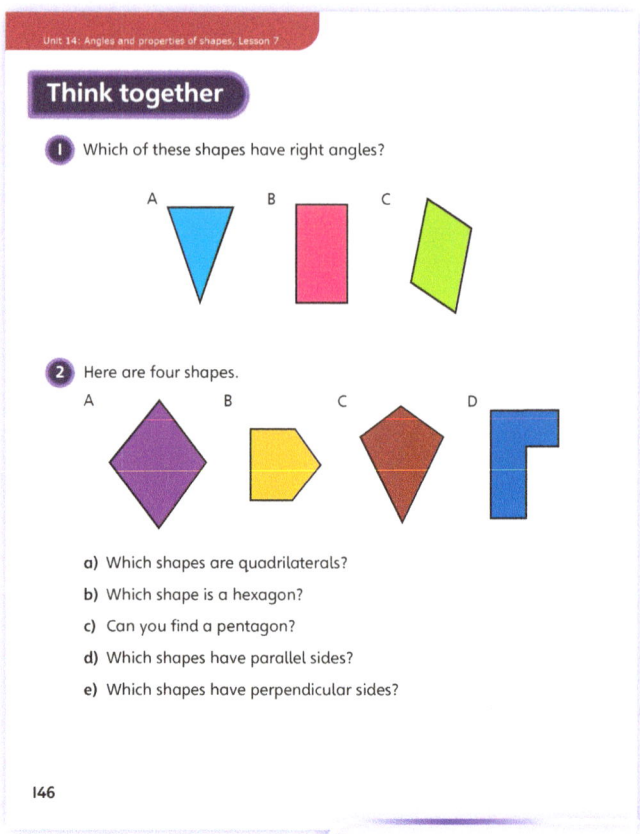

PUPIL TEXTBOOK 3C PAGE 146

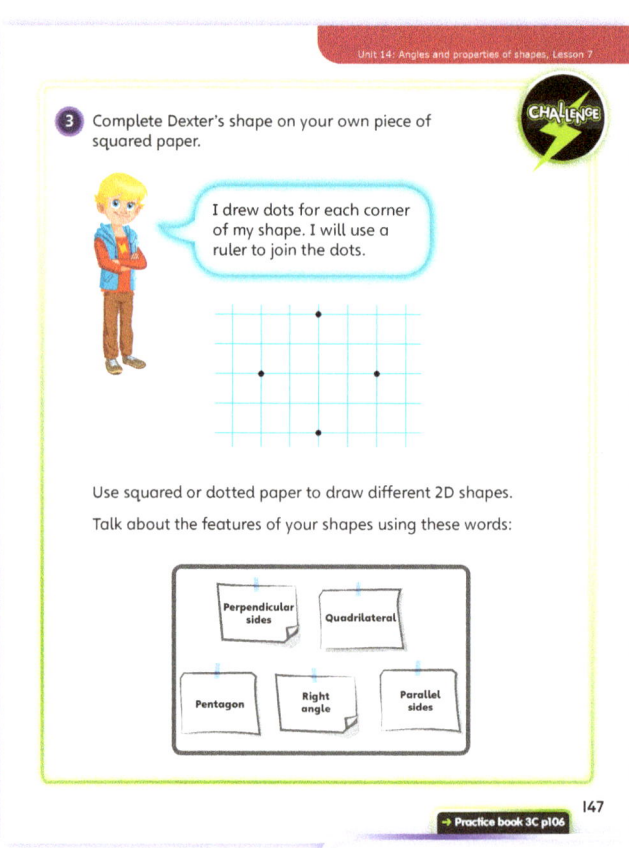

PUPIL TEXTBOOK 3C PAGE 147

184

Unit 14: Angles and properties of shapes, Lesson 7

Practice

WAYS OF WORKING Independent thinking

IN FOCUS Questions ❶ and ❷ focus on recognition of the names of common 2D shapes in different orientations and proportions, with a focus on quadrilaterals in question ❷.

Question ❸ focuses on the properties of pentagons. As children are asked to draw three different examples, at least two of their shapes will be irregular. The key thing they should identify is that each has five sides and five angles.

Question ❺ requires logical reasoning using knowledge of the properties of quadrilaterals.

STRENGTHEN Explore the properties of shapes using geoboards or string to form the outline of the shapes.

DEEPEN Challenge children to invent their own properties puzzle. They should write clues and ask a partner to identify the shape.

THINK DIFFERENTLY Question ❹ requires children to think about the properties of shapes as they design suitable but different shapes that will fulfil the given criteria. Ask pairs of children to check each other's work. Ask: *Are the shapes different or has the same one been turned? Do they all match the properties required?*

ASSESSMENT CHECKPOINT Children should be able to name many of the common 2D shapes covered in this unit and describe some of them in relation to their angle and line properties. Questions ❶ to ❸ will support this assessment.

ANSWERS Answers for the **Practice** part of the lesson can be found in the *Power Maths* online subscription.

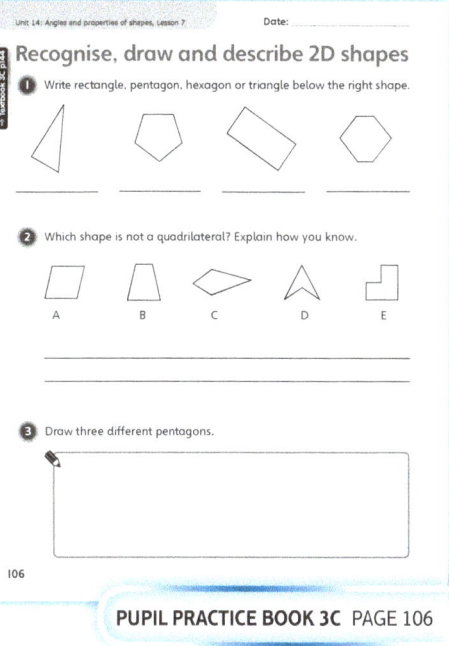

PUPIL PRACTICE BOOK 3C PAGE 106

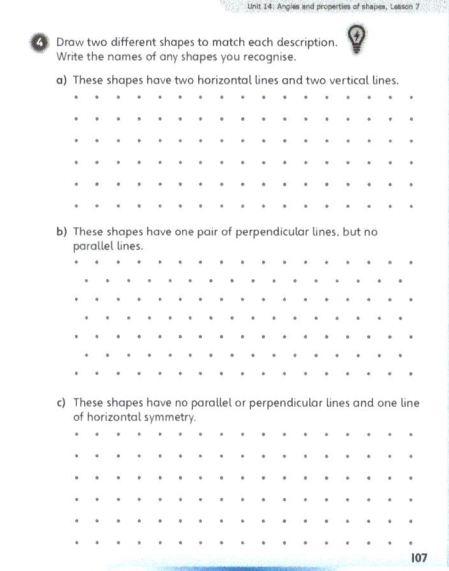

PUPIL PRACTICE BOOK 3C PAGE 107

Reflect

WAYS OF WORKING Independent thinking

IN FOCUS Children should use this opportunity to show the depth of their knowledge of the properties of shapes, including reference to three of these properties: parallel lines, perpendicular lines, number of right angles, vertical/horizontal symmetry or the fact that a rectangle is a quadrilateral.

ASSESSMENT CHECKPOINT Can children apply their knowledge of types of lines and angles, and the correct names, to the properties of a rectangle?

ANSWERS Answers to the **Reflect** part of the lesson can be found in the *Power Maths* online subscription.

After the lesson

- Can children apply their knowledge of types of line, types of angle and symmetry to recognition and description of 2D shapes?
- How could you help children to reinforce the names of 2D shapes and their properties in curriculum areas such as art, DT and computing?

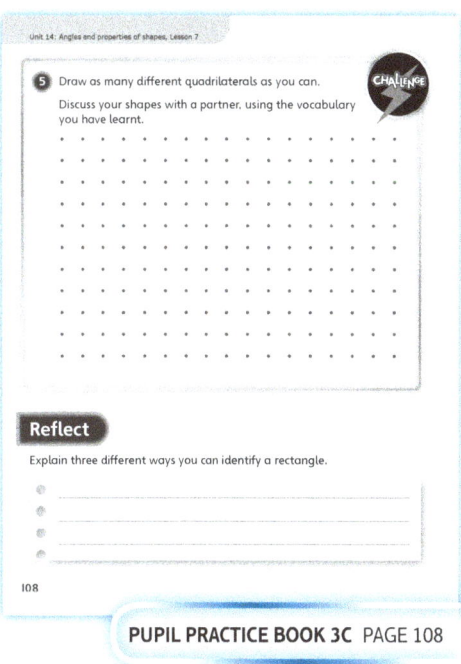

PUPIL PRACTICE BOOK 3C PAGE 108

Unit 14: Angles and properties of shapes, Lesson 8

Recognise and describe 3D shapes

Learning focus
In this lesson, children identify and sort 3D shapes based on properties of faces, vertices and edges. They deepen their understanding of cubes and cuboids, and also describe the shapes and dimensions of faces of different 3D shapes.

Before you teach
- Do children know the names of any 3D shapes?
- Can children find the faces/edges/vertices of a shape?

NATIONAL CURRICULUM LINKS

Year 3 Geometry – properties of shapes

Draw 2D shapes and make 3D shapes using modelling materials; recognise 3D shapes in different orientations and describe them.

ASSESSING MASTERY

Children can describe and visualise the faces of 3D shapes from 2D representations. They can name and describe a range of 3D shapes, including prisms and cuboids. Children can sort and classify shapes according to their properties.

COMMON MISCONCEPTIONS

Children may confuse prisms with pyramids. Ask:
- *What is the same and what is different about a triangular-based pyramid and a triangular prism?*

Children may find it difficult to visualise individual faces from a 2D representation of a 3D shape. Ask:
- *What shape is the face opposite to this one? What shape are the faces on a cuboid?*

STRENGTHENING UNDERSTANDING

Handling models of 3D shapes, and exploring them in different orientations, is essential. Use equipment to build 3D shapes that will unfold to reveal the faces. Explore prisms: all have rectangular faces and two identical end faces that give the prism its name. Define 3D as an object with three dimensions (such as height, width and length), as opposed to 2D, which has only two dimensions (width and length usually).

GOING DEEPER

Challenge children to describe similarities as well as differences when comparing different 3D shapes. Use a range of sorting circles or sorting tables to prompt reasoning about properties of 3D shapes.

KEY LANGUAGE

In lesson: cuboid, cube, prism, pyramid, vertices, face, edge, square, rectangle, shape, size, opposite, sorting circles, parallel, perpendicular, symmetrical

Other language to be used by the teacher: vertex, sphere, square-based pyramid, triangular-based pyramid, triangular prism, cone, cylinder, rectangular

STRUCTURES AND REPRESENTATIONS

3D shapes

RESOURCES

Mandatory: 3D shapes to handle (cuboid, cube, prisms, pyramids, sphere, cone, cylinder)

Optional: range of cardboard boxes of different proportions, 3D shapes that open out into nets, 3D shapes represented as solid shapes and as wireframe models

 In the eTextbook of this lesson, you will find interactive links to a selection of teaching tools.

Quick recap

Ask children to write down as many 3D shapes as they can. Ask children to think of some real-life objects that have these shapes. You might want to take the opportunity to remind children of the key words 'face' and 'edge'.

Unit 14: Angles and properties of shapes, Lesson 8

Discover

WAYS OF WORKING Pair work

ASK
- Question 1 a): *How could you describe the faces of a cube precisely?*
- Question 1 a): *What is the difference between a cube and a cuboid?*
- Question 1 b): *How can you check the properties of a 3D shape accurately?*

IN FOCUS Children explore the properties of cubes and cuboids in terms of the lengths of their edges and the shapes of their faces. They use careful measuring to identify these properties and find out whether a shape is a cube or a cuboid.

PRACTICAL TIPS Children could explore the faces of cardboard boxes that have different proportions. This could be done with eyes closed, so children try to judge whether a box is cube or cuboid without looking. They could then check by measuring the edges of any faces that could be square. Measuring opposite faces will enable children to understand that opposite faces of a cuboid are identical.

ANSWERS

Question 1 a): Ambika can measure all the edges of her gift to find out if it is a cube.
The edges are not all the same so Ambika's gift is not a cube.

Question 1 b): All the faces of Ambika's gift are rectangles. The opposite faces are exactly the same shape and size. Ambika's gift is a cuboid.

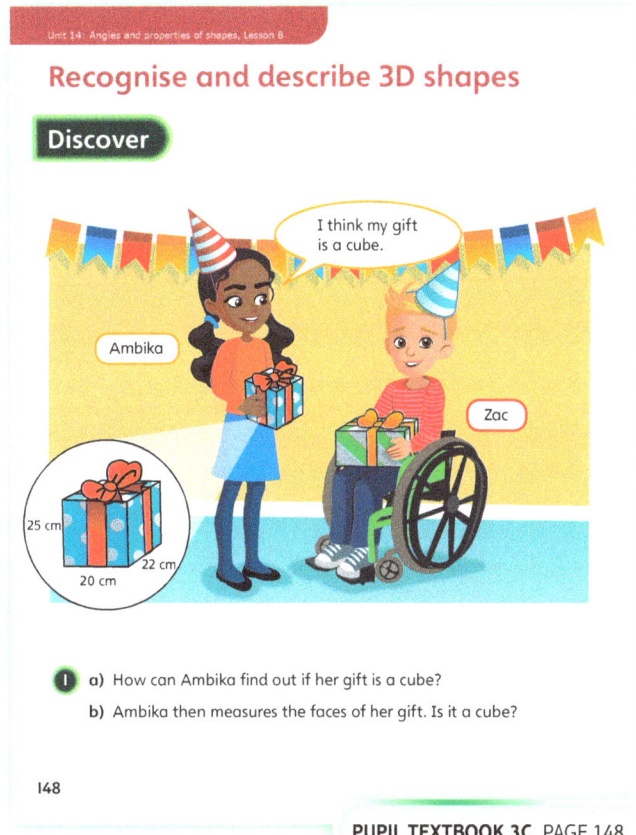

PUPIL TEXTBOOK 3C PAGE 148

Share

WAYS OF WORKING Whole class teacher led

ASK
- Question 1 a): *Why is it important to measure the edges?*
- Question 1 a): *What do you know about the length of the edges of a cube?*
- Question 1 b): *What shape are the faces of a cube/cuboid?*
- Question 1 b): *What is special about the opposite faces of a cuboid?*

IN FOCUS In question 1 a), children are deepening their understanding of a cube as having all edges the same length and all faces as squares.

In question 1 b), children explore how to describe individual faces of a cuboid and recognise that opposite faces are identical.

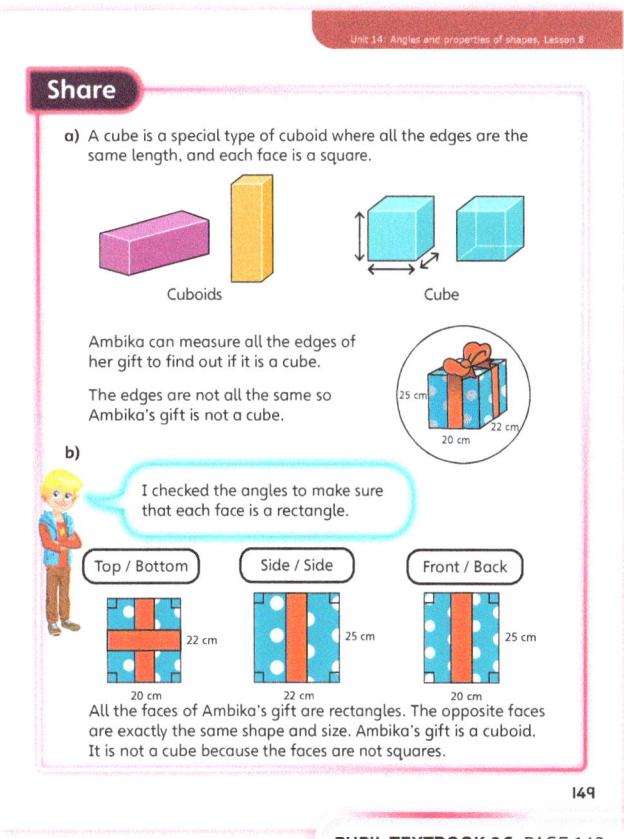

PUPIL TEXTBOOK 3C PAGE 149

187

Unit 14: Angles and properties of shapes, Lesson 8

Think together

WAYS OF WORKING Whole class teacher led (I do, We do, You do)

ASK
- Question ❶: *Do only cubes have square faces?*
- Question ❶: *What other shapes could have one or more square faces?*
- Question ❷: *How can you tell a prism from a pyramid?*
- Question ❸: *What is true of all prisms?*

IN FOCUS Question ❶ focuses on the faces of a cuboid where two opposite faces are squares. This tackles a misconception that only cubes have square faces.

Question ❸ covers a range of common 3D shapes, and prompts children to sort them in to the sorting circles based on their properties. Children should be taught that shapes that do not fit into either circle must be put outside, but still within the rectangle that surrounds the sorting circles.

STRENGTHEN Children should handle models of the common 3D shapes in order to fully explore the properties in a concrete way, but should be encouraged to do so after reasoning from the 2D representations. Some children may need to use stickers to keep track of the faces/edges/vertices as they count them. Some children will need support in question ❸ to realise that cubes, cuboids and cylinders are also classed as prisms.

DEEPEN Challenge children to sort shapes in different ways by altering the headings of the sorting circles in question ❸, using properties such as 'Has an odd number of rectangular faces' or 'Has no parallel edges'.

ASSESSMENT CHECKPOINT Responses to question ❸ should indicate a good level of understanding of a range of 3D shapes and their properties, recognising that cubes, cuboids and cylinders are also classed as prisms.

ANSWERS

Question ❶: The faces of Zac's gift measure: 24 cm × 12 cm, 24 cm × 12 cm and 12 cm × 12 cm.

Question ❷: The tent is a triangular prism. It has two equilateral triangles at either end, with sides of 2 m, and it has two rectangular sides (and one base) measuring 2 m × 3 m 50 cm.

Question ❸:

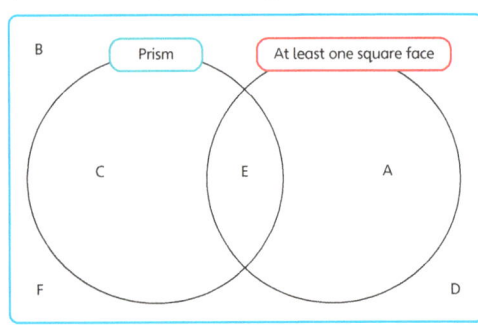

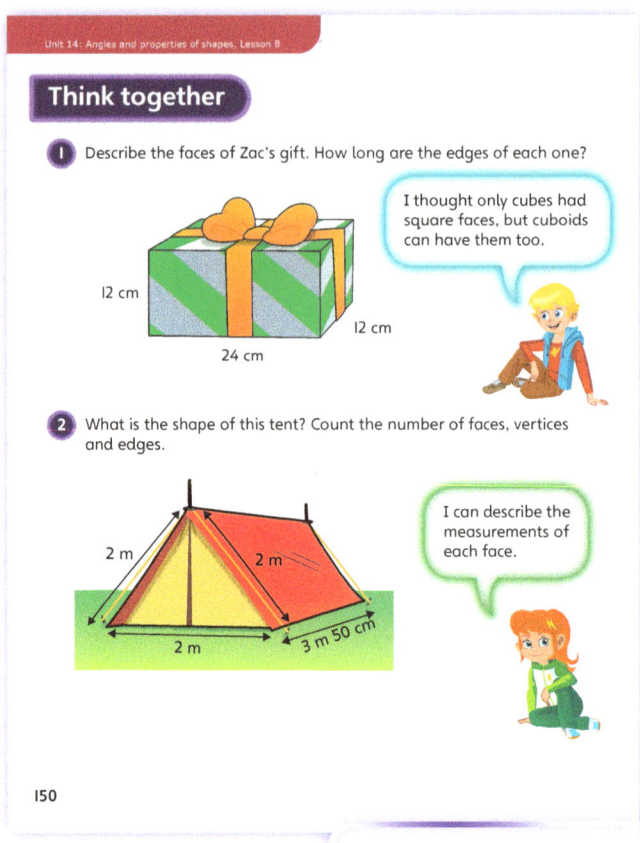

PUPIL TEXTBOOK 3C PAGE 150

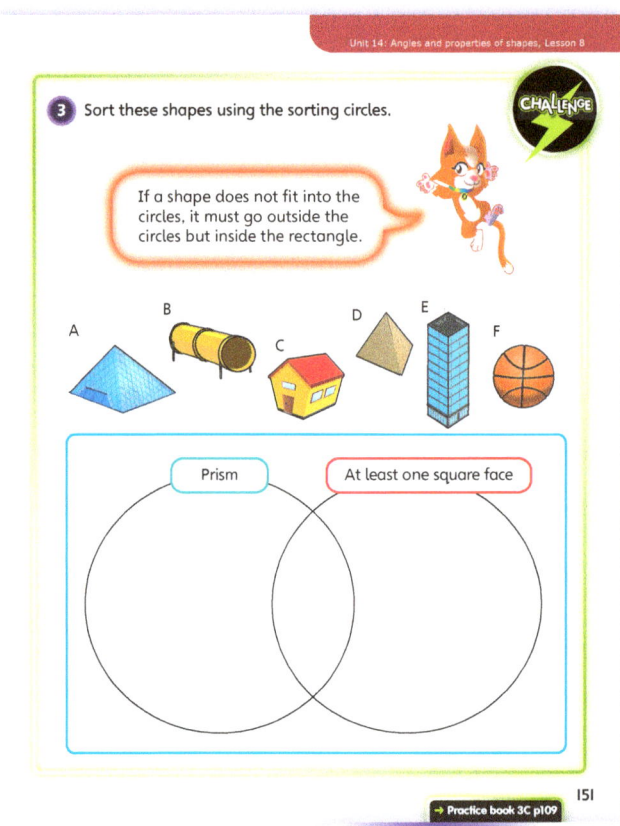

PUPIL TEXTBOOK 3C PAGE 151

188

Unit 14: Angles and properties of shapes, Lesson 8

Practice

WAYS OF WORKING Independent thinking

IN FOCUS Question 4 requires children to complete a table based on properties of prisms. It should show that all prisms have more than one rectangular face (the exception being cylinders). Take the opportunity to discuss the differences between prisms and pyramids, as both non-prisms in this question are actually pyramids. Only a truncated pyramid has more than one rectangular face, and not every pyramid has a rectangular face. Children may need reminding that a square is a special type of rectangle. None of the common shapes would fit into the 'Not a prism' and 'More than one rectangular face' section.

STRENGTHEN Encourage children to make reasoned predictions about given faces and properties of shapes, before checking the predictions by handling 3D models or by measuring edges.

DEEPEN Challenge children to sort shapes into different sorting circles or tables by choosing their own headings. Extend question 5 to compare other shapes. Ask: *What is the same? What is different?*

THINK DIFFERENTLY Question 3 requires children to use reasoning skills to match the cuboids to the correct set of faces.

ASSESSMENT CHECKPOINT Responses to these questions will show whether children can name common 3D shapes, describe them using the shape and number of their faces, and sort them according to their properties.

ANSWERS Answers for the **Practice** part of the lesson can be found in the *Power Maths* online subscription.

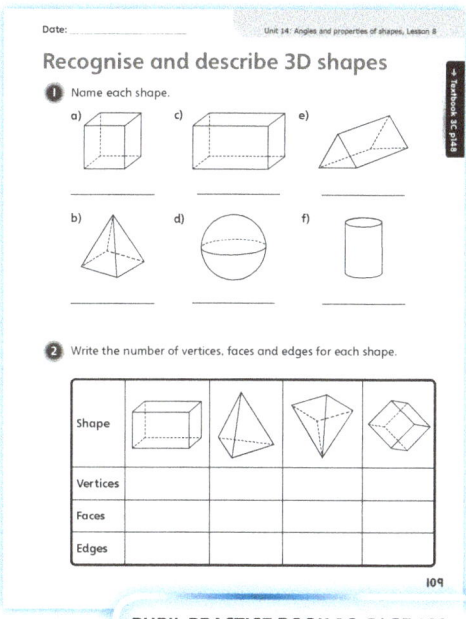

PUPIL PRACTICE BOOK 3C PAGE 109

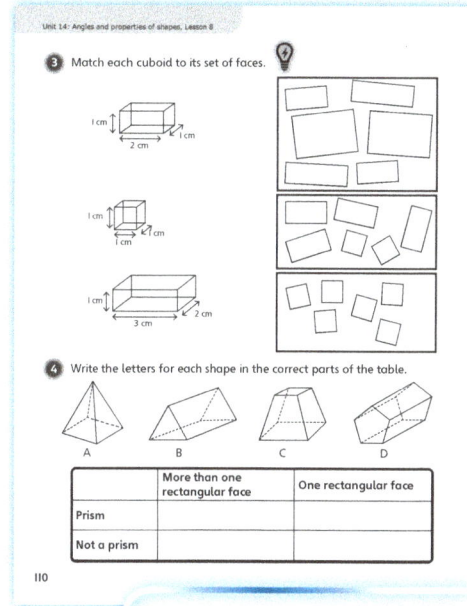

PUPIL PRACTICE BOOK 3C PAGE 110

Reflect

WAYS OF WORKING Independent thinking

IN FOCUS This task looks simple, but children will need to make sure that their checklist includes enough information so that it does not also encompass other shapes. For example, a shape with six faces and edges all the same length could be a pentagon-based pyramid.

ASSESSMENT CHECKPOINT Can children use the properties of a cube to identify it definitively?

ANSWERS Answers to the **Reflect** part of the lesson can be found in the *Power Maths* online subscription.

After the lesson ⏸

- Are children able to describe individual faces of a 3D shape?
- Can children explain the properties of a cuboid that is not a cube, including reference to opposite faces?
- Are children confident in sorting 3D shapes according to different criteria relating to their properties?

PUPIL PRACTICE BOOK 3C PAGE 111

189

Unit 14: Angles and properties of shapes, Lesson 9

Make 3D shapes

Learning focus
In this lesson, children learn to construct 3D shapes by considering their properties in relation to different construction materials.

Before you teach
- Which 3D shapes can children construct?
- What materials have children used before to make 3D shapes?

NATIONAL CURRICULUM LINKS

Year 3 Geometry – properties of shapes

Draw 2D shapes and make 3D shapes using modelling materials; recognise 3D shapes in different orientations and describe them.

ASSESSING MASTERY

Children can describe different ways to construct 3D shapes from different construction materials, by reasoning about their properties.

COMMON MISCONCEPTIONS

Children may not recognise that the same cuboid could be represented in different orientations. Ask:
- *Which cuboid is the same as this one, but has been turned around?*

Children may not understand that different construction materials are based on different properties. Ask:
- *What features of a 3D shape are sticks and marshmallows good at representing? Why is it not possible to make a cone, cylinder or sphere with these construction materials?*

STRENGTHENING UNDERSTANDING

All children would benefit from having access to different construction materials in order to build the different shapes. Children should be shown how to make connecting edges with sticks and vertices with the marshmallows.

GOING DEEPER

Challenge children to think of innovative ways to construct a 3D shape – for example, a cylinder. Children should not be pushed to formal consideration of nets, but their inventiveness should be encouraged.

KEY LANGUAGE

In lesson: cube, length, direction, angle, features, vertices, faces, edges, sphere, pyramid

Other language to be used by the teacher: cuboid, prism, cone, cylinder, sphere, vertex

STRUCTURES AND REPRESENTATIONS

3D shapes

RESOURCES

Mandatory: multilink cubes, construction materials

Optional: sticks and marshmallows, modelling clay, snap-together construction materials, wireframe models

 In the eTextbook of this lesson, you will find interactive links to a selection of teaching tools.

Quick recap

Ask children to close their eyes and visualise a large cube. Ask: *How many faces does a cube have? What shape are the faces? How many edges does the shape have?* See if children can answer these questions only by visualising the shape.

Unit 14: Angles and properties of shapes, Lesson 9

Discover

WAYS OF WORKING Pair work

ASK
- Question 1 a): *How can you identify a cube?*
- Question 1 a): *How many cubes does each child have?*
- Questions 1 a) and b): *What do you need to think about before you start using the multilink cubes?*
- Question 1 b): *Would this be the same shape if you turned it around like this?*

IN FOCUS Question 1 a) challenges children to form a larger cube from the limited numbers available to Lee or Bella. Some children may mistakenly identify a 1 × 4 arrangement as a cube.

Question 1 b) pushes children to consider how the same cuboid can be represented in different orientations, and to distinguish which properties of two cuboids make them distinct or identical.

PRACTICAL TIPS Ideally, all children should have access to multilink cubes, so that they can test out their ideas. However, in this case, it is also important that children have a chance to consider their ideas before trying out the materials.

ANSWERS

Question 1 a): Bella can make a 2 × 2 × 2 cube using all her smaller cubes.

Question 1 b): Lee is not correct. He can only make two different cuboids with his six cubes.

PUPIL TEXTBOOK 3C PAGE 152

Share

WAYS OF WORKING Whole class teacher led

ASK
- Question 1 a): *Can you prove that Bella's shape is definitely a cube?*
- Question 1 b): *Why are some of Lee's shapes not really different?*

IN FOCUS Question 1 a) focuses on the property that a cube has edges of equal length.

Question 1 b) focuses on how cuboids can be represented in different orientations.

Some children may argue that Lee could make many different cuboids if he did not have to use all six cubes every time. This is an interesting point to clarify and a good extension for children to explore.

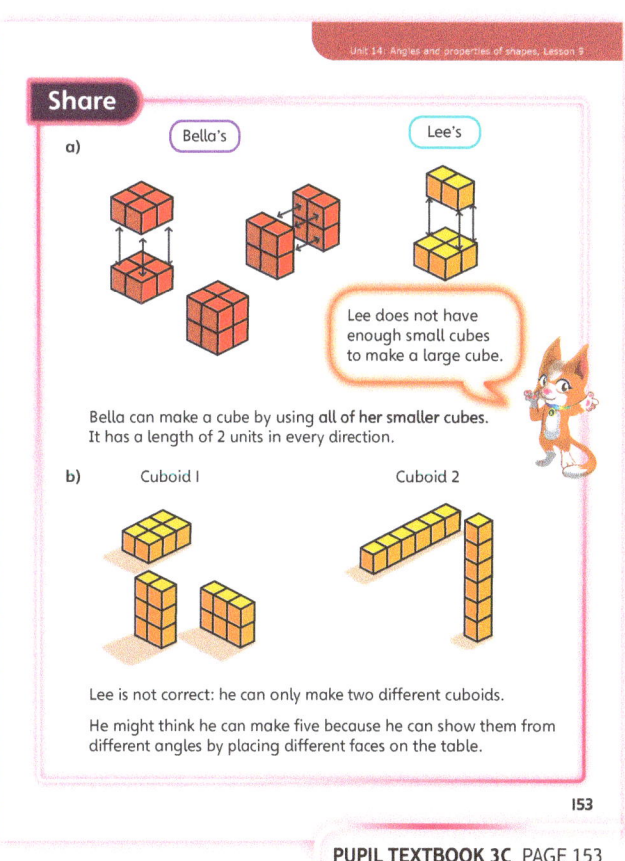

PUPIL TEXTBOOK 3C PAGE 153

Think together

WAYS OF WORKING Whole class teacher led (I do, We do, You do)

ASK

- Questions ❶ and ❷: *What is the same and what is different about these sets of construction materials?*
- Questions ❶ to ❸: *What is the same about the shapes that you cannot make from these materials?*

IN FOCUS All questions focus on the different ways of building shapes using different materials. Children are asked to consider which materials are appropriate and which are not when constructing specific 3D shapes. The marshmallows are used to connect the sticks to form edges and vertices. Small balls of modelling clay could replace the marshmallows.

STRENGTHEN Give children access to different types of construction materials to build and experiment with.

DEEPEN Challenge children to produce a list or table identifying which 3D shapes can and cannot be made from the different construction materials shown or available in class.

ASSESSMENT CHECKPOINT Are children able to explain which materials can be used to construct a prism and why cubes cannot be used for this purpose? Can they explain the difficulty in making shapes that have any curved surfaces?

ANSWERS

Question ❶: You need 6 squares to make the faces of a cube.
You need 12 sticks to make the edges of a cube.
You need 8 marshmallows to make the corners (vertices) of a cube.

Question ❷ a): Children can make a cube from the sticks and marshmallows, from the squares and from the linking cubes. They can make a pyramid from the sticks and marshmallows and from the squares and triangles.

Question ❷ b): A sphere cannot be made using the shapes as a sphere is not made using edges and vertices, or from flat-shaped faces.

Question ❸ a): Shape A cannot be made using the sets of material given as there is nothing which can be used to make the circles for the ends. Shape B can be made using the sticks and marshmallows or the squares and triangles.

Question ❸ b): You would need 2 triangles and 3 squares to make the prism.
You could also join 2 squares together to make each rectangular face of the prism, therefore using 2 triangles and 6 squares to make a different triangular prism.
You would need 9 sticks and 6 marshmallows to make the prism.

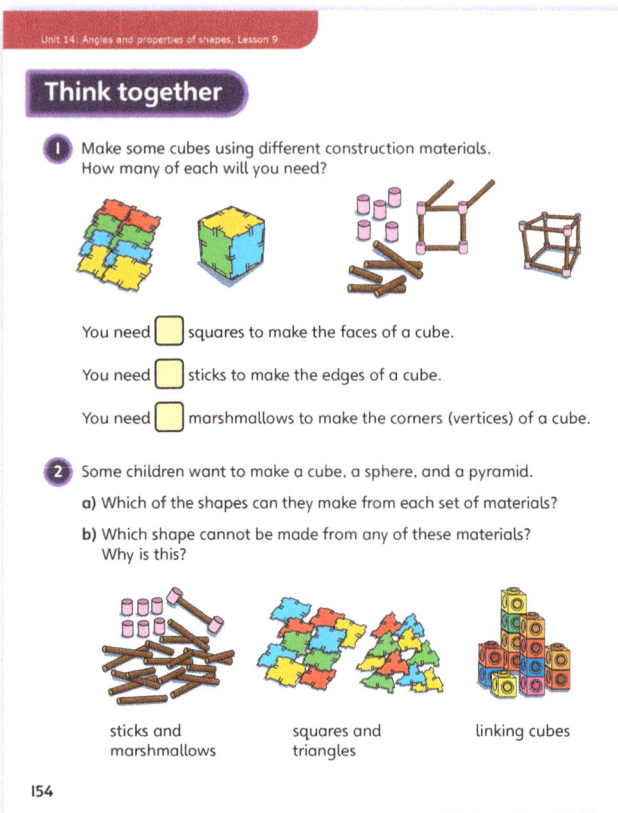

PUPIL TEXTBOOK 3C PAGE 154

PUPIL TEXTBOOK 3C PAGE 155

Unit 14: Angles and properties of shapes, Lesson 9

Practice

WAYS OF WORKING Independent thinking

IN FOCUS Question **1** focuses on recognising how many cubes have been used to build a 3D shape given as a 2D representation.

Question **3** requires children to link the number of edges and vertices to the amount of construction materials they will need.

STRENGTHEN Encourage children to form an idea of how to build shapes before trialling their ideas with construction materials.

DEEPEN Question **6** challenges children to explain the pattern that appears in the table, based on a comparison of the shapes. The answers will depend on the length of each prism, but the number of sticks should be a multiple of the number of sides on the end piece, assuming each end piece has equal length sides.

THINK DIFFERENTLY Question **5** requires children to use the language and terminology of edges and vertices to describe how to make a triangular prism. As all sides are not the same length, they will have to think about how many short and how many long sticks they will need.

ASSESSMENT CHECKPOINT Are children able to explain which materials can and cannot be used to make specific shapes? Can they explain the difficulty in constructing shapes that have any curved surfaces?

ANSWERS Answers to the **Practice** part of the lesson can be found in the *Power Maths* online subscription.

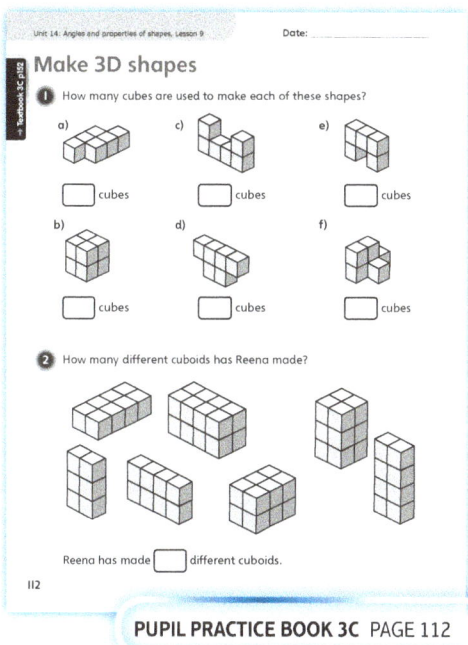

PUPIL PRACTICE BOOK 3C PAGE 112

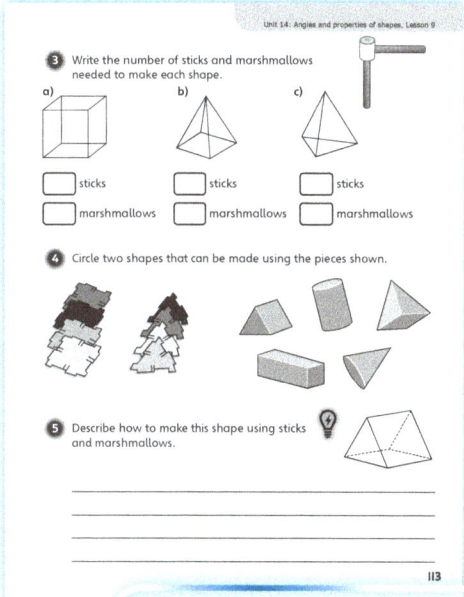

PUPIL PRACTICE BOOK 3C PAGE 113

Reflect

WAYS OF WORKING Individual study

IN FOCUS Children should discuss and compile a list by looking back through the unit. Children should then decide individually which three things they feel are the most important.

ASSESSMENT CHECKPOINT Are children able to make basic 3D shapes using a variety of materials (cubes, 2D shapes that fit together, or sticks and marshmallows)? Can they identify concepts they understood well and concepts they may need to do more work on?

ANSWERS Answers to the **Reflect** part of the lesson can be found in the *Power Maths* online subscription.

After the lesson ⏸

- Can children justify decisions about how to construct 3D shapes, based on the properties of the shapes and of the materials?
- Are children able to recognise shapes when they are presented in different orientations?
- What did children learn in this unit that could be useful in DT, science or art?

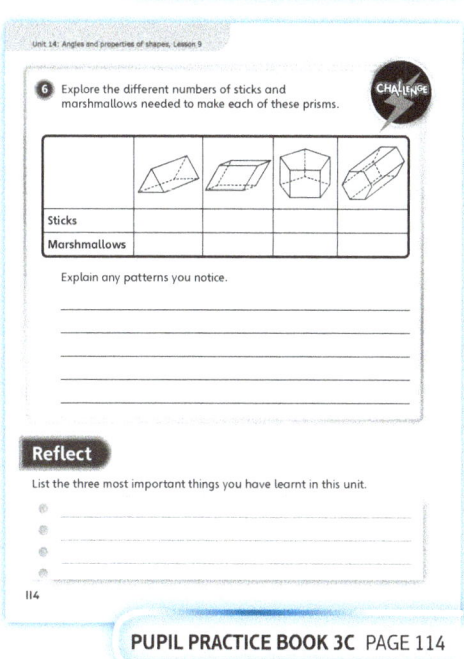

PUPIL PRACTICE BOOK 3C PAGE 114

Unit 14: Angles and properties of shapes

End of unit check

Don't forget the unit assessment grid in your *Power Maths* online subscription.

WAYS OF WORKING Group work adult led

IN FOCUS
- Question ❶ assesses children's ability to identify a shape with just one right angle.
- Question ❷ assesses children's ability to identify an obtuse angle shown using the hands of a clock face.
- Question ❸ assesses children's ability to identify the horizontal lines in letter shapes.
- Question ❹ assesses children's ability to identify a shape made up of perpendicular lines, recognising that these are lines that meet at a right angle.
- Question ❺ assesses children's ability to identify the shape with no pairs of parallel sides.
- Question ❻ is a SATs-style question relating to the faces of a cuboid.

ANSWERS AND COMMENTARY

Children who have mastered the concepts in this unit will be able to define these terms: right angle, parallel lines, perpendicular lines, vertical and horizontal. Children will be able to use these terms to describe some of the properties of 2D and 3D shapes.

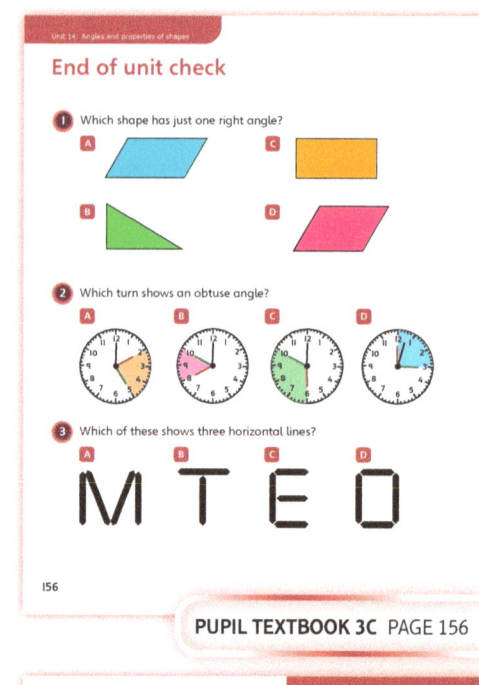

PUPIL TEXTBOOK 3C PAGE 156

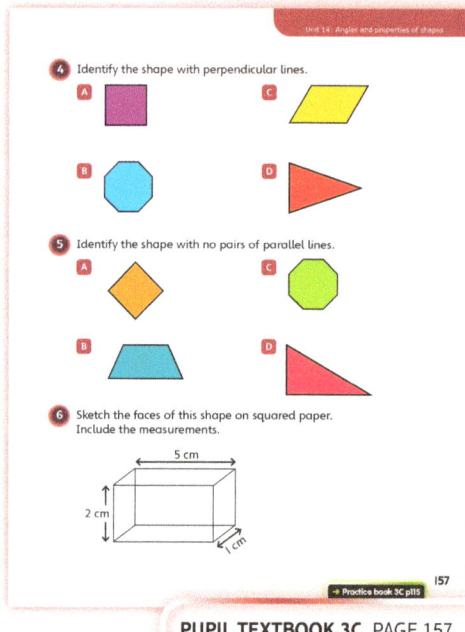

PUPIL TEXTBOOK 3C PAGE 157

Q	A	WRONG ANSWERS AND MISCONCEPTIONS	STRENGTHENING UNDERSTANDING
1	B	A or D suggest that the child is not sure how to identify right angles; C suggests that the child did not read the question carefully as C has more than one right angle.	Help children to remember key vocabulary with these tips: • Acute, obtuse or right angle? *Acute* contains the word *cute* which usually describes something small, so it is the smallest type of angle; right angles can stand up*right* • Horizontal or vertical? *Horizont*al is flat like the *horizon*; *vert*ical stands up straight like *vert*ebrae in your back • Parallel or perpendicular? The word *parallel* has double *ll* in the middle, which looks like a pair of parallel lines; parallel lines are like train tracks – they will never touch or cross; perpendicular lines can be made from a pair of parallel lines by turning one line a quarter turn.
2	C	A or D suggest that the child does not know that the angle between three numbers on a clock face is exactly 90 degrees. B suggests confusion between acute and obtuse angles.	
3	C	A, B or D suggest confusion between horizontal and vertical lines, and that the child has not counted properly.	
4	A	B, C or D suggest that the child is not able to recognise perpendicular lines.	
5	D	B or C suggest confusion between parallel and perpendicular lines.	
6		Have children drawn six faces in three pairs? Have they included the measurements on their sketches?	

Unit 14: Angles and properties of shapes

My journal

WAYS OF WORKING Independent thinking

ANSWERS AND COMMENTARY

Questions 1 a) and b): Ensure children think about which shape to start with and whether they could make some shapes on a different rotation to the ones shown. Ensure children explain how they lined up the ruler.

Question 2 a): Children could use the dots on the grid to draw diagonal parallel lines.

Question 2 b): Children could use the vertical and horizontal grid lines to create a right angle or simply use the end of their ruler.

Question 2 c): Children should draw a right-angled trapezium.

Question 2 d): Children should draw a pentagon with one right angle.

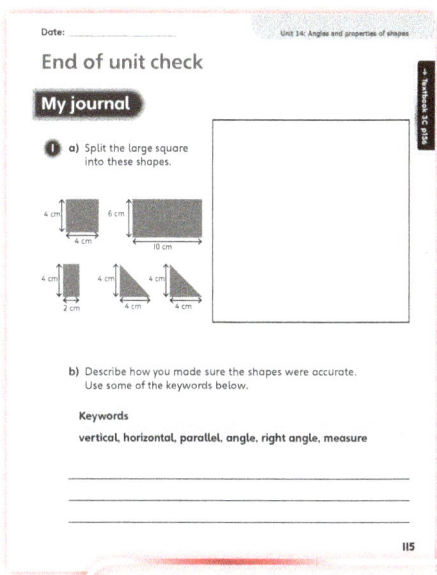

PUPIL PRACTICE BOOK 3C PAGE 115

Power check

WAYS OF WORKING Independent thinking

ASK
- Can you explain the difference between parallel and perpendicular lines?
- Are you confident describing 2D and 3D shapes?
- Can you draw a right-angled triangle?

Power play

WAYS OF WORKING Pair work

IN FOCUS Children should explore whether a shape can be split physically into two identical halves. All the designs shown in the **Power play** cannot be split into symmetric halves where each half is a whole number of cubes. Children may suggest slicing through the design on the left to make identical top and bottom halves, which would represent a plane of symmetry. This is acceptable, and an interesting discussion point, but this concept is not introduced until KS3.

ANSWERS AND COMMENTARY This **Power play** is open-ended as there are many ways of making a 3D shape with no symmetry. All the designs shown here cannot be split into symmetric halves where each half is a whole number of cubes. Try to elicit key shapes which *do* have symmetry, in order to understand which shapes *do not*. Most T shapes cannot be physically split, but they are all vertically symmetrical. A squared C or U shape can only be physically split where there is an even number of cubes in the middle section, but, again, they are still symmetrical. Asymmetrical shapes are easily constructed where there is a different number of cubes on each layer, as the two designs in the centre and the design on the right show.

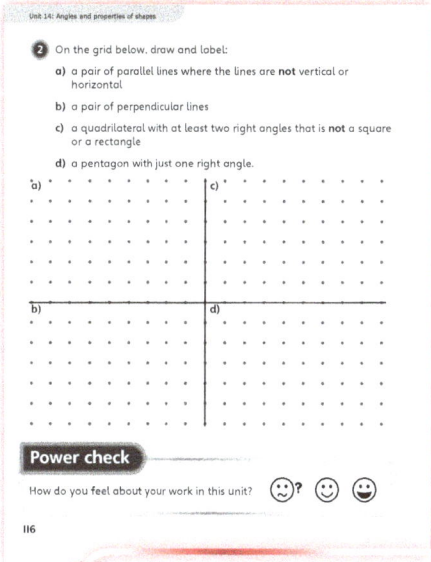

PUPIL PRACTICE BOOK 3C PAGE 116

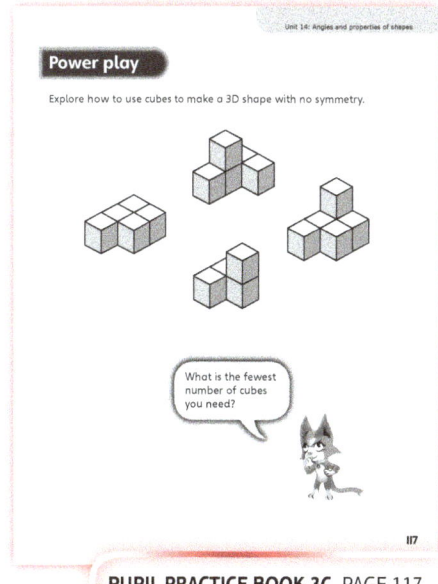

PUPIL PRACTICE BOOK 3C PAGE 117

After the unit

- How well did children respond to the new words introduced in this unit? How can you continue reinforcing these words in future lessons?
- Based on children's responses, are there things that you would do differently next time you teach this unit?

Strengthen and **Deepen** activities for this unit can be found in the *Power Maths* online subscription.

Unit 15
Statistics

Mastery Expert tip! 'My class really liked it when I replaced some of the data and contexts given in the **Discover** sections of this unit with data from our class or school – they found this really gave the data and their interpretation of it meaning. I used data from a science experiment and PE lessons.'

Don't forget to watch the Unit 15 video!

WHY THIS UNIT IS IMPORTANT

This unit presents children with a range of ways in which information and data can be presented and interpreted. Children will explore pictograms in detail, including exploring the use of keys where symbols represent more than 1. Children will be introduced to data presented in bar charts and more complex tables. The unit also provides a good opportunity for children to practise and apply their calculation and reasoning skills, including addition and subtraction and counting in multiples of 2, 5 and 10.

WHERE THIS UNIT FITS

→ Unit 14: Angles and properties of shapes
→ **Unit 15: Statistics**

This unit builds on Year 2 Statistics, where children were introduced to basic pictograms and block charts. It develops their understanding further, and encourages them to explore the range of information that they can get from the data presented to them.

Before they start this unit, it is expected that children:
- know how to interpret a basic pictogram
- are confident in carrying out addition and subtraction calculations
- know how to count in multiples of 2, 5 and 10.

ASSESSING MASTERY

Children who have mastered this unit can interpret data presented in pictograms, bar charts and tables, and use this data to answer a range of questions, including comparison, ordering and total questions. They can also make their own statements based on the data presented to them and they are beginning to be able to compare linked data that is presented in more than one diagram or table to answer questions.

COMMON MISCONCEPTIONS	STRENGTHENING UNDERSTANDING	GOING DEEPER
Children may miscount the number of pictogram symbols (and their value) or misread a value on the vertical axis of a bar chart.	Represent the pictogram and/or bar chart physically using counters, cubes or other objects. Encourage children to count each object physically, and to use the key in a pictogram to work out the total value.	Encourage children to make their own increasingly complex statements based on data, including data presented across multiple styles of charts and tables. Ask: *What else can you tell me based on this data? How do you know?*
Children may use the incorrect operation when answering questions and carrying out calculations based on the data presented to them.	Draw children's attention to the structure of the question. Ask: *What is the question asking you to do? What operation could this involve?* For comparison and total questions, representing the pictogram or bar chart using objects can also be beneficial.	

Unit 15: Statistics

UNIT STARTER PAGES

Give children a few minutes to read the unit starter pages of the Textbook. Then read it together as a class. In Year 2, children only encountered block diagrams, which are similar to bar charts, but each unit in each bar is represented by a single block. Ask: *Do you recognise the diagram next to Ash? In what ways is it similar to a block diagram? In what ways is it different?* Then, look at the key words next to Flo. Ask: *Which of the key words did you understand? What do they mean?*

STRUCTURES AND REPRESENTATIONS

Pictograms:

Sport	Number of people
skiing	😊 😊 😊 😊
snowboarding	😊 😊 😊 😊 😊

Key: Each 😊 represents 2 people.

Bar charts:

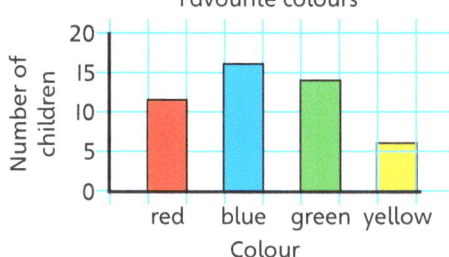

KEY LANGUAGE

There is some key language that children will need to know as part of the learning in this unit:
- pictogram, key, symbol
- compare, least, most, altogether, total
- bar chart, horizontal axis, vertical axis, scale
- half-way between
- table, row, column
- order, smallest, largest

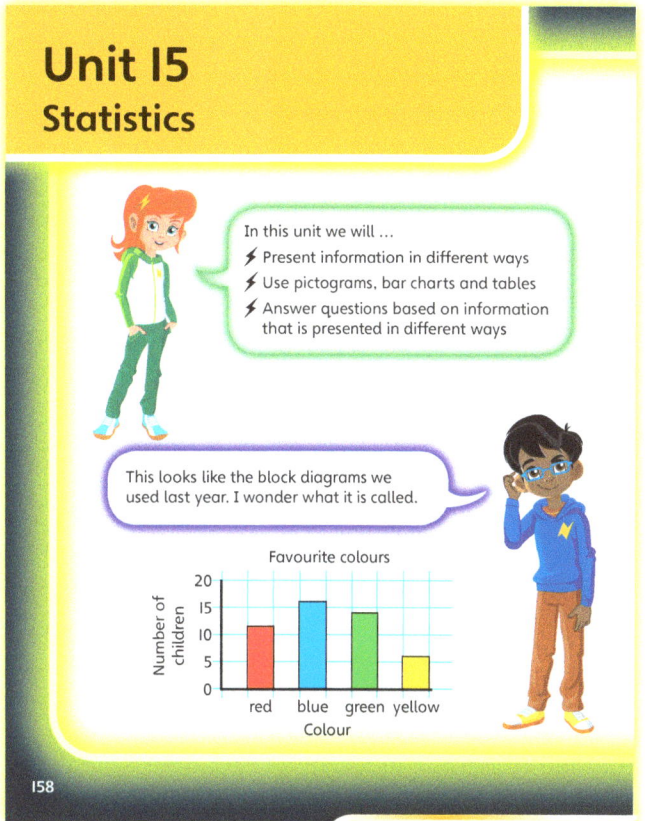

PUPIL TEXTBOOK 3C PAGE 158

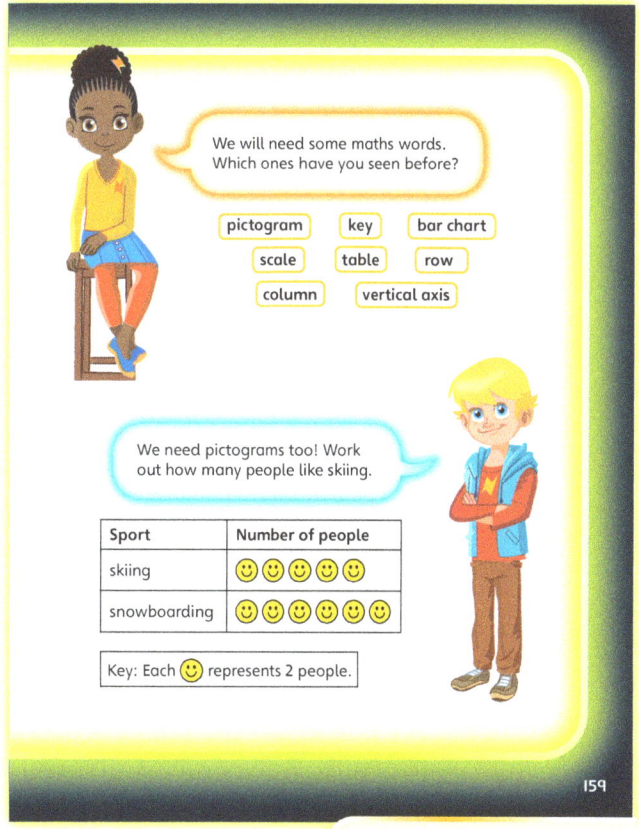

PUPIL TEXTBOOK 3C PAGE 159

197

Unit 15: Statistics, Lesson 1

Interpret pictograms

Learning focus
In this lesson, children will interpret pictograms where each symbol is worth more than 1.

Before you teach
- Are children secure with their multiplication facts from the 2, 5 and 10 times-tables?
- Can you relate the learning in this lesson to work across the curriculum, for example using real data from a science experiment?

NATIONAL CURRICULUM LINKS

Year 3 Statistics

Interpret and present data using bar charts, pictograms and tables.

Solve one-step and two-step questions [for example, 'How many more?' and 'How many fewer?'] using information presented in scaled bar charts, pictograms and tables.

ASSESSING MASTERY

Children can interpret a range of pictograms, including those whose keys are based on known multiplication facts and where half symbols are used. Children can also construct their own pictograms, using information that is provided and employing inverse operations to work out missing information.

COMMON MISCONCEPTIONS

Children may believe that each pictogram symbol always represents '1'. Draw attention to the key and ask:
- Where can you look on the pictogram to find out what each symbol is worth?

Children may count half symbols as full symbols. To help address this, identify a half symbol and ask:
- What do you notice about this symbol? Is it the same size as the symbol in the key?

Children may not understand that a half symbol represents half as many things as a full symbol represents. Draw a whole symbol and a half symbol on the board. Ask:
- If the full symbol represents 6 people, how many people does the half symbol represent?

STRENGTHENING UNDERSTANDING

Encourage children to represent pictograms using practical equipment, such as cubes or counters. Manipulating each item physically will help children to do calculations, such as repeated addition.

GOING DEEPER

Encourage children to think about the purpose of pictograms, and how and where they could be useful for presenting data. Also challenge children to consider some of the possible drawbacks of using pictograms.

KEY LANGUAGE

In lesson: pictogram, key, symbol

Other language to be used by the teacher: inverse, worth, whole symbol, half symbol

STRUCTURES AND REPRESENTATIONS

Number lines, pictograms

RESOURCES

Optional: number lines, cubes or counters

 In the eTextbook of this lesson, you will find interactive links to a selection of teaching tools.

Quick recap

Go over multiplication and division facts from the 2, 5 and 10 times-tables. Ask children questions and have them respond on mini-whiteboards. Include examples such as 10 × ? = 60.

Unit 15: Statistics, Lesson 1

Discover

WAYS OF WORKING Pair work

ASK
- Question 1 a): *What type of chart is this? How does it work?*
- Question 1 a): *What can you tell from the pictogram?*
- Question 1 b): *How much is one symbol worth?*

IN FOCUS Pictograms are revisited here for the first time since they were covered in Year 2. Review with children the important points of a pictogram such as what the symbols mean and how to read the key.

PRACTICAL TIPS Consider replacing the data shown here with data from your own mini-beast hunt, or a similar activity, which may link to your science teaching.

ANSWERS

Question 1 a): Amelia found 8 beetles.

Question 1 b): Amelia found 3 spiders.

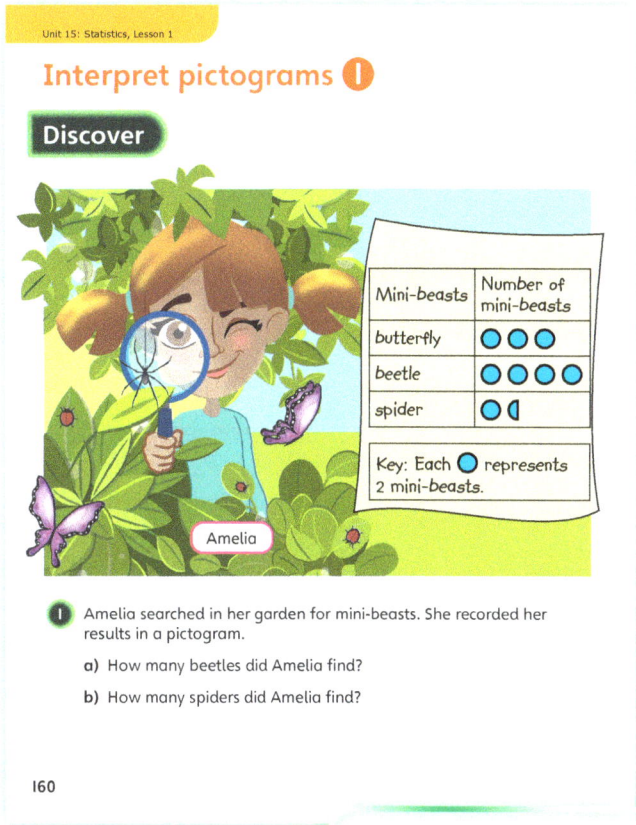

PUPIL TEXTBOOK 3C PAGE 160

Share

WAYS OF WORKING Whole class teacher led

ASK
- Question 1 a): *How many symbols are there for beetles?*
- Question 1 a): *How much does each symbol represent?*
- Question 1 a): *How can you work out the total number of beetles?*
- Question 1 b): *How can you work out what the half symbol is worth?*

IN FOCUS It is important that children realise the importance of the key, and that pictograms are not always based on a 1-to-1 correspondence. For example, 5 symbols do not always mean 5 'things'. The value of each symbol is linked to known multiplication facts, in this case the 2 times-table. Discuss question 1 a) and revisit the link between multiplication and repeated addition.

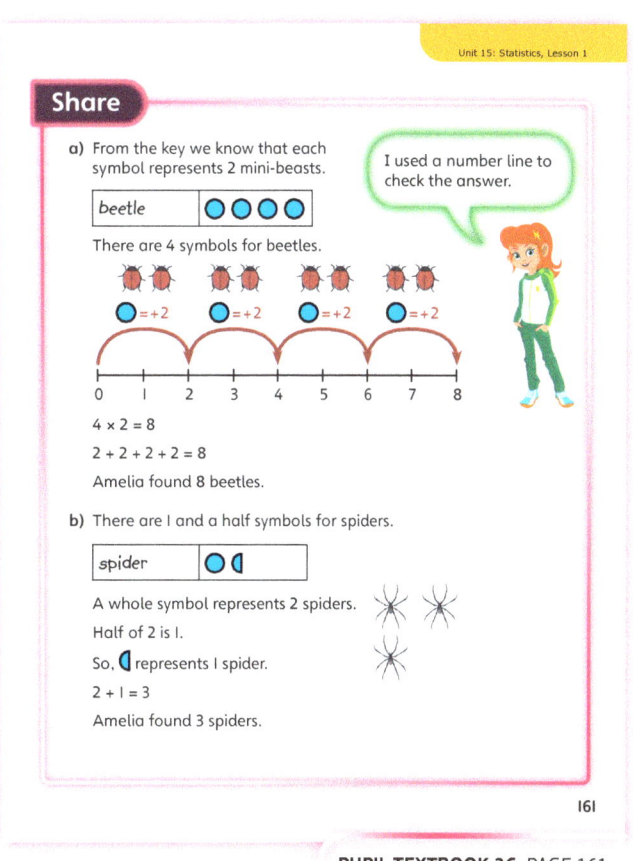

PUPIL TEXTBOOK 3C PAGE 161

199

Unit 15: Statistics, Lesson 1

Think together

WAYS OF WORKING Whole class teacher led (I do, We do, You do)

ASK
- Question 1: *Did Amelia find more bluebells or honeysuckles? How can you tell?*
- Question 1 a): *How can you work out how much each symbol is worth?*
- Question 1 b): *How can you work out how much each half symbol is worth?*
- Question 2: *What operation do you need to use to work out how much each symbol is worth?*
- Question 3: *What is different about Lee's pictogram compared to the other pictograms you have looked at? Why is Lee's pictogram confusing?*

IN FOCUS Draw children's attention to the value of the symbol in question 1 compared to the value of the symbols in **Discover** and **Share**. Draw children's attention to the benefits of pictograms, including how they can be used to compare data visually.

STRENGTHEN Encourage children to use a number line or counters to help with the division calculation in question 2.

DEEPEN Children should begin to develop a deeper understanding of the purpose and benefits of a pictogram, and how these benefits can only be achieved if the same key value is used for all items on the same pictogram. Discuss question 3 and how the pictogram has lost most of its usefulness. This is because the key changes for each symbol and so does not allow the data to be compared visually. Encourage children to create their own pictograms. Ask: *What would it look like if each symbol represented 2 mini-beasts?*

ASSESSMENT CHECKPOINT Use question 1 b) to assess if children can work out the value of half symbols. Do children understand that these do not have the same value as whole symbols?

ANSWERS

Question 1 a): Amelia found 50 honeysuckles.

Question 1 b): Amelia found 25 daffodils.

Question 1 c): There are two symbols for bluebells which represent 20 bluebells and $\frac{1}{2}$ a symbol for primroses which represents 5 primroses. Therefore there are more bluebells than primroses.

Question 1 d): There are 20 bluebells and 5 primroses. Therefore there are 15 more bluebells than primroses.

Question 2 a): Each symbol is worth 5 trees.

Question 2 b): Amelia found 20 conifer trees.

Question 2 c): Amelia found 45 trees in total.

Question 3: You could improve the pictogram by using the same symbol for all the insects and giving that symbol the same value so that it is easier to compare.

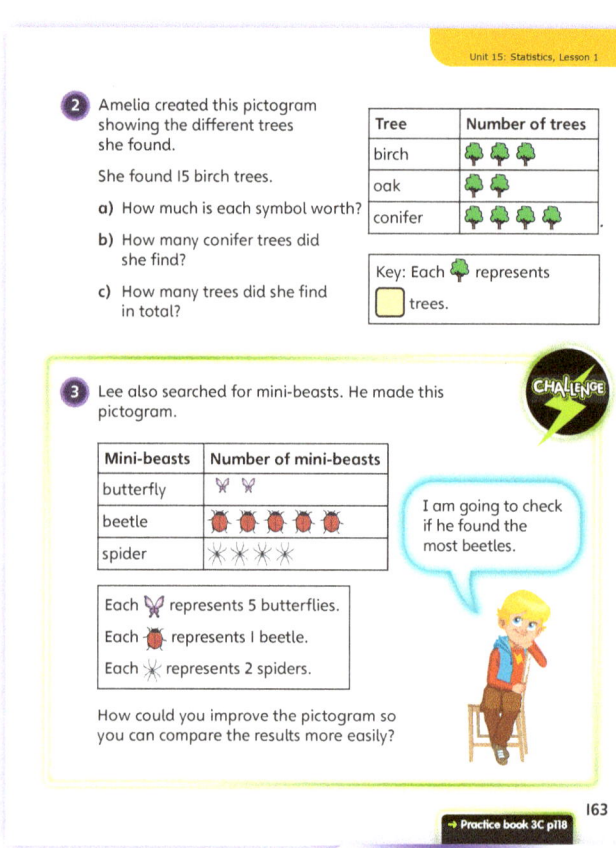

PUPIL TEXTBOOK 3C PAGE 162

PUPIL TEXTBOOK 3C PAGE 163

Unit 15: Statistics, Lesson 1

Practice

WAYS OF WORKING Independent thinking

IN FOCUS Questions ❶ and ❷ ask children to interpret information from a pictogram. Children should take care when working with half symbols. Look out for children's common misconceptions, such as not reading the key and assuming how many children each symbol represents. Question ❸ presents children with more of a multi-step problem. Encourage them to be efficient with their calculations – for example, adding and then multiplying (or multiplying then adding). Ask children to consider both methods. Question ❹ is the first example where a symbol represents 4 items, and where there are quarter, half and three-quarter symbols.

STRENGTHEN Check that children do not have the common misconception that one symbol always represents one item. Ensure they read all the information given (including the key) before they tackle the problem. If children are struggling with their multiplication facts, you might want to encourage them to skip count in order to get the answers. They could also use a number line for support.

DEEPEN Question ❸ again asks children to calculate totals where the number of symbols in a pictogram does not tally 1:1 for the number of objects (in this case, the number of farm animals). For question ❺, ask children to draw their own correct pictogram without the mistakes.

ASSESSMENT CHECKPOINT Use questions ❶ to ❹ to check that children can answer simple questions based on pictograms using their knowledge of multiplication facts. Use question ❺ to check that children understand common mistakes that people make when drawing pictograms.

ANSWERS Answers for the **Practice** part of the lesson can be found in the *Power Maths* online subscription.

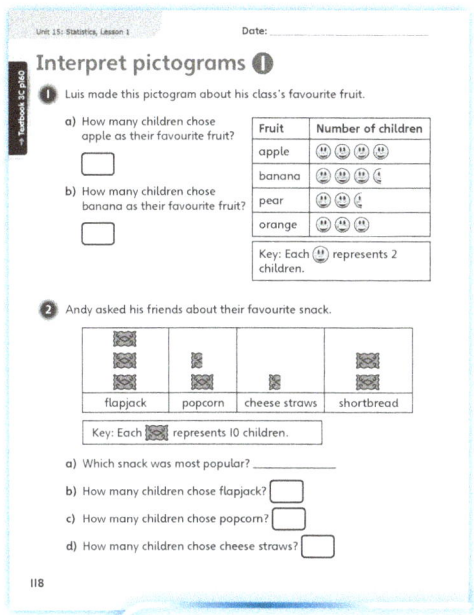

PUPIL PRACTICE BOOK 3C PAGE 118

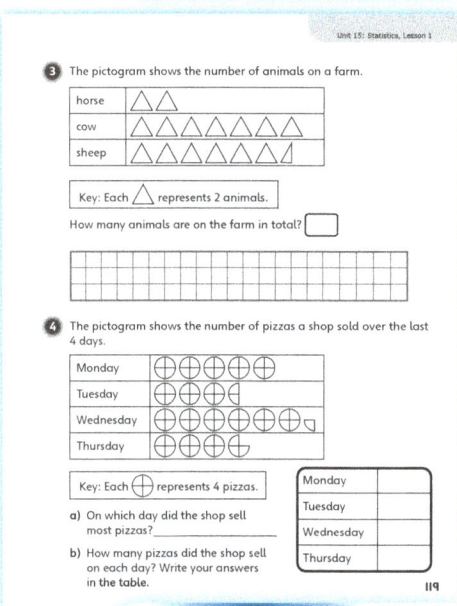

PUPIL PRACTICE BOOK 3C PAGE 119

Reflect

WAYS OF WORKING Pair work

IN FOCUS Children should discuss the **Reflect** question in pairs. They should then rehearse their answers before the teacher asks for feedback. To help children with this, you could draw any pictogram on the board where a symbol represents 2 or more people, but do not give children a key. You could then ask children a question about the pictogram (which they will not be able to answer because they do not have a key). Ask: *Can you explain to a partner why you can't answer this question?*

ASSESSMENT CHECKPOINT Use this activity to assess if children are able to explain the importance of the key in pictograms. Listen for children giving answers such as if the key was not there then they would not be able to work out the answer accurately.

ANSWERS Answers for the **Reflect** part of the lesson can be found in the *Power Maths* online subscription.

After the lesson

- Are children secure at interpreting pictograms where symbols do not equal 1?
- Are children able to spot mistakes on pictograms?

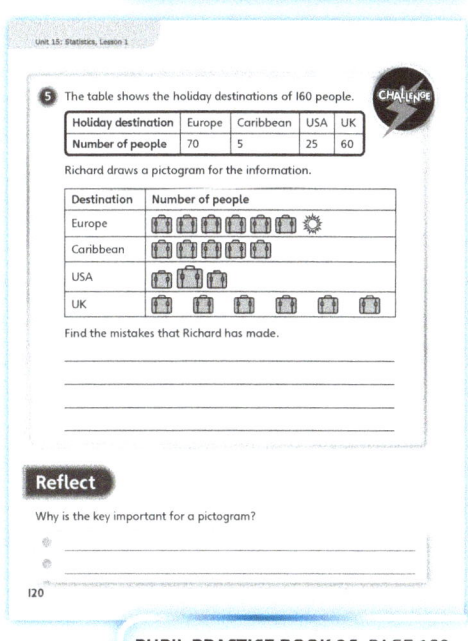

PUPIL PRACTICE BOOK 3C PAGE 120

201

Unit 15: Statistics, Lesson 2

Interpret pictograms ❷

Learning focus
In this lesson, children will solve 1- and 2-step problems based on information that is presented in pictograms.

Before you teach
- Are children secure with interpreting pictograms, including those with half symbols?
- Are there any common misconceptions from Lesson 1 that you need to address?

NATIONAL CURRICULUM LINKS

Year 3 Statistics

Solve one-step and two-step questions [for example, 'How many more?' and 'How many fewer?'] using information presented in scaled bar charts, pictograms and tables.

ASSESSING MASTERY

Children can answer a range of 1- and 2-step problems using information that is presented in a pictogram. Children can answer questions involving the analysis of more than one pictogram and make comparisons based on this data.

COMMON MISCONCEPTIONS

When comparing and using data across more than one pictogram, children may assume the key is the same. To help address this, draw attention to the key on each pictogram and ask:
- *Where do you look on a pictogram to find out what each symbol is worth? Are the symbols worth the same on both pictograms?*

STRENGTHENING UNDERSTANDING

Continue to encourage children to represent pictograms using practical equipment. By doing so, children will be able to physically manipulate each item, which will assist them in calculations, especially when they are using data across more than one pictogram. It is also important to continue to encourage children to use number lines and other representations to support their calculations.

GOING DEEPER

Encourage children to make their own statements based on the pictograms that are presented. These statements should go beyond the questions asked. Ask: *What else can you tell me based on this pictogram?*

KEY LANGUAGE

In lesson: pictogram, key, symbol, compare, least, most, altogether

Other language to be used by the teacher: whole symbol, half symbol, inverse

STRUCTURES AND REPRESENTATIONS

Number lines, pictograms

RESOURCES

Optional: number lines, cubes or counters

 In the eTextbook of this lesson, you will find interactive links to a selection of teaching tools.

Quick recap

On the board show 3 packs of 10 pens (or something similar). Ask: *How many pens are there? How do you know?* Then, next to this, show 5 packs of 10 pens. Ask: *How many pens are there now?* Discuss with children the different ways to work out how many pens there are in total.

Unit 15: Statistics, Lesson 2

Discover

WAYS OF WORKING Pair work

ASK
- Question 1 a): *What type of chart is this? How does it work?*
- Question 1 a): *What can you tell from the pictogram?*
- Question 1 a): *Can you think of any other questions that you could ask about this pictogram?*
- Question 1 b): *How can you compare data from the pictogram?*

IN FOCUS In this lesson, children are asked to use information that is presented in pictograms to answer 1- and 2-step questions about the data. Draw children's attention to the pictorial nature of pictograms and how this means that sometimes they do not need to calculate the value for each item in their calculations.

PRACTICAL TIPS Encourage children to use physical representations, such as blocks or counters, to represent the pictogram in question 1.

ANSWERS

Question 1 a): 8 more children chose to play in midfield than goalkeeper.

Question 1 b): 16 children chose defender or forward as their favourite position.

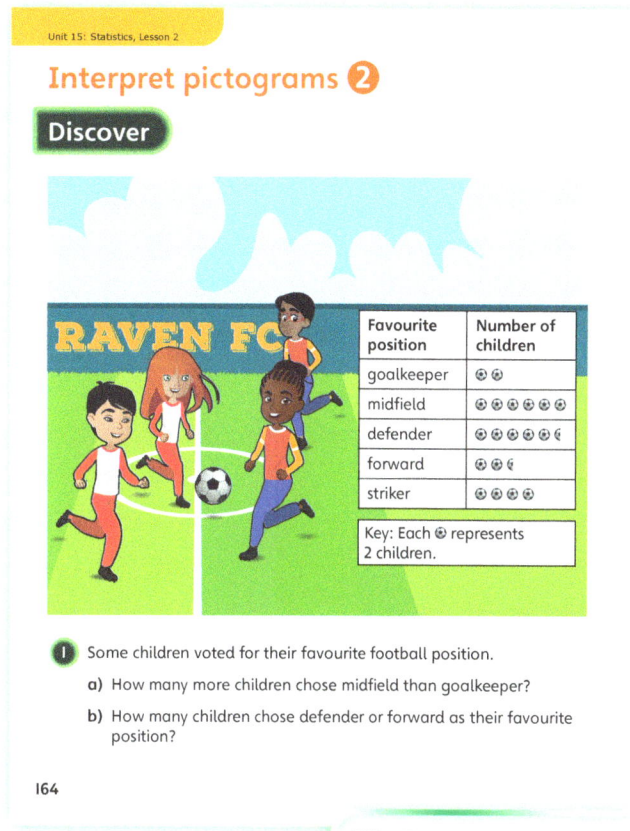

PUPIL TEXTBOOK 3C PAGE 164

Share

WAYS OF WORKING Whole class teacher led

ASK
- Question 1: *How did you work out the answer?*
- Question 1: *Did anyone use a different method? Which method is the most efficient?*

IN FOCUS There are a number of ways of arriving at the answers for both parts of this question. For questions 1 a) and b), two different methods are modelled. Draw children's attention to these different ways of working, and use these as a stimulus for discussing which method is most efficient. Also discuss how children can use the pictorial nature of the pictogram to help them in their calculations.

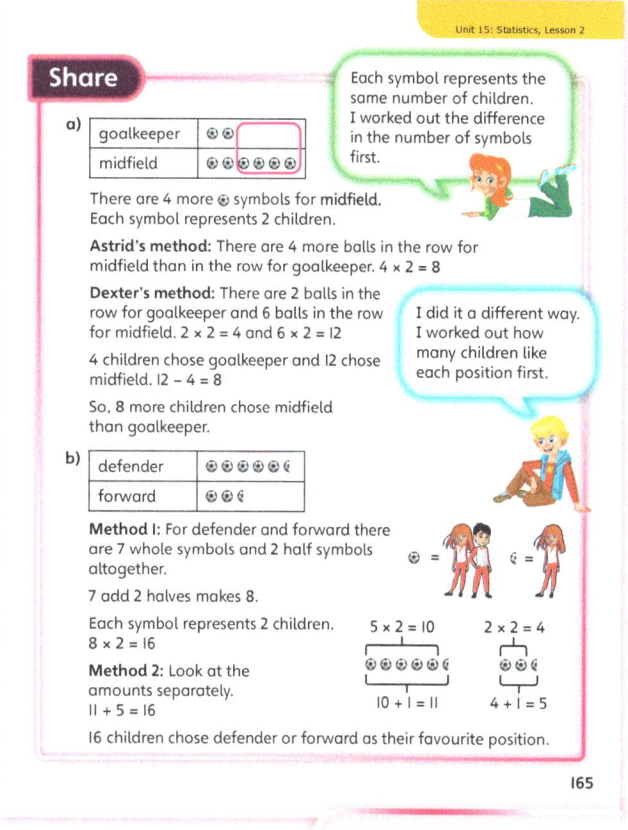

PUPIL TEXTBOOK 3C PAGE 165

203

Unit 15: Statistics, Lesson 2

Think together

WAYS OF WORKING Whole class teacher led (I do, We do, You do)

ASK
- Question ❶ a): *How can you work this out? Is there more than one way of doing this?*
- Question ❶ a): *How can you work out how many seven-year-olds there are? How can you work out how many six-year-olds they are?*
- Question ❶ b): *How can you work out the total number of seven- and eight-year-olds? Can you think of a different way?*
- Question ❶ c): *Is this information given on the pictogram? How can you use the information you have to answer the question?*
- Question ❷: *What steps do you need to take to find the solution?*

IN FOCUS Question ❶ draws children's attention to the range of information that they can find by analysing a pictogram, with part c) focusing on children finding the total number in the set (the total number of children who train for Raven FC) by totalling the number of symbols and working out the total value of these.

Question ❷ uses a different key from the other questions in this section. Ensure that children remember the importance of checking the key for each pictogram they analyse.

STRENGTHEN Question ❷ also requires two steps and it may be useful to model the thinking out loud (for example, ask: *I think I need to work out who is the top goal scorer first. How can I do this? What do I need to work out next?*)

DEEPEN Question ❸ provides a good opportunity to begin to practise complex reasoning, as it requires the use of the given information together with the relationships between pieces of information in order to solve the problems presented. Ask: *What information is missing? How can you find the missing information?*

ASSESSMENT CHECKPOINT Use question ❶ to assess if children can answer simple analysis questions based on the information presented to them in a pictogram.

ANSWERS

Question ❶ a): There are 4 more seven-year-olds than six-year-olds.

Question ❶ b): There are 19 seven- and eight-year-olds altogether.
(8 seven-year-olds + 11 eight-year-olds)

Question ❶ c): There are 40 players altogether.

Question ❷: The top goal scorer has scored 15 more goals than the next highest goal scorer.

Question ❸ a): 23 children have played 5 or more games.

Question ❸ b): 11 children have played 3 games.

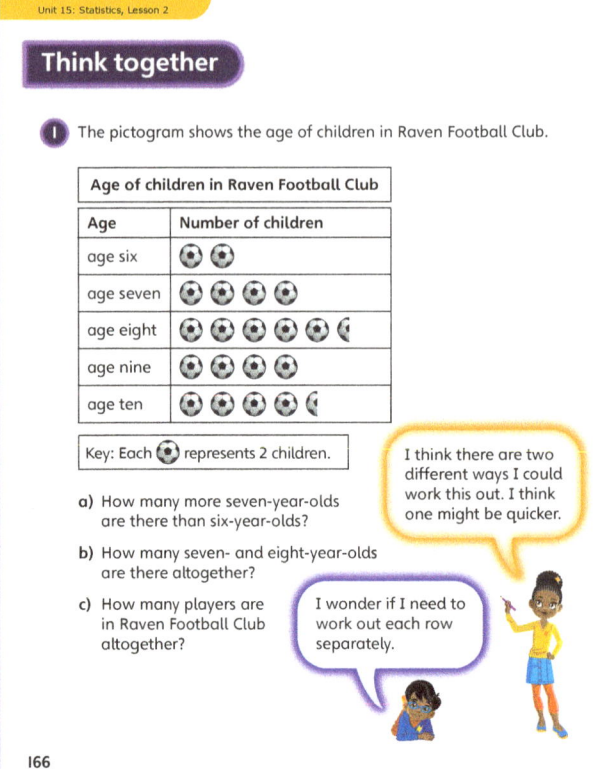

PUPIL TEXTBOOK 3C PAGE 166

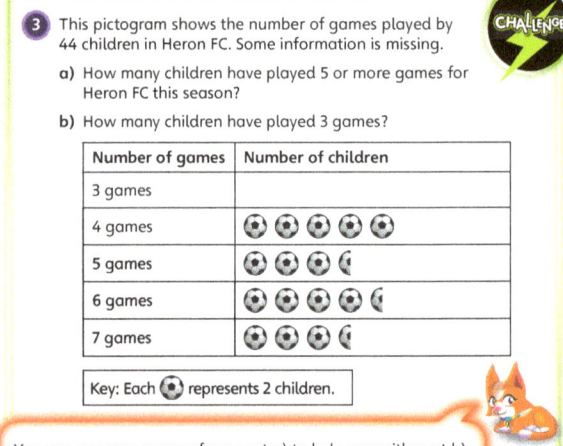

PUPIL TEXTBOOK 3C PAGE 167

Unit 15: Statistics, Lesson 2

Practice

WAYS OF WORKING Independent thinking

IN FOCUS Whilst only one method is scaffolded in questions ❶ and ❷, continue to draw children's attention to the different ways of working and continue to discuss which ways are the most efficient.

Questions ❷ and ❸ involve children interpreting data from across different pictograms and tables, and answering questions that involve data from both. Ensure that children notice that in question ❷ the symbols in the two pictograms have different values.

STRENGTHEN To support children with question ❶, encourage them to cross off each case symbol once they have counted it. This will help them to avoid double counting.

Children may need to practise answering questions about each individual pictogram in question ❷, before moving on to answering the question given, which involves using information from across both pictograms.

DEEPEN Children should be able to give their own statements about what they can tell through the interpretation of one or more pictograms.

THINK DIFFERENTLY In question ❷, children are required to choose the correct pictogram from which to extract the information for the first time. Discuss with children what the different pictograms show, before tackling the question.

ASSESSMENT CHECKPOINT Use questions ❶ and ❷ to assess if children can solve more complicated problems involving pictograms.

ANSWERS Answers for the **Practice** part of the lesson can be found in the *Power Maths* online subscription.

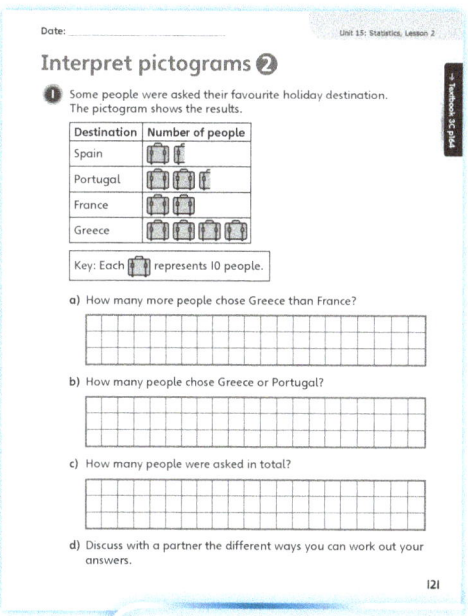

PUPIL PRACTICE BOOK 3C PAGE 121

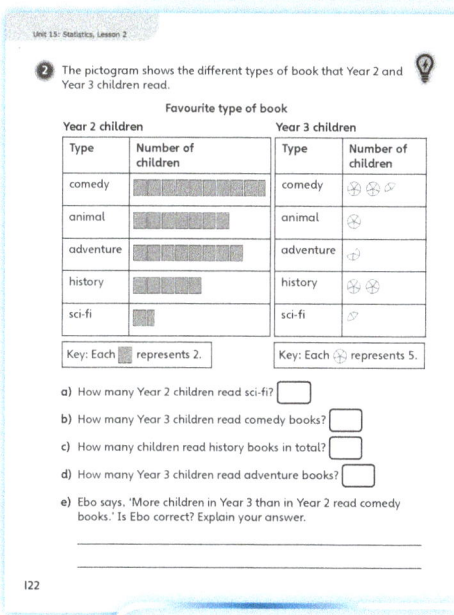

PUPIL PRACTICE BOOK 3C PAGE 122

Reflect

WAYS OF WORKING Independent thinking

IN FOCUS This **Reflect** question focuses on children understanding the things they need to do to create a pictogram which can be easily understood.

ASSESSMENT CHECKPOINT Use this question to assess if children are able to identify the key features of a pictogram and explain why they are necessary.

ANSWERS Answers for the **Reflect** part of the lesson can be found in the *Power Maths* online subscription.

PUPIL PRACTICE BOOK 3C PAGE 123

After the lesson
- Are children secure at making comparisons using pictograms?
- Can you incorporate pictograms into your day-to-day classroom routine to provide more practice at interpreting pictograms?

Unit 15: Statistics, Lesson 3

Draw pictograms

Learning focus
In this lesson, children will construct pictograms from a table of data.

Before you teach
- Check that children know the key features of a pictogram.
- Recap multiplication and division facts for the 2, 5 and 10 times-tables.

NATIONAL CURRICULUM LINKS

Year 3 Statistics

Interpret and present data using bar charts, pictograms and tables.

Solve one-step and two-step questions [for example, 'How many more?' and 'How many fewer?'] using information presented in scaled bar charts, pictograms and tables.

ASSESSING MASTERY

Children can represent data on a pictogram, using a symbol that represents 1, 2, 5 or 10 objects. The pictograms that children draw are accurate.

COMMON MISCONCEPTIONS

Children may believe that each pictogram symbol always represents '1'. Encourage children to think carefully about whether this is the case for every pictogram. Ask:
- *What happens if you are trying to represent 45? Would it be better to use a symbol that represents 5 or 10? Why?*

Children may only draw full symbols instead of part symbols. Discuss with children why sometimes they may need only part of a symbol to represent a number. Remind children of the work in previous lessons where half symbols were used and what each meant and how the value depended on the key. Explain why it is useful to have a symmetrical symbol.

Children may use different symbols when drawing a pictogram, or they may not line the symbols up. Emphasise the importance of using one consistent symbol that is easy to draw, that remains the same size throughout and that is correctly lined up with the other symbols. These are often the most common errors children make when drawing a pictogram.

STRENGTHENING UNDERSTANDING

Before they start drawing, encourage children to represent pictograms using practical equipment, such as cubes or counters. Have children look at examples of pictograms from previous lessons as well. Recapping what a good pictogram looks like will help children when they come to draw their own.

GOING DEEPER

Children collect their own data from the class and represent the information on a poster, writing down five summary points of what the data shows them. Challenge children to also consider some of the possible drawbacks of using pictograms.

KEY LANGUAGE

In lesson: pictogram, key, symbol

Other language to be used by the teacher: half symbol, part symbol, represent, draw, construct

STRUCTURES AND REPRESENTATIONS

Pictograms, tally charts

RESOURCES

Mandatory: rulers

Optional: number lines, connecting cubes

 In the eTextbook of this lesson, you will find interactive links to a selection of teaching tools.

Quick recap

Show children an accurate pictogram from a previous lesson and ask them to write down what elements they will always see on every pictogram. Then ask them to write down a fact about the pictogram you are displaying.

Unit 15: Statistics, Lesson 3

Discover

WAYS OF WORKING Pair work

ASK
- Question ① a): *Do you remember what we call this table? What information does the table show?*
- Question ① b): *What do you need to draw for the pictogram? What is the key? What symbol is being used and what does it represent? How many circles do you need to draw for walk? Why?*

IN FOCUS Ensure that children fully understand the situation. Set the context of a class working out how each person gets to school. Take time to discuss with children what the table tells them and that, for question ①, they only need to find the number that does *not* match the tally marks. Question ① b) asks children to represent the data on a pictogram themselves. Let children have a go and see if they can replicate ones they have seen earlier in this unit. Some children may want to make the pictogram using counters first. Watch out for children making any common errors, such as using symbols of different sizes or symbols that are not correctly aligned. If children are struggling, you could give them a pre-drawn table onto which they can add the symbols and key.

PRACTICAL TIPS You could do your own survey instead with the children and represent the children's results in your class.

ANSWERS

Question ① a): The number of children travelling by car should be 5.

Question ① b):

How children travel to school		
Method	Tally	Number
walk	⚫⚫⚫⚫⚫⚫⚫⚫⚫⚫⚫⚫	12
car	⚫⚫⚫⚫⚫	5
bike	⚫⚫⚫⚫⚫⚫⚫⚫⚫	9
bus	⚫⚫	2

Key: Each ⚫ represents 1 child

Share

WAYS OF WORKING Whole class teacher led

ASK
- Question ① a): *How did you check? How did you count the tallies quickly?*
- Question ① b): *What do you need to draw first? What does each symbol represent? How many symbols do you need for 'walk' and why?*

IN FOCUS In question ① a), remind children how to count the tallies efficiently by first counting in 5s. In question ① b), take children step by step through the process of constructing the pictogram starting with the table and the different rows and columns. You might want to construct it slowly on a whiteboard. Remind children that the symbols should be the same size and lined up. Talk through each row and the number of symbols in each one.

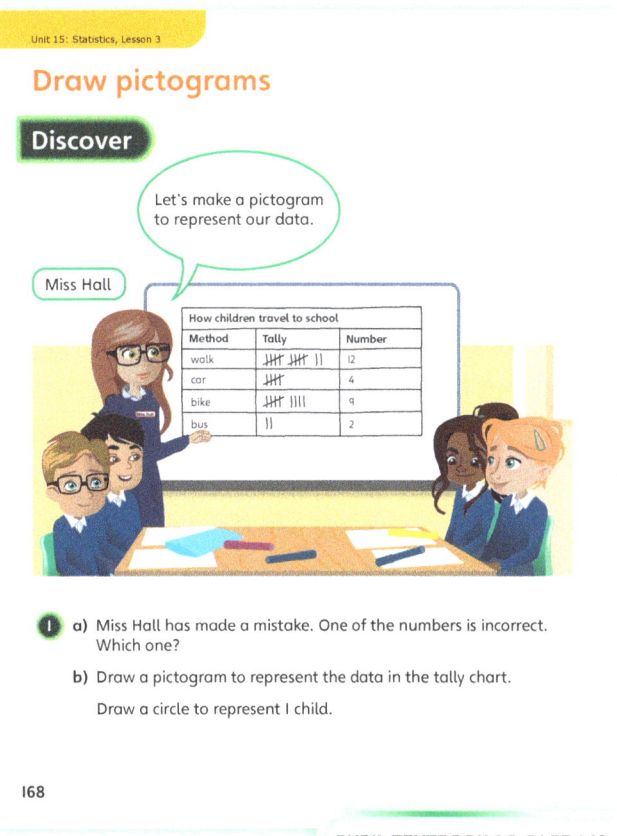

PUPIL TEXTBOOK 3C PAGE 168

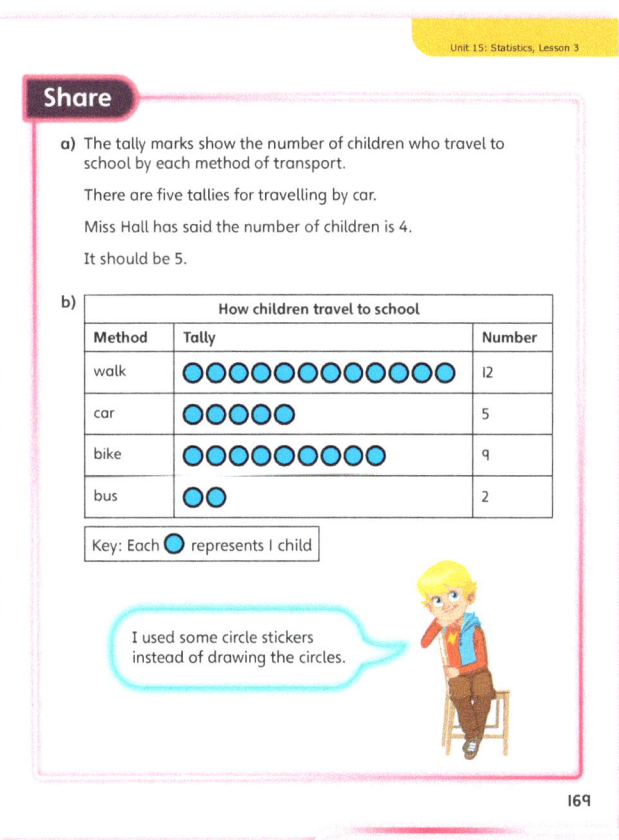

PUPIL TEXTBOOK 3C PAGE 169

Unit 15: Statistics, Lesson 3

Think together

WAYS OF WORKING Whole class teacher led (I do, We do, You do)

ASK
- Questions ① c) and d): *What would be the best symbol to use for the pictogram? Why are the others not useful? Why is it important the symbols are lined up and all about the same size?*
- Question ② a): *How can you work out how many symbols you need to draw for each line?*

IN FOCUS Question ① guides children through the thinking process involved in drawing a pictogram, encouraging them to consider why it might be better to use a symbol to represent 2 children instead of 1 this time. Explain that it will make drawing the pictogram more efficient and also make interpreting it easier. In question ① c), discuss the symbols that are most appropriate to use. Children should settle for a symmetrical symbol that they can easily replicate. In question ① d), work through the drawing of the pictogram slowly alongside the children, asking how many symbols will be needed for each line of information.

STRENGTHEN Before they start drawing, encourage children to represent pictograms using practical equipment, such as cubes or counters.

DEEPEN Look at question ② b) with children. Establish that the value of the symbol should be a number that will divide into all the numbers in the table. However, this is not possible for the numbers given in question ②, so it is best to choose a number that means drawing $\frac{1}{2}$ or $\frac{1}{4}$ of a symbol, rather than $\frac{1}{5}$ or $\frac{1}{6}$, for example. Then ask children to draw a pictogram to show this data where the symbol represents a different value (that is, not 4).

ASSESSMENT CHECKPOINT Check that children can work through the steps of drawing a pictogram, picking appropriate symbols and using an efficient key.

ANSWERS

Question ① a): Some children may have more than or fewer than one pet.

Question ① b): So that the pictogram has fewer symbols.

Question ① c): Children should choose the stick man, star or circle, as they are shapes that are easy to draw half of.

Question ① d): Check children's pictograms show 7 symbols for dog, 2 symbols for cat, $1\frac{1}{2}$ for rabbit, $2\frac{1}{2}$ for other and 5 symbols for none.

Question ② a):

Class	Number of children in class
3A	●●●●●●●
4A	●●●●●●●◖
5A	●●●●●●◖
6A	●●●●◖

Key: Each ● represents 4 children

Question ② b): Not all the numbers are divisible by 5 and it would be difficult to show $\frac{1}{5}$ of the circle.

Question ③: Emma should have used the same symbol, , for each flavour of ice cream.

Unit 15: Statistics, Lesson 3

Think together

① Some children were asked what type of pets they have.

What pets children have		
Type of pet	Tally	Number
dog	|||| |||| ||||	14
cat	||||	4
rabbit	|||	3
other	||||	5
none	|||| ||||	10

a) There are 28 children in the class.
Explain why the tally chart might not total to 28.

b) Miss Hall wants the class to use a symbol to represent 2 children instead of 1.
Why might she want to do this?

c) Which of these symbols could be used?
Which symbols would you not use?

d) Make the pictogram for the data in the table using a symbol to represent 2 children.

② Here is some data about the number of children in four classes.

Draw a circle to represent 4 children.

Class	Number of children in class
3A	28
4A	30
5A	25
6A	18

a) Draw a pictogram to represent this data.

b) Why might it not be a good idea to draw a circle to represent 5 children?

③ **CHALLENGE** What mistakes has Emma made in drawing this pictogram?

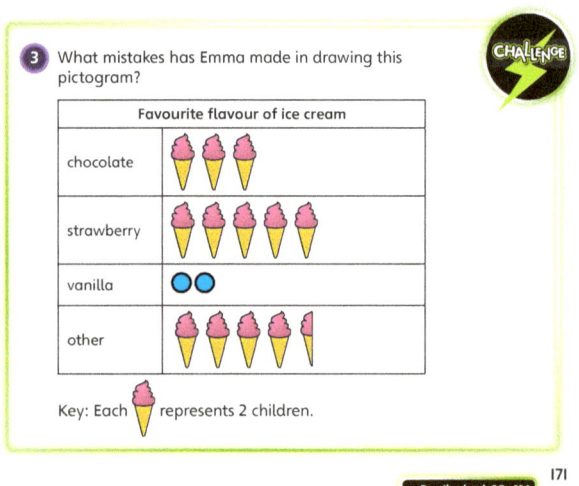

Key: Each 🍦 represents 2 children.

Unit 15: Statistics, Lesson 3

Practice

WAYS OF WORKING Independent thinking

IN FOCUS In question ①, children draw a simple pictogram where one symbol represents 1. Look for children making an effort to keep their symbols equal in size and lined up. These are really important skills for them to develop when they draw pictograms. Question ② asks children to complete a pictogram where the symbol represents 10. In this case, children will need to use half symbols. In question ③, children are given part of a table and part of a pictogram and they have to complete both using the cross-over information.

STRENGTHEN Before they start drawing, encourage children to represent pictograms using practical equipment, such as cubes or counters. Have them also look at examples of pictograms from previous lessons. Recapping what a good pictogram looks like will help children when they come to draw their own.

DEEPEN Children collect their own data from the class and represent the information on a poster, writing down five summary points of what the data shows them. Challenge children to also consider some of the possible drawbacks of using pictograms.

THINK DIFFERENTLY In question ③, children first need to work out the value of the symbol. They should realise that the chocolate bar is the only one where they are given information on both charts. They should then use this to work out that each symbol is worth 5.

ASSESSMENT CHECKPOINT Use questions ① and ② to assess that children can draw pictograms accurately.

ANSWERS Answers for the **Practice** part of the lesson can be found in the *Power Maths* online subscription.

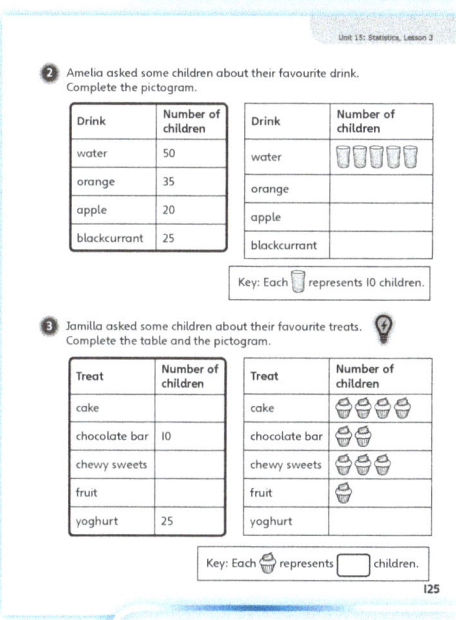

PUPIL PRACTICE BOOK 3C PAGE 124

PUPIL PRACTICE BOOK 3C PAGE 125

Reflect

WAYS OF WORKING Pair work or independent thinking

IN FOCUS Children should discuss the statement in pairs, or think about it by themselves. Children may be able to think of some advantages and disadvantages of doing as Kate suggests.

ASSESSMENT CHECKPOINT Check that children know that, for problems involving large numbers, it would be inefficient always to draw pictograms where 1 item is represented by 1 symbol.

ANSWERS Answers for the **Reflect** part of the lesson can be found in the *Power Maths* online subscription.

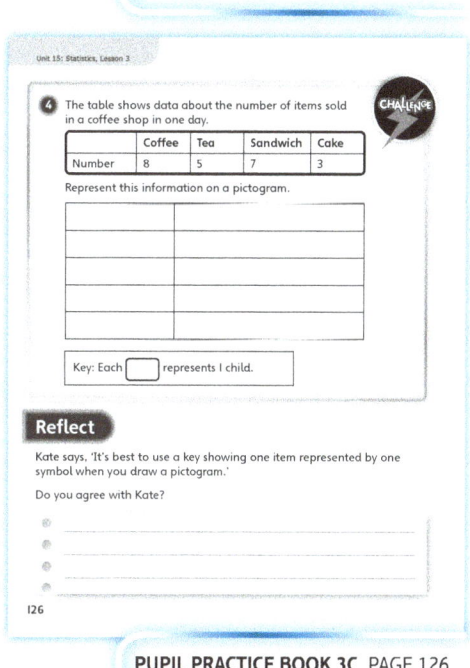

PUPIL PRACTICE BOOK 3C PAGE 126

After the lesson

- Can children draw a pictogram when given information in a table?
- Do children draw pictograms with correctly aligned and sized symbols?
- Do children know they need to include a key on their pictograms?

Unit 15: Statistics, Lesson 4

Interpret bar charts

Learning focus
In this lesson, children will read and interpret bar charts that have a range of scales.

Before you teach
- Are children confident counting in multiples of 2, 5 and 10?
- Are children confident finding numbers that lie half-way between two different numbers?

NATIONAL CURRICULUM LINKS

Year 3 Statistics
Interpret and present data using bar charts, pictograms and tables.

ASSESSING MASTERY
Children can read data and values from a range of bar charts, including those that have scales that increase in multiples of 2, 5 and 10. Children can read values where the height of the bar falls in between two marked divisions and can answer simple comparison questions by comparing the heights of the bars.

COMMON MISCONCEPTIONS
Children may misread the scales on a bar chart, assuming that each division always represents 1. To help address this, draw attention to the scale by asking:
- *What do you notice about the scale on the vertical axis of this chart? What do the numbers increase by each time?*

Children may work out comparison questions by working out the value of each bar, rather than by comparing the height of each bar. Draw children's attention to the structure of a bar chart. Ask:
- *What does it mean if the bar for one category is higher than the bar for another? Will this always be the case? Do you need to work out the value of each bar in order to compare them?*

STRENGTHENING UNDERSTANDING
To help children interpret the scale on a bar chart, liken it to a number line. Rotate the bar chart so that the vertical axis runs horizontally, with 0 to the left hand side, to help children see this connection. Ask: *What does this look like? How is it similar to a number line? How is it different?* Children may also benefit from recreating bar charts using connecting cubes, in order to help them compare the height of each bar.

GOING DEEPER
Encourage children to begin to make statements based on the data presented to them in bar charts. For example, you could ask: *What can you tell based on this bar chart?*

KEY LANGUAGE
In lesson: **bar chart**, horizontal axis, **vertical axis**, scale, half-way between

Other language to be used by the teacher: inverse, most, least, altogether

STRUCTURES AND REPRESENTATIONS
Number lines, bar charts

RESOURCES
Mandatory: rulers
Optional: number lines, connecting cubes

 In the eTextbook of this lesson, you will find interactive links to a selection of teaching tools.

Quick recap
Ask children what they remember about block diagrams from Year 2. Ask: *What information can you put on a block diagram? How can you get information from a block diagram?*

210

Unit 15: Statistics, Lesson 4

Discover

WAYS OF WORKING Pair work

ASK

- Question 1 a): *What type of chart is this? How does it work?*
- Question 1 a): *Can you come up with questions that could be answered with this bar chart?*
- Question 1 a): *Does it look similar to any other type of chart you have seen before?*
- Question 1 b): *What can you tell from the chart?*

IN FOCUS Children are familiar with block diagrams from Year 2, but this is the first time that children have met bar charts. Introduce the vocabulary 'bar chart' and discuss any initial similarities children can see between block diagrams and bar charts.

PRACTICAL TIPS Provide children with interlocking cubes and encourage them to replicate the bar charts.

ANSWERS

Question 1 a): The children found 6 clams and 10 barnacles.
The children found 16 clams and barnacles altogether.

Question 1 b): The children found 11 limpets.

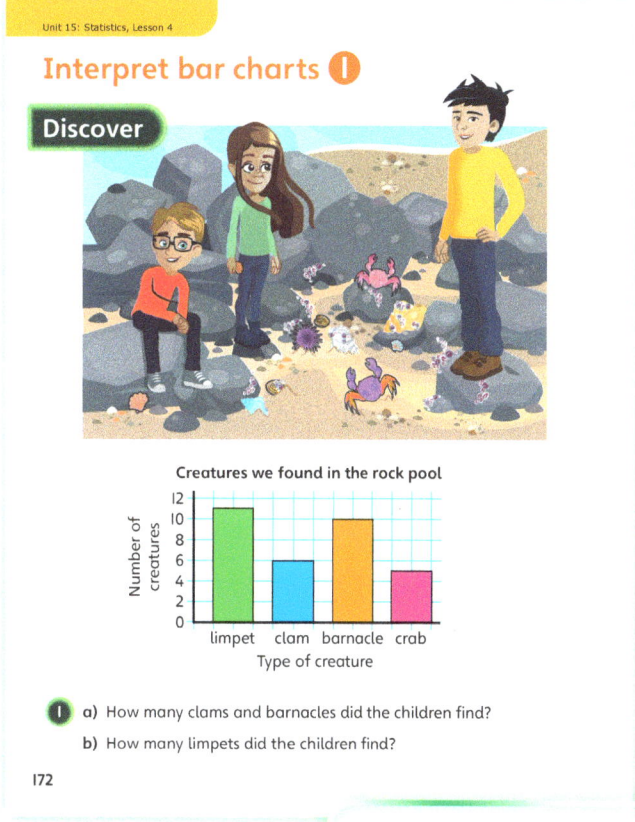

PUPIL TEXTBOOK 3C PAGE 172

Share

WAYS OF WORKING Whole class teacher led

ASK

- Question 1 a): *How did you work out the answer? How do you know how many they collected?*
- Question 1 a): *How can you make sure you read the correct value on the vertical axis?*
- Question 1 b): *What do you notice about the height of the bar for limpets?*
- Question 1 b): *How can you work out the value if the bar is half-way between two numbers on the scale?*

IN FOCUS The focus in this section is on children accurately reading values from the chart. It is suggested that children use a ruler, lining up the top of the ruler with the top of the bar, in order to help them read across to the scale on the vertical axis. Through discussion in this section, draw children's attention to the different features of a bar chart, including how the scale always increases in steps of the same size.

Question 1 b) involves a bar with a height half-way between two marked values on the scale. Through questioning, help children to make the connection with the number that is half-way between 10 and 12.

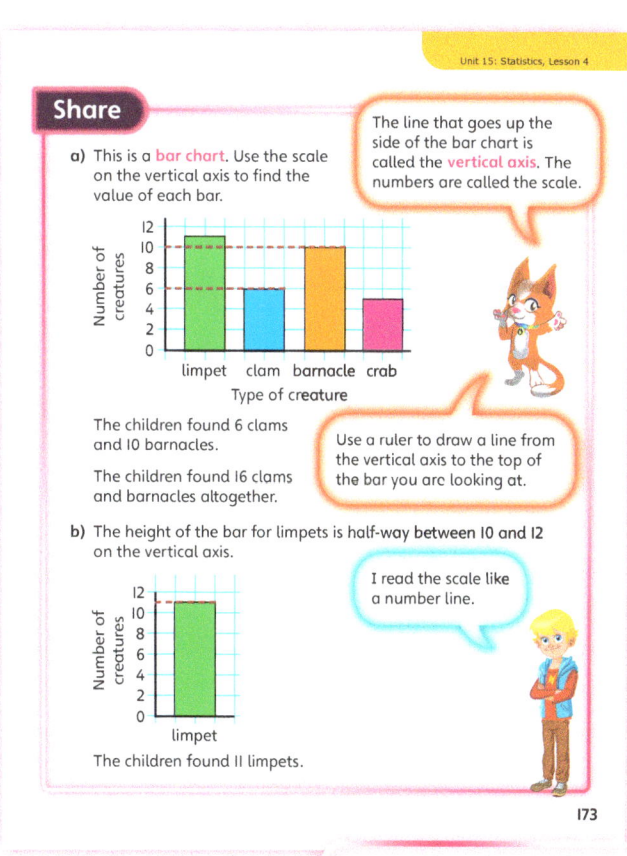

PUPIL TEXTBOOK 3C PAGE 173

211

Unit 15: Statistics, Lesson 4

Think together

WAYS OF WORKING Whole class teacher led (I do, We do, You do)

ASK
- Question ① a): *How can you work out the value for gutweed? What number does it line up with on the vertical axis?*
- Question ① b): *How can you work out the value for sea lettuce? What number is half-way between 6 and 8?*
- Question ②: *What do you notice about the scale on the vertical axis for this question?*
- Questions ② a) and b): *Is there more than one way to solve this question? Do you need to work out the value for each child first?*
- Question ③: *What was the most popular sea animal?*

IN FOCUS In this section, children are introduced to bar charts where the scales increase in multiples of 2 or 10 and where each division is a new number on the scale. In question ②, children should discuss how they can use the heights of the bars in order to compare the values, rather than needing to read the values for each bar individually. Question ③ involves children writing their own news article based on the data. You may want to make it real by having children collect data about their own class.

STRENGTHEN Children may need support to work out the values of bars where the height is between two marked numbers on the scale. It may be helpful to: discuss what the bar chart would look like if the scale went up in 1s; and to re-draw the chart in question ① so that the scale has one division per number, but only multiples of 2 marked.

DEEPEN Children should begin to answer simple questions that involve interpreting data from bar charts. Question ③ provides an opportunity to do this. You could ask the class the same question and then compare Class 3A's data with the data collected from your own class. Encourage children to think about statements that they could make that compare the two sets of data.

ASSESSMENT CHECKPOINT Use question ① to assess if children can independently read values from a bar chart, including where the height of the bar is half-way between two marked numbers on the scale.

ANSWERS

Question ① a): The children found 8 gutweeds.

Question ① b): They found 7 sea lettuces.

Question ① c): They found 13 pieces of coral weed and 1 sea oak.

Question ② a) Emma found the most shells.

Question ② b): Andy found the fewest shells.

Question ② c): Andy found 45 shells, Yugo found 60 shells and Emma found 75 shells.

Question ③: Children's news summaries will vary. For example: Class 3A has a total of 32 votes. The sea animals voted for in order of most favourite to least favourite were as follows: dolphin – 9; sea horse – 8; shark – 7; starfish – 6; jellyfish – 2.

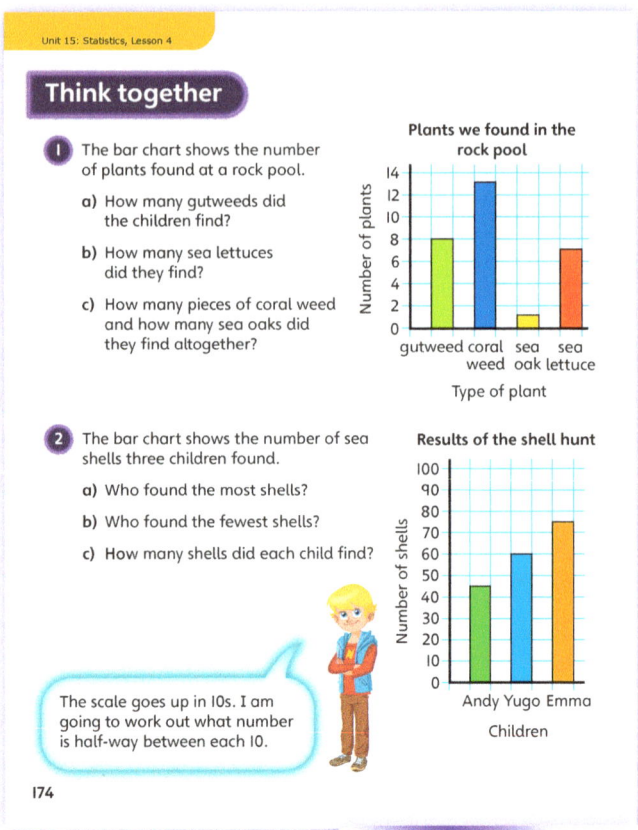

PUPIL TEXTBOOK 3C PAGE 174

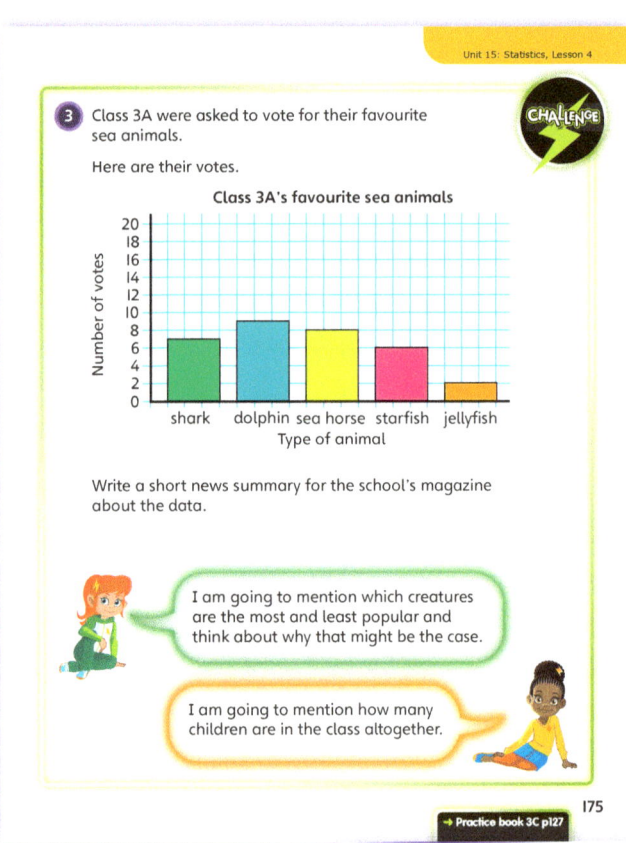

PUPIL TEXTBOOK 3C PAGE 175

Unit 15: Statistics, Lesson 4

Practice

WAYS OF WORKING Independent thinking

IN FOCUS The bar charts used in the questions in this section have scales that go up in 2s or 10s. Ensure that children pay attention to the scale of each chart and therefore read the values correctly.

In question ❷, draw children's attention to the fact that they can work out the most and least popular days without finding the values of all the bars.

Question ❸ involves a bar chart where the scale on the vertical axis is missing.

STRENGTHEN To support children with completing the bar chart in question ❸, encourage them to notice that the height of the bar for '2 weeks' is $5\frac{1}{2}$ squares high. Agree that the table shows this bar should represent 55 animals. Help children to see that this means each square represents 10 animals, so they can label the intervals on the vertical axis. From this point, children can complete the table by reading off the values for '1 week' and '4 weeks'. Lastly, they should use a ruler to draw a bar $3\frac{1}{2}$ squares high for '3 weeks'.

DEEPEN Children should begin to solve more complex problems involving a bar chart, including where they need to compare different sources of data in order to complete the chart, and where there are missing pieces of information. Question ❸ encourages children to do this, and relies on children having a deep understanding of how bar charts are constructed. Ask: *How can you work out what the scale should be on the vertical axis? What information can you use to help?*

ASSESSMENT CHECKPOINT Use question ❷ to assess if children understand the structure and features of a bar chart. Can they explain why Emma is incorrect?

ANSWERS Answers for the **Practice** part of the lesson can be found in the *Power Maths* online subscription.

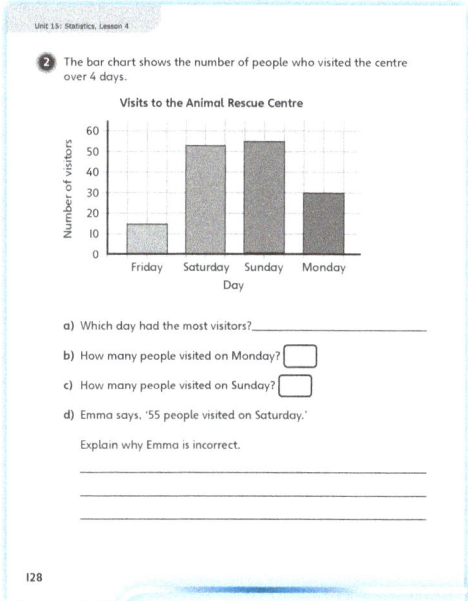

PUPIL PRACTICE BOOK 3C PAGE 127

PUPIL PRACTICE BOOK 3C PAGE 128

Reflect

WAYS OF WORKING Pair work

IN FOCUS This **Reflect** question focuses on children being able to read the value of each bar from the scale and also on their understanding that one of the benefits of a bar chart is the ability to compare information visually, without having to work out the value of each bar.

ASSESSMENT CHECKPOINT Use this question to assess if children are secure in their understanding of the concept of a scale. Children who have mastered this lesson will be able to reason that the scale goes up in 10s and that the next closest value to that of Baxter is half a division less on the scale. Therefore it cannot mean 10 people less.

ANSWERS Answers for the **Reflect** part of the lesson can be found in the *Power Maths* online subscription.

After the lesson
- Are children secure at reading data from bar charts?
- Can children make comparisons by comparing the heights of the bars?

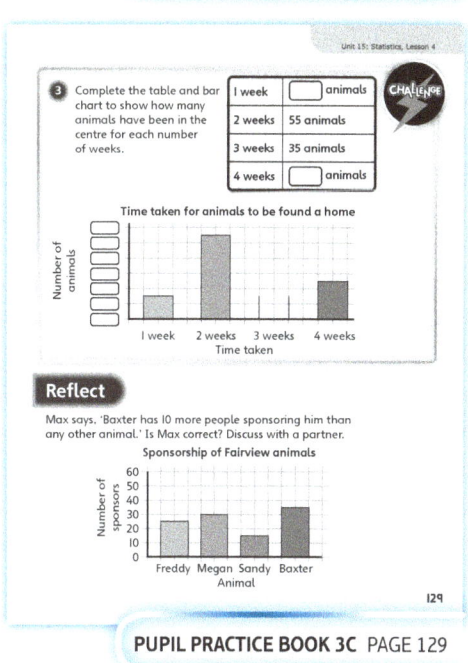

PUPIL PRACTICE BOOK 3C PAGE 129

Unit 15: Statistics, Lesson 5

Interpret bar charts 2

Learning focus
In this lesson, children will solve a range of 1- and 2-step problems based on the interpretation of bar charts.

Before you teach
- Are children secure at reading information from bar charts?
- Are there any common misconceptions from Lesson 4 that need to be addressed during this lesson?

NATIONAL CURRICULUM LINKS

Year 3 Statistics

Interpret and present data using bar charts, pictograms and tables.

Solve one-step and two-step questions [for example, 'How many more?' and 'How many fewer?'] using information presented in scaled bar charts, pictograms and tables.

ASSESSING MASTERY

Children can answer a wide range of 1- and 2-step problems using the information that is presented to them in bar charts. They can answer problems that involve comparisons and the finding of the total value of groups, as well as comparing the largest groups. Children can also answer simple logical reasoning questions based on bar charts.

COMMON MISCONCEPTIONS

Children may misinterpret the operation that is required to answer questions presented in words. To help address this, draw attention to the structure of the question, by asking:
- *What is the question asking you to do? Will this involve addition or subtraction? Do you have to carry out more than one step?*

STRENGTHENING UNDERSTANDING

Support children with the calculations required in this lesson by using the structures and representations from their earlier work on addition and subtraction. Help children to read the values of the different bars correctly by continuing to encourage the use of a ruler in order to ensure the correct value is read off the vertical axis.

GOING DEEPER

Encourage children to create their own bar chart with missing information and to come up with a set of clues to help their partner complete the chart.

KEY LANGUAGE

In lesson: bar chart, scale, horizontal axis, vertical axis, half-way between

Other language to be used by the teacher: inverse, most, altogether

STRUCTURES AND REPRESENTATIONS

Number lines, bar charts

RESOURCES

Optional: number lines, interlocking cubes, rulers

 In the eTextbook of this lesson, you will find interactive links to a selection of teaching tools.

Quick recap

On the board, display a bar chart that children met in the previous lesson. Ask them to write down five things about the data that it shows. Share the results as a class. You might want to encourage the use of words such as 'most', 'fewest', 'more' and 'less'.

Unit 15: Statistics, Lesson 5

Discover

WAYS OF WORKING Pair work

ASK

- Question 1 a): *What information can you tell from the bar chart? How many people like superhero films? How many people like comedy films?*
- Question 1 a): *What method could you use to find the difference? Do you prefer to count on from 12 to 20 using a number line? Or would you prefer to find 20 – 12 using column subtraction?*
- Question 1 b): *What operation do you need to use in order to work out the number of people who like either comedy or animated films best?*

IN FOCUS In this activity, children use the skills and knowledge of bar charts gained in Lesson 4 to answer problems based on data. In question 1 a), children should realise that they need to find the difference between 20 and 12. Talk about the different approaches that children could take: they could use column subtraction; they could use a number line to count on or back; or they could simply do the calculation mentally. In question 1 b), help children to read the question carefully and work out what they are being asked to find. The number of children who like *either* comedy *or* animated films is the sum of the 'comedy' and 'animated' categories.

ANSWERS

Question 1 a): 8 more people like superhero films than like comedy films.

Question 1 b): 27 people like comedy or animated films best.

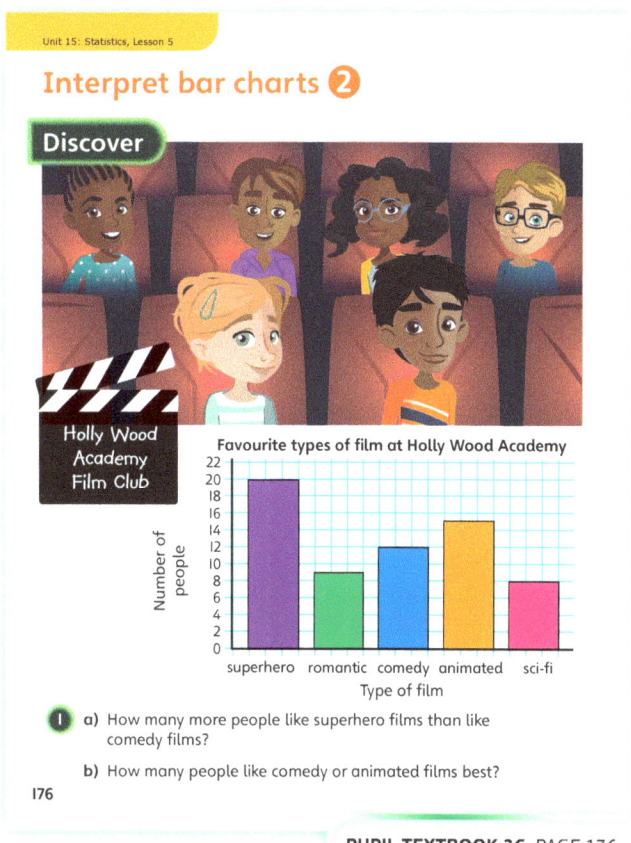

PUPIL TEXTBOOK 3C PAGE 176

Share

WAYS OF WORKING Whole class teacher led

ASK

- Question 1 a): *How did you work out the answer? What steps did you take?*
- Question 1 a): *Is there any other way you could have worked out the answer?*
- Question 1 b): *How did you work out how many people liked animated films best?*
- Question 1 b): *Which addition method is most efficient for this calculation?*

IN FOCUS In question 1 a), in order to work out the difference between the number of people who like superhero films best and the number who like comedy films best, children will need to use their subtraction skills. They can calculate this answer by finding the difference (between 12 and 20) by counting on, or by subtraction (20 subtract 12). It would be beneficial to discuss both methods with children and to discuss which is more efficient here.

Question 1 b) involves reading a value for a bar where the height is half-way between two marked values on the *y*-axis. This is a skill children developed in Lesson 4, but it is important to check that they can still accurately read values which do not line up with a marked value on the *y*-axis.

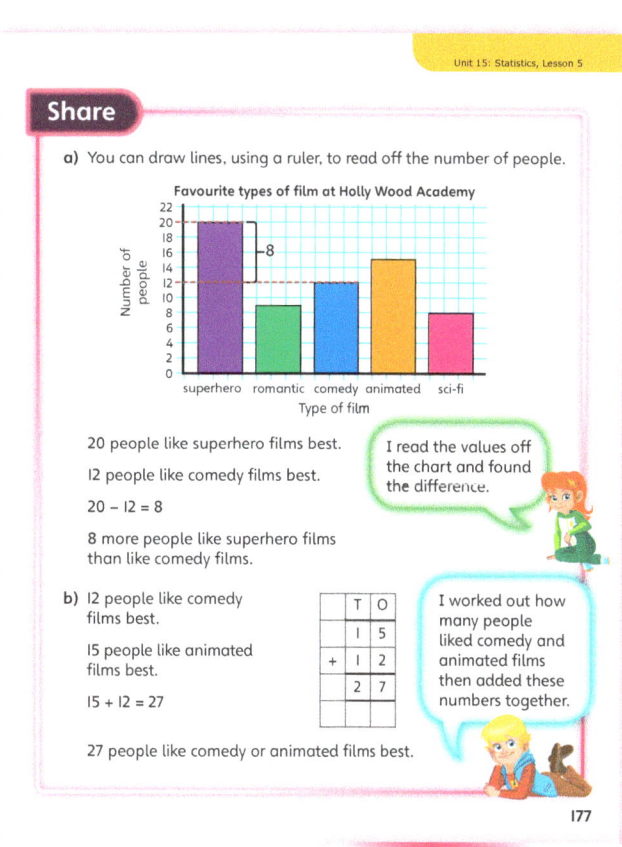

PUPIL TEXTBOOK 3C PAGE 177

215

Unit 15: Statistics, Lesson 5

Think together

WAYS OF WORKING Whole class teacher led (I do, We do, You do)

ASK
- Question ① a): *How can you work out the difference between the number of portions of sandwiches and the number of portions of nachos? What calculations do you need to carry out?*
- Question ① b): *How can you work out the most and least popular foods without working out the values of all the bars?*
- Question ③: *Which days are during the week? How can you work out the total for during the week?*
- Question ③: *What is the difference between the number of people who prefer to go to the cinema during the week and the number who prefer to go at the weekend?*
- Question ③: *How can you use the information you have in order to help you work out the answer?*

IN FOCUS In question ① b), ensure that children can recall how they can find the most and least popular by comparing the heights of the bars, rather than working out the value of each bar. Because question ③ is a multi-step question, ask children to describe their method to you before they begin. If children are finding it difficult to know what to do, ask: *How can you find the total number of people who went to the cinema during the week? How can you find the total number who went to the cinema at the weekend? Which of the two totals is greater?*

STRENGTHEN Support children with the calculations in this section by using a range of representations and models from their earlier work on addition and subtraction.

DEEPEN Children should begin to find solutions to more complex problems, including using the information provided to help deduce missing information and values. Question ③ provides a good opportunity to practise this. Encourage children to consider different ways of finding the solutions to problems.

ASSESSMENT CHECKPOINT Use question ① to assess if children can independently answer simple comparison and total questions based on bar charts.

ANSWERS

Question ① a): 50 − 35 = 15
15 more sandwiches were sold than nachos.

Question ① b): The most popular food was popcorn.
The least popular food was chocolate.
The bar was the tallest for popcorn and the shortest for chocolate.
The difference between the most popular and least popular food is 45 portions.

Question ②: The vertical axis should be labelled 0, 10, 20, 30, 40; the bars for student and senior need to be swapped.

Question ③: The number of people who prefer going during the week is 100.
The number of people who prefer going at the weekend is 105.
More people prefer to go to the cinema at the weekend.

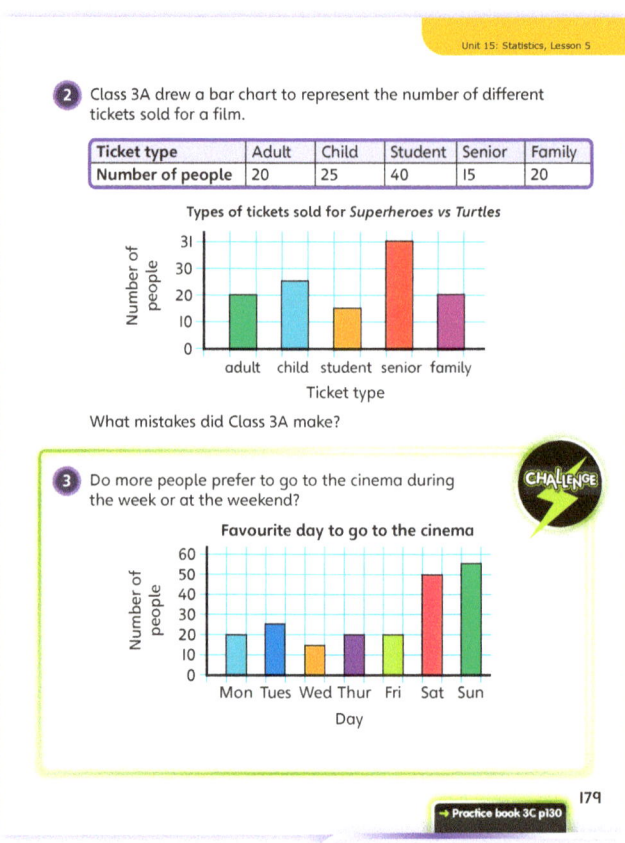

PUPIL TEXTBOOK 3C PAGE 178

PUPIL TEXTBOOK 3C PAGE 179

Unit 15: Statistics, Lesson 5

Practice

WAYS OF WORKING Independent thinking

IN FOCUS Question ❷ involves children using the information provided in order to work out missing information. Children need to be secure in interpreting the value of bars half-way between two divisions on the scale. In question ❷ a), children need to use the information that there were 6 Hanover kings and queens to work out that the scale increases in 2s. In question ❷ b), children need to add 1 (0·5 divisions) to the value of the Hanover bar to find the value of the Stuart bar.

Question ❸ introduces children to answering questions by comparing data from more than one bar chart for the first time.

The activities in this section use scales that go up in 2s or 10s. Ensure that children pay attention to the scale of each chart and therefore read the values correctly.

STRENGTHEN To support children with question ❷, encourage them to imagine the scale on the *y*-axis as a number line. They can draw on the jumps required to get from 0, which is the bottom of the *y*-axis, to the top of the Hanover bar, which is 6. Children will see that each jump must be 2, so the scale goes up in 2s. Then, children can draw the jump required from Hanover to Stuart and will see that $\frac{1}{2}$ a division has a value of 1.

DEEPEN Children should begin to find solutions to problems that require an element of logical reasoning. Question ❹ provides a good opportunity to practise this. Encourage children to consider the order in which they need to use the clues in order to find the missing information and construct the chart.

ASSESSMENT CHECKPOINT Use question ❷ to assess if children understand the structure and features of a bar chart and can therefore complete the chart.

ANSWERS Answers for the **Practice** part of the lesson can be found in the *Power Maths* online subscription.

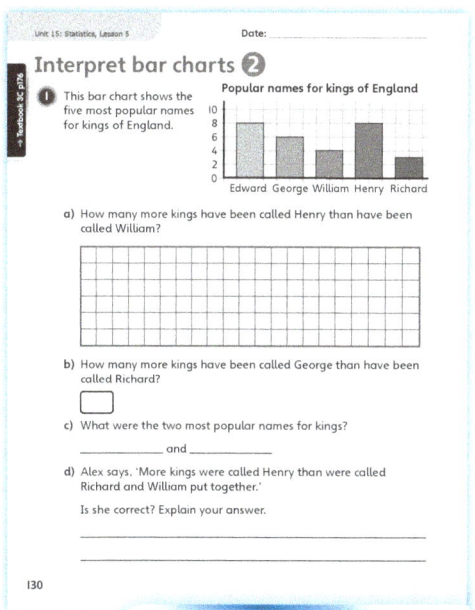

PUPIL PRACTICE BOOK 3C PAGE 130

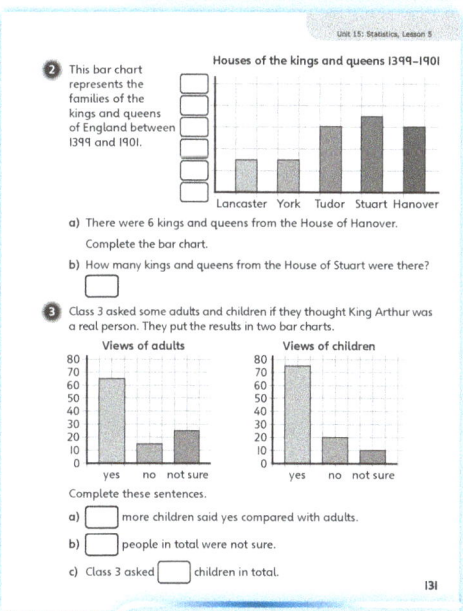

PUPIL PRACTICE BOOK 3C PAGE 131

Reflect

WAYS OF WORKING Pair work

IN FOCUS This **Reflect** question helps children begin to compare and contrast the different ways of presenting data that they have come across so far.

ASSESSMENT CHECKPOINT Use this activity to assess if children understand the features of pictograms and bar charts and if they are able to compare their advantages and disadvantages.

ANSWERS Answers for the **Reflect** part of the lesson can be found in the *Power Maths* online subscription.

After the lesson

- Are children secure answering questions based on bar charts?
- Can you incorporate bar charts into your day-to-day classroom activities so children have continued practice at interpreting them?

PUPIL PRACTICE BOOK 3C PAGE 132

Unit 15: Statistics, Lesson 6

Collect and represent data in a bar chart

Learning focus
In this lesson, children will solve a range of 1- and 2-step problems based on the interpretation of bar charts.

Before you teach
- Check that children understand how to collect data using a tally chart.
- Check that children understand how to read data from a bar chart.
- Ensure that children understand the key features of a bar chart.

NATIONAL CURRICULUM LINKS

Year 3 Statistics

Interpret and present data using bar charts, pictograms and tables.

ASSESSING MASTERY

Children can represent data from a table in a bar chart. They can draw accurate bar charts, including ensuring that the bars are the same width and there are gaps between the bars.

COMMON MISCONCEPTIONS

Children draw bars of different widths. Explain to children that each of the bars on a bar chart should be the same width.

Children do not put gaps between bars. Explain to children that there should be gaps between the bars and also between the axes and the first bar. The gaps should be of equal width, but don't need to be the same width as the bars.

Children label the axes incorrectly. Children should label the bars directly underneath the bar. For the vertical axes, the numbers should increase by an equal amount each time.

STRENGTHENING UNDERSTANDING

Before children start drawing bar charts, encourage them to represent the data using practical equipment, such as cubes. Ask children to look at examples of bar charts from previous lessons as well. Ask them to identify common features. Recapping what a good bar chart looks like will help children when they come to draw their own.

GOING DEEPER

Ask children to represent some of the pictograms in Lessons 1 and 2 of this unit as bar charts. Ask children what is the same and what is different about the pictograms and bar charts. Ask: *What are the advantages and disadvantages of each diagram? Which do you prefer?*

KEY LANGUAGE

In lesson: bar chart, bar, width, height, represent, axis, scale

Other language to be used by the teacher: most, fewest

STRUCTURES AND REPRESENTATIONS

Bar charts

RESOURCES

Optional: cubes, squared paper

 In the eTextbook of this lesson, you will find interactive links to a selection of teaching tools.

Quick recap

Children should look back at the last two lessons and identify the common features of the bar charts they have seen. Ask children questions such as: *What goes on each axis? What do the heights of the bars tell you? Is there always a gap between the bars? Are the widths of the bars all the same?*

218

Discover

WAYS OF WORKING Pair work

ASK

- Question 1 a): *What does the bar chart show? What have the children done? How can you work out how many bottles were picked up?*
- Question 1 b): *Where would the bars for crisp packets and cans start and end? Why? Should there be a gap between the bars?*

IN FOCUS In question 1 a), children use their knowledge from previous lessons to read off how many bottles were picked up. For question 1 b), children should think about the bar they need to draw. You could provide squared paper for children to draw the bar chart. Children also need to discuss in pairs where the bars should be drawn. Remind them that they need to make sure the bars are the same width and that there is a gap between the bars.

PRACTICAL TIPS You could collect your own data and draw a bar chart to represent that data.

ANSWERS

Question 1 a): 15 bottles were picked up.

Question 1 b):

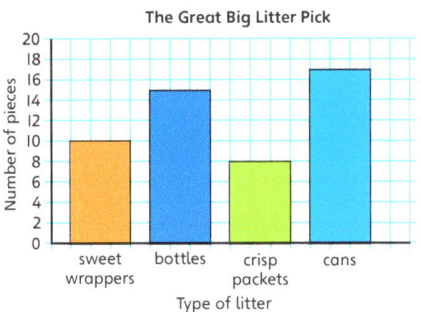

Share

WAYS OF WORKING Whole class teacher led

ASK

- Question 1 a): *Where does the bar go up to? What intervals does the axes go up in? How many bottles were picked up?*
- Question 1 b): *Where should the heights for the other items go? Why? What do you notice about the width of the bars? What do you notice about the gaps between the bars?*

IN FOCUS In question 1 a), explain that the bar for bottles goes up to half-way between 14 and 16. Discuss with children that this would be 15 bottles. In question 1 b), you may want to cover up the bars before children see where they have been drawn. Discuss with children that the bars need to be same width as the other ones and that there also needs to be a gap between the bars. They should note that the gap between the bars is the same as the gaps between the other bars.

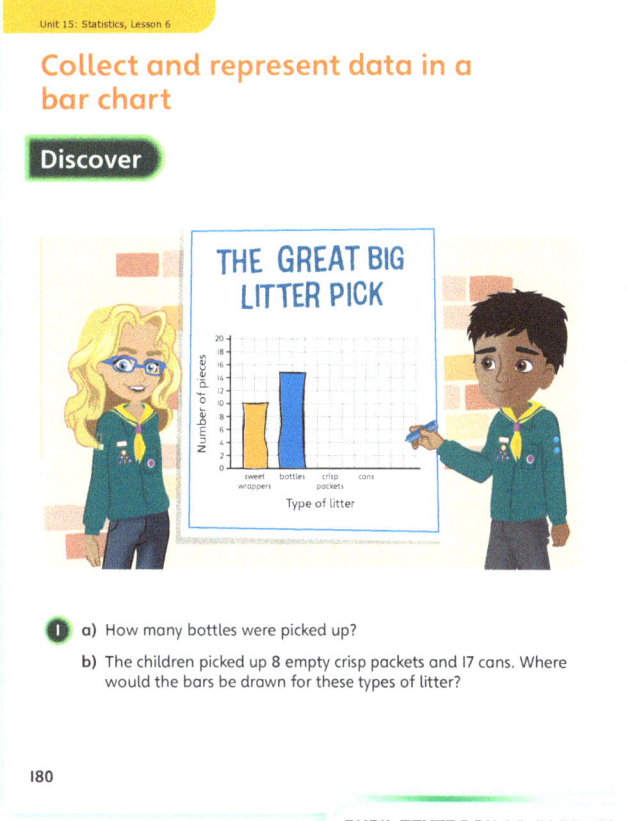

PUPIL TEXTBOOK 3C PAGE 180

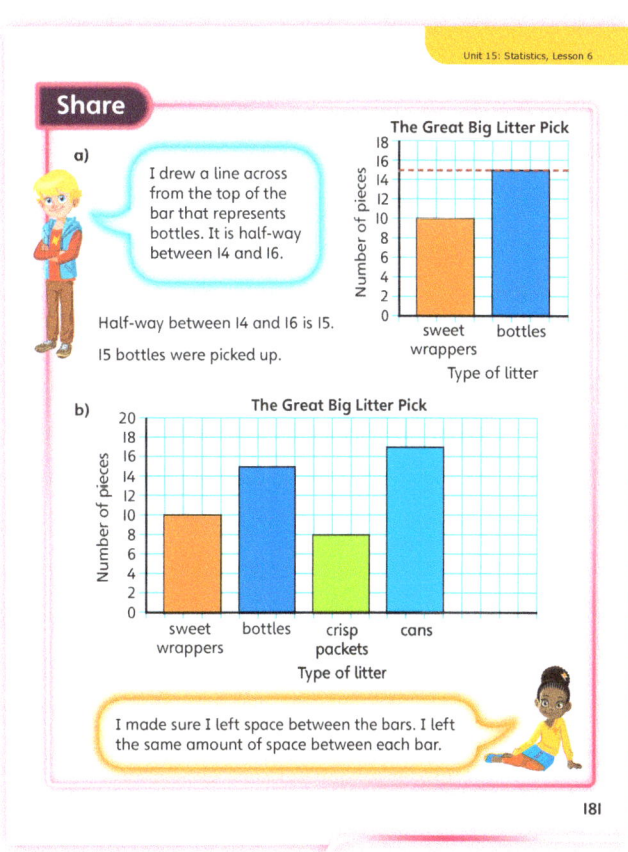

PUPIL TEXTBOOK 3C PAGE 181

Think together

WAYS OF WORKING Whole class teacher led (I do, We do, You do)

ASK

- Question ❶: *What interval does the y-axis on the bar chart go up by each time? Where would you draw the bar for 75? Where would the bar for cans go?*
- Question ❷: *What mistakes can you see? What is wrong with the numbers on the axes? What things are wrong with the bars?*
- Question ❸: *What do you want to find out about? How could you collect the data? How could you use a tally chart? How could you use a table? How will you represent this in a bar chart?*

IN FOCUS In question ❶, children complete a bar chart. This time the chart goes up in 10s, and they must find where values, such as 75 and 98, go. In question ❷, children work in pairs and discuss the things that are wrong with the chart. Focus their attention on the axes first and then on the bars. You might want to ask children to draw a correct version of the bar chart on a sheet of squared paper.

Question ❸ brings together all the learning from the last few lessons. As a class, choose something children would like to find out and then ask them to collect the data. Children should then represent this data on both a bar chart and pictogram and write a short summary. You might want to get children to do this as a poster or presentation.

STRENGTHEN Before children start drawing bar charts, encourage them to represent the data using practical equipment, such as cubes. Have children look at examples of bar charts from previous lessons as well. Ask them to identify common features. Recapping what a good bar chart looks like will help children when they come to draw their own.

DEEPEN For question ❷, ask children to draw a correct bar chart.

ASSESSMENT CHECKPOINT Questions ❶ and ❷ will help you check whether children can correctly draw a bar chart. Look to ensure that they do not make any common errors when drawing the bars.

ANSWERS

Question ❶:

Question ❷: The vertical axis needs to go up by the same increments. There should be a space between A and B. Each bar needs to be the same width.

Question ❸: Children are likely to want to record their data in a tally chart, then summarise their data in a table. Finally, they should draw a bar chart to represent their data.

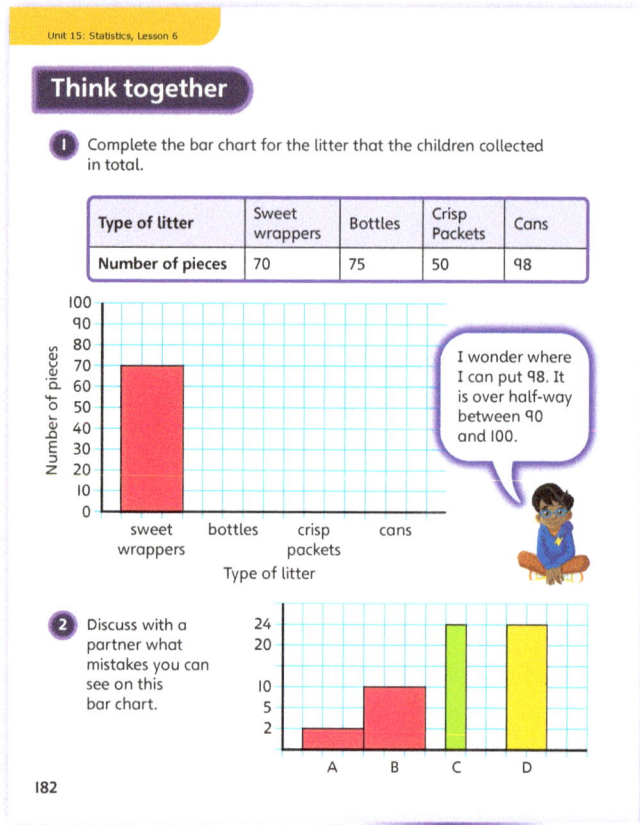

PUPIL TEXTBOOK 3C PAGE 182

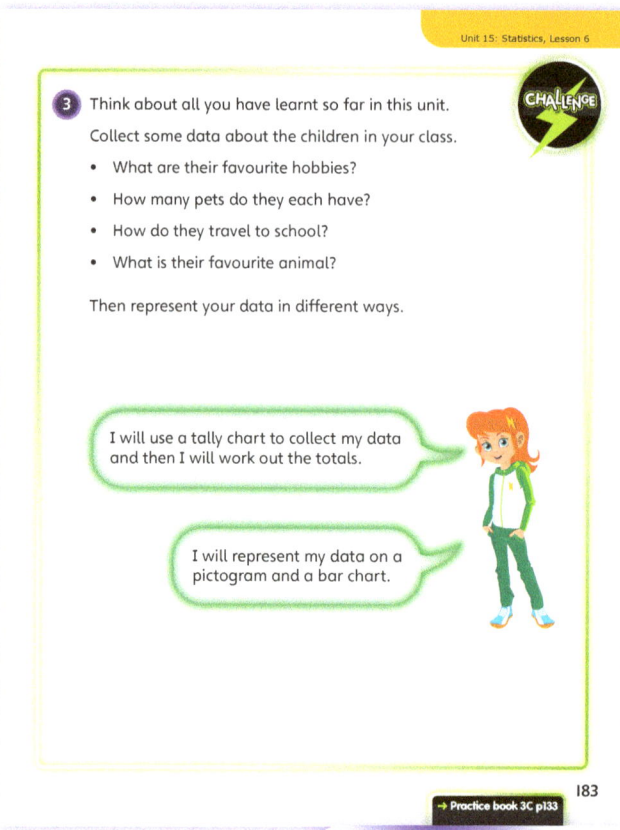

PUPIL TEXTBOOK 3C PAGE 183

Unit 15: Statistics, Lesson 6

Practice

WAYS OF WORKING Independent thinking

IN FOCUS In question ❶, children draw a bar chart showing data about favourite wild birds. Parts of the bars are already drawn for them for support. In question ❷, children draw bars to show the number of children in a class, with the axes going up in 2s rather than 1s. Therefore, they need to put the height of the bar half-way between 2 squares. In question ❸, children first tally the results and find the total, and then they represent this data in a bar chart. Ask children to double check that they have tallied all the values (they may cross out the values as they tally). Also, children can double check at the end by adding the numbers. In question ❹, children draw their own bar chart, including the axes.

STRENGTHEN Before they start drawing bar charts, encourage children to represent the data using practical equipment, such as cubes. Have children look at examples of bar charts from previous lessons as well. Ask them to identify common features. Recapping what a good bar chart looks like will help children when they come to draw their own.

DEEPEN Ask children to create a survey to find information about their class's favourite ice cream flavours. Ask children to create a presentation that includes tables, graphs and charts.

ASSESSMENT CHECKPOINT Use questions ❶ to ❸ to check that children can draw bar charts and identify any common errors.

ANSWERS Answers for the **Practice** part of the lesson can be found in the *Power Maths* online subscription.

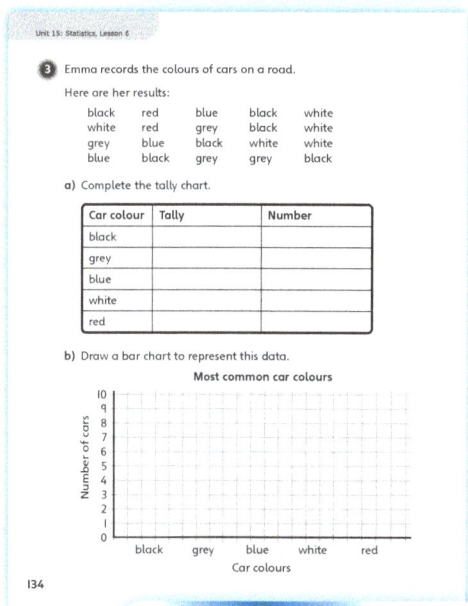

PUPIL PRACTICE BOOK 3C PAGE 133

PUPIL PRACTICE BOOK 3C PAGE 134

PUPIL PRACTICE BOOK 3C PAGE 135

Reflect

WAYS OF WORKING Pair work

IN FOCUS Children discuss with a partner what they need to remember when drawing a bar chart. They may want to look back at their work from earlier in the lesson to help them. They may want to draw a bar chart to exemplify their understanding.

ASSESSMENT CHECKPOINT Children should be able to talk about the key features of bar charts, such as the axes, and that the height of a bar represents a value. They should also be able to talk about why the bars should be same width and why there should be gaps between bars.

ANSWERS Answers for the **Reflect** part of the lesson can be found in the *Power Maths* online subscription.

After the lesson

- Can children construct a bar chart from given data?
- Can children find common errors in bar charts?

221

Unit 15: Statistics, Lesson 7

Simple two-way tables

Learning focus
In this lesson, children will interpret data that is presented in tables, and use this data to answer 1- and 2-step problems.

Before you teach
- How will you help children to understand that a table lets you compare two sets of data systematically?
- Are you able to use real data, such as from a PE lesson, to complement some of the examples given in this lesson?

NATIONAL CURRICULUM LINKS

Year 3 Statistics

Interpret and present data using bar charts, pictograms and tables.

ASSESSING MASTERY

Children can interpret data that is presented in tables. They can identify the correct row and column needed to find the required data. They can use this data to answer a range of 1- and 2-step problems, including questions involving comparison and ordering.

COMMON MISCONCEPTIONS

Children may incorrectly identify the row and column needed to answer a question. Encourage children to make a decision about the column first, identifying this with a star, and then to look at the row. Ask:
- *Which column will you find the information in? Which row will you find the information in?*

STRENGTHENING UNDERSTANDING

Support children with correctly identifying the row and column that they need to refer to by encouraging the use of two rulers, one laid to the right of the column they need and the other laid under the row they need, so that the data they require is where the rulers cross. Support children with the calculations required in this lesson by using the structures and representations from their earlier work on addition and subtraction.

GOING DEEPER

Encourage children to answer questions that involve more than one data source. For example, ask them to use linked data presented in a pictogram, a bar chart and a table to make comparisons and draw conclusions from the data.

KEY LANGUAGE

In lesson: table, rows, columns, order, smallest, largest, total

Other language to be used by the teacher: data

STRUCTURES AND REPRESENTATIONS

Number lines

RESOURCES

Optional: rulers

 In the eTextbook of this lesson, you will find interactive links to a selection of teaching tools.

Quick recap
Check that children can do simple 2-digit additions and subtractions. This skill will help them to do calculations using the data presented in two-way tables in this lesson.

Unit 15: Statistics, Lesson 7

Discover

WAYS OF WORKING Pair work

ASK

- Question 1 a): *Do you think you can answer the question using the **Discover** artwork? Where is the information you need?*
- Question 1 a): *How do you know which child threw the ball 23 m?*
- Question 1 b): *Which row and column do you need to look at to find out the distance Alex threw the discus?*

IN FOCUS In this activity, children are introduced to reading data presented in more complex tables where, for the first time, they need to identify both the row and the column in order to find a particular piece of data. Encourage children to explore the structure of the table, and to identify where information for each event and each child is found.

ANSWERS

Question 1 a): Ebo came in 1st place, Richard came in 2nd place, Alex came in 3rd place and Ambika came in 4th place for the ball throw.

Question 1 b): Alex threw the discus 8 m further than Ambika.

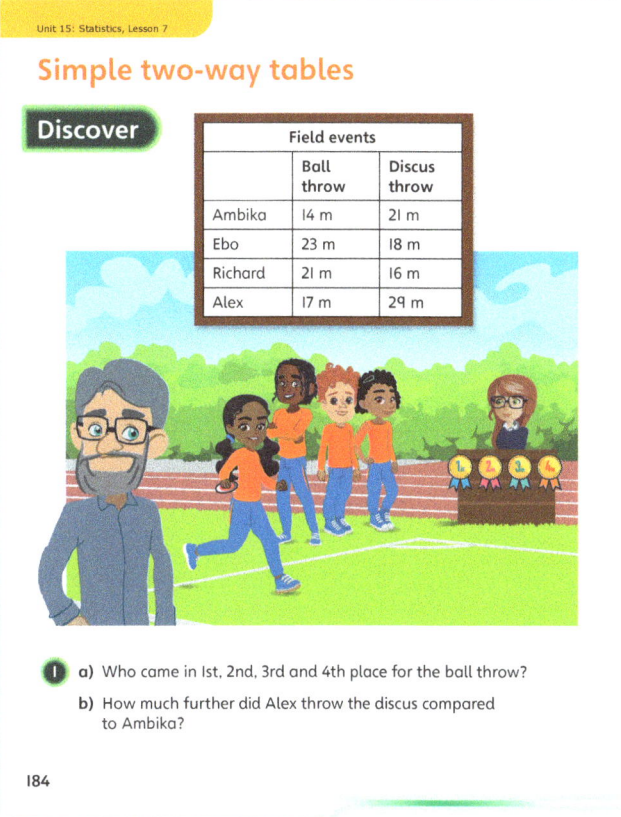

PUPIL TEXTBOOK 3C PAGE 184

Share

WAYS OF WORKING Whole class teacher led

ASK

- Question 1 a): *How did you work out the answer? What steps did you take?*
- Question 1 a): *Which column did you look at?*
- Question 1 a): *How did you know which child threw the ball for each distance?*
- Question 1 b): *Where did you find the information about Ambika's and Alex's discus throws?*
- Question 1 b): *Which operation and method did you use to help you work out the difference?*

IN FOCUS This is the first time children have been introduced to the formal vocabulary of rows and columns. Draw their attention to Sparks' comment which will help them identify rows and columns within a table.

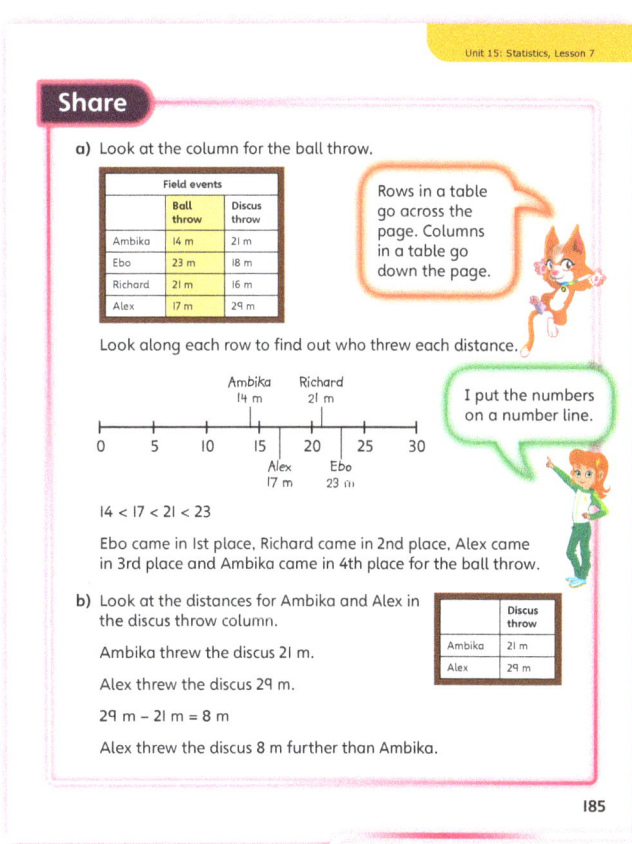

PUPIL TEXTBOOK 3C PAGE 185

Think together

WAYS OF WORKING Whole class teacher led (I do, We do, You do)

ASK
- Question ① a): *Where can you find the times that the children took for the 100 m race?*
- Question ① a): *How do you know which child took each time?*
- Question ① e): *How are you going to find the times for Richard and Ambika?*
- Question ① e): *What calculation method are you going to use to find the difference between the times?*
- Question ③: *What do you think needs to go in the total column?*

IN FOCUS Question ① involves times in a race, where, unlike many other circumstances, the lower the value (the quicker the time), the better. The context of question ② is house points, so if your school has its own house point or similar system, consider replacing or repeating this question with actual data from your school. Help children to see that they can complete the missing value in each row using the other two values given in that row. Ask: *To complete this row, do you need to use an addition or a subtraction?*

STRENGTHEN The key to children answering question ① correctly is to first determine the correct column to look at in each question part. Once they have done that, children should identify the row or rows that they need to use to find the answer. You could ask them to highlight the cell or cells in the table that they need to use. Then ask: *What operation do you need to do?*

DEEPEN Children should begin to be able to exploit the structure of a table in order to work out missing data. Question ③ gives children an opportunity to practise this. Extend children further by asking them to explain how they have worked out each missing value. Children should also begin to make their own statements based on the data presented in the table. Ask: *What else can you tell from this table?*

ASSESSMENT CHECKPOINT Use question ① to assess if, working independently, children can identify the correct row and column in which to find the data they require and use this to carry out simple calculations.

ANSWERS

Question ① a): Alex took 26 seconds to run the 100 m race.

Question ① b): Ebo took 50 seconds to run the 200 m race.

Question ① c): Ambika ran the 200 m race the fastest.

Question ① d): Ambika 1st; Alex 2nd; Richard 3rd; Ebo 4th.

Question ① e): Ambika ran the 200 m race 2 seconds faster than Richard.

Question ②: Ash House, Total: 80; Oak House, Field: 35; Maple House, Field: 45.

Question ③ a): Year 3, Total: 36; Year 4, 1st, 2nd or 3rd: 33; Year 5, Commended: 17; Year 6, 1st, 2nd or 3rd: 22.

Question ③ b): Children's answers will vary. Any three pieces of information using data from the table, for example, more Year 3 medals were for 1st, 2nd or 3rd than for commended.

Think together

① These are the results from the running events.

	Running event times	
	100 m race	200 m race
Ambika	22 seconds	49 seconds
Ebo	31 seconds	50 seconds
Richard	27 seconds	51 seconds
Alex	26 seconds	53 seconds

a) How long did Alex take to run the 100 m race?
b) How long did Ebo take to run the 200 m race?
c) Who ran the 200 m race the fastest?
d) Who came 1st, 2nd, 3rd and 4th in the 100 m race?
e) How much faster did Ambika run the 200 m race compared to Richard?

I think the fastest time will be the lowest number and the slowest time will be the highest number.

PUPIL TEXTBOOK 3C PAGE 186

② This table shows how many points each house won on sports day. Complete the table.

House	Running	Field	Total
Ash House	30	50	
Oak House	45		80
Maple House	40		85

③ This table shows the number of medals won by each year group.

Year	1st, 2nd or 3rd	Commended	Total
Year 3	21	15	
Year 4		9	42
Year 5	20		37
Year 6		26	48

a) Use the information to complete the table.
b) Write three sentences about the information in the table.

I think I can work out this question based on the information I know.

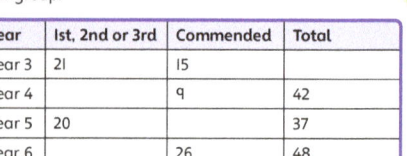

PUPIL TEXTBOOK 3C PAGE 187

Unit 15: Statistics, Lesson 7

Practice

WAYS OF WORKING Independent thinking

IN FOCUS Question ❷ involves more complex calculations than have been experienced so far within this lesson. Remind children of the calculation methods for addition, subtraction and division.

In question ❺, both the pictogram and the bar chart show the amount spent on non-food items for Morgan, Tan and Agg. Children should use the pictogram and bar chart to fill in the 'Non-food' column in the table. Once they have done that, they can subtract 'Non-food' from the total in each row to work out the amount spend on food.

DEEPEN Encourage children to solve problems that involve data presented in more than one place – for example, two different pictograms or, as in question ❺, a table, a bar chart and a pictogram.

Encourage children to consider what is the same and what is different about the ways of presenting data shown in question ❺, and to consider which is most useful.

THINK DIFFERENTLY Ensure children read question ❹ carefully and encourage them to find a range of possible numbers.

ASSESSMENT CHECKPOINT Use question ❷ to assess if children are able to carry out calculations based on the data presented in a table with accuracy.

ANSWERS Answers for the **Practice** part of the lesson can be found in the *Power Maths* online subscription.

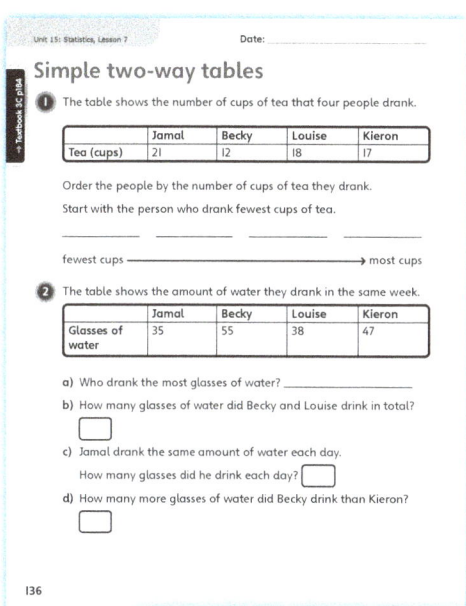

PUPIL PRACTICE BOOK 3C PAGE 136

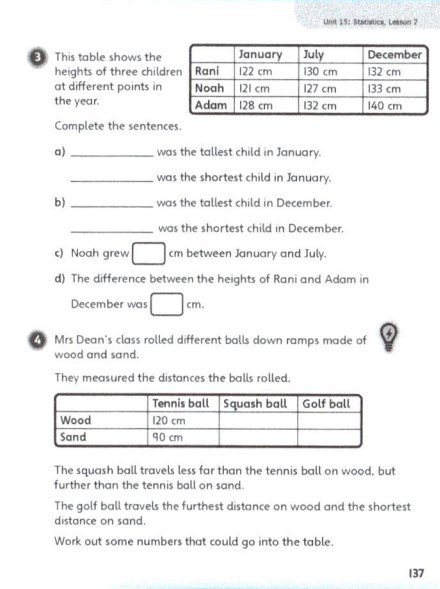

PUPIL PRACTICE BOOK 3C PAGE 137

Reflect

WAYS OF WORKING Pair work

IN FOCUS This question encourages children to make connections between the different ways of presenting data that they have explored within this unit.

ASSESSMENT CHECKPOINT Use this question to assess if children understand the features of tables, pictograms and bar charts and if they are able to compare their advantages and disadvantages.

ANSWERS Answers for the **Reflect** part of the lesson can be found in the *Power Maths* online subscription.

PUPIL PRACTICE BOOK 3C PAGE 138

After the lesson ⏸

- Are children secure answering questions based on tables?
- Can you incorporate tables, pictograms and bar charts into your day-to-day classroom or school environment?

Unit 15: Statistics

End of unit check

Don't forget the unit assessment grid in your *Power Maths* online subscription.

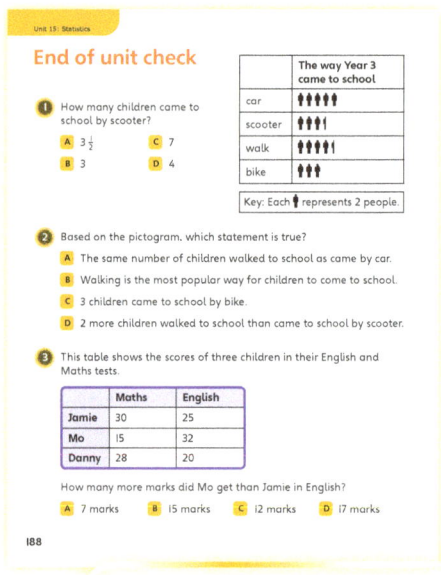

PUPIL TEXTBOOK 3C PAGE 188

WAYS OF WORKING Group work adult led

IN FOCUS

The questions in the **End of unit check** will give you a good insight into which children can read and interpret information presented in pictograms, bar charts and tables.

Question ❻ is a SATs-style question. Children should determine the correct column and/or row to look at in each question part. You could ask them to highlight the cell or cells in the table that they need to use. Then ask: *What operation do you need to do?*

ANSWERS AND COMMENTARY

Children who have mastered this unit can interpret data presented in pictograms, bar charts and tables, and use this data to answer a range of questions, including comparison, ordering and total questions. They can also make their own statements based on the data presented to them and are beginning to be able to compare linked data that is presented in more than one diagram or table to answer questions.

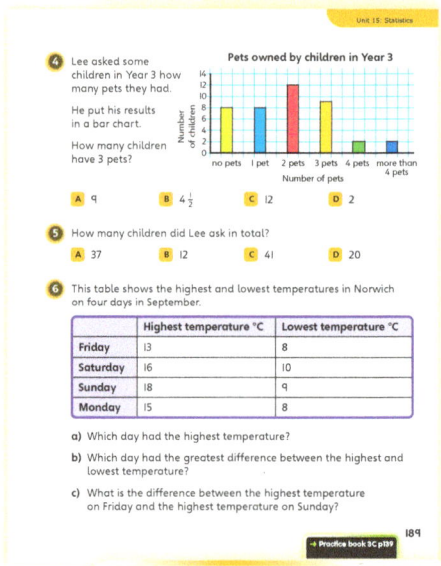

PUPIL TEXTBOOK 3C PAGE 189

Q	A	WRONG ANSWERS AND MISCONCEPTIONS	STRENGTHENING UNDERSTANDING
1	C	A suggests that the child is interpreting each symbol as having a value of 1.	To help children interpret bar charts and pictograms, represent them physically using counters, cubes or other objects. Encourage children to count each object, and to use the key in a pictogram or the scale on the vertical axis in a bar chart to work out the total value. Remind children that they can use a ruler to help them identify the value on the vertical axis of a graph and/or a pair of rulers to help them identify the correct row and column needed in a table.
2	D	A and B suggest that the child is not interpreting half symbols correctly. C suggests that the child is interpreting each symbol as having a value of 1, rather than 2 as identified in the key.	
3	A	B, C and D suggest that the child is selecting the wrong values from the table to compare.	
4	A	B suggests the child is counting the number of divisions, rather than using the scale on the vertical axis. C and D suggest that the child is incorrectly identifying which bar to read.	
5	C	B suggests the child has read the value of one bar only, rather than totalling all the bars. A and D suggest the child has made arithmetic errors in their calculations.	
6	Sunday Sunday 5 °C	Are children reading across the columns to access the correct data?	

Unit 15: Statistics

My journal

WAYS OF WORKING Independent thinking

ANSWERS AND COMMENTARY

Question **1**: Children should identify that Izzy's statement is incorrect, and their explanation should refer to the fact that there is a total of 1 more cone for caramel than for vanilla, and that each cone symbol is worth 10 cones of ice cream.

Question **2**: Children should provide a range of statements based on the pictogram, for example: 'Caramel is the most popular flavour.' 'Raspberry is the least popular flavour.' 'Izzy sold 145 cones of ice cream altogether.'

Support children in creating their own statements by providing them with gapped sentences to use. For example: ☐ is the ☐ popular flavour. Izzy sold ☐ cones of ice cream altogether. Izzy sold ☐ more/fewer cones of ☐ ice cream than ☐.

Power check

WAYS OF WORKING Independent thinking

ASK

- What do you know now that you did not know at the start of this unit?
- Are there some charts or tables that you feel more confident about than others?

Power puzzle

WAYS OF WORKING Pair work

IN FOCUS Use this **Power puzzle** to identify whether children can use logical reasoning in order to complete a bar chart. Encourage them to first complete the scale and then to work through the clues step by step, identifying which bar could refer to which type of fruit.

This **Power puzzle** also helps you identify whether children are able to transfer data between two different representations, in this case from a bar chart to a pictogram, and whether children are able to identify an appropriate key for the pictogram.

ANSWERS AND COMMENTARY If children can complete question **1** of the **Power puzzle**, it suggests they are able to use logical reasoning to interpret clues about a chart. If they are not able to complete this, support them by asking specific questions about each clue. For example: *You know raspberry is the most popular fruit. Which bar shows the most popular fruit?*

If children can complete the question **2** of the **Power puzzle**, it suggests they are able to make links between different ways of presenting data, and are confident with creating pictograms. If they cannot, it suggests they may not be confident with the structure of bar charts and pictograms. Support children by encouraging them to first record the value for each fruit and then to consider what key they need to use. Before using the key children should draw the appropriate number of symbols.

After the unit

- How can you continue to show children a range of statistics and the different ways in which these are presented through your day-to-day classroom activities?
- What links can you make between the work in this unit and your work in other subjects, such as science and PE?

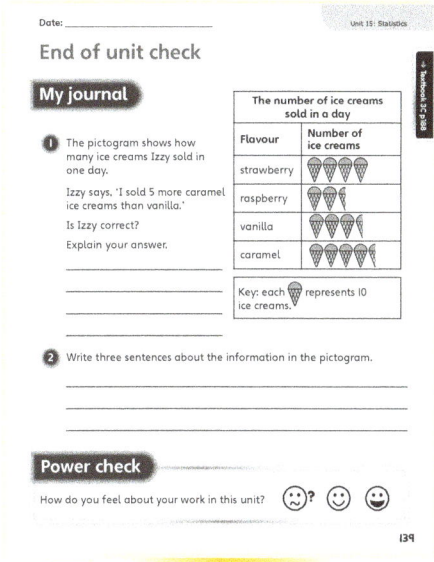

PUPIL PRACTICE BOOK 3C PAGE 139

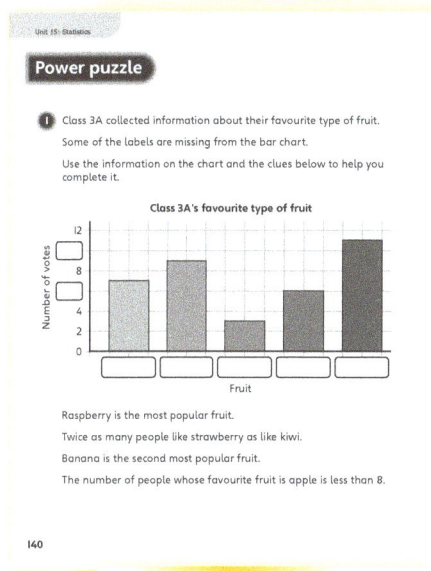

PUPIL PRACTICE BOOK 3C PAGE 140

PUPIL PRACTICE BOOK 3C PAGE 141

Strengthen and **Deepen** activities for this unit can be found in the *Power Maths* online subscription.

Published by Pearson Education Limited, 80 Strand, London, WC2R 0RL.

www.pearsonschools.co.uk

Text © Pearson Education Limited 2018, 2023
Edited by Pearson and Florence Production Ltd
First edition edited by Pearson, Little Grey Cells Publishing Services and Haremi Ltd
Designed and typeset by Pearson and PDQ Digital Media Solutions Ltd
First edition designed and typeset by Kamae Design
Original illustrations © Pearson Education Limited 2018, 2023
Illustrated by Fran and David Brylewski, Virginia Fontanabona, Adam Linley and Paul Moran at Beehive Illustration; Emily Skinner at Graham-Cameron Illustration; and Kamae Design
Images: The Royal Mint, 1971, 1982, 1990, 1992, 1998, 2017, 2023: 64, 68, 81, 83–85, 87–89, 91–93, 95–97, 99–102; Bank of England: 81, 83–85, 87–89, 91–93, 95–97, 99–102
Images Power Up: The Royal Mint, 1971, 1982, 1990, 1992, 2017, 2023: Y3 Unit 11 Slides: Lesson 8, Y3 Unit 12 Slides: Lesson 2, Y3 Unit 12 Slides: Lesson 3, Y3 Unit 12 Slides: Lesson 4
Cover design by Pearson Education Ltd
Back cover illustration © Diego Diaz and Nadene Naude at Beehive Illustration

Series editor: Tony Staneff; Lead author: Josh Lury
Authors (first edition): Tony Staneff and Josh Lury
Consultants (first edition): Professor Liu Jian and Professor Zhang Dan

The rights of Tony Staneff and Josh Lury to be identified as authors of this work have been asserted by them in accordance with the Copyright, Designs and Patents Act 1988.

This publication is protected by copyright, and permission should be obtained from the publisher prior to any prohibited reproduction, storage in a retrieval system, or transmission in any form or by any means, electronic, mechanical, photocopying, recording, or otherwise. For information regarding permissions, request forms and the appropriate contacts, please visit https://www.pearson.com/us/contact-us/permissions.html Pearson Education Limited Rights and Permissions Department.

First published 2018
This edition first published 2023

27 26 25 24 23
10 9 8 7 6 5 4 3 2 1

British Library Cataloguing in Publication Data
A catalogue record for this book is available from the British Library

ISBN 978 1 292 45055 1

Copyright notice
All rights reserved. No part of this publication may be reproduced in any form or by any means (including photocopying or storing it in any medium by electronic means and whether or not transiently or incidentally to some other use of this publication) without the written permission of the copyright owner, except in accordance with the provisions of the Copyright, Designs and Patents Act 1988 or under the terms of a licence issued by the Copyright Licensing Agency, Barnards Inn, 86 Fetter Lane, London EC4A 1EN (http://www.cla.co.uk). Applications for the copyright owner's written permission should be addressed to the publisher.

Printed in the UK by Ashford Press Ltd

For Power Maths online resources, go to:
www.activelearnprimary.co.uk

Note from the publisher
Pearson has robust editorial processes, including answer and fact checks, to ensure the accuracy of the content in this publication, and every effort is made to ensure this publication is free of errors. We are, however, only human, and occasionally errors do occur. Pearson is not liable for any misunderstandings that arise as a result of errors in this publication, but it is our priority to ensure that the content is accurate. If you spot an error, please do contact us at resourcescorrections@pearson.com so we can make sure it is corrected.